# GEOMETRY
# OF CURVES

# CHAPMAN & HALL/CRC MATHEMATICS

OTHER CHAPMAN & HALL/CRC MATHEMATICS TEXTS:

**Functions of Two Variables, Second edition**
S. Dineen

**Network Optimization**
V. K. Balakrishnan

**Sets, Functions, and Logic: An introduction to abstract mathematics, Third edition**
Kevin Devlin

**Algebraic Numbers and Algebraic Functions**
P. M. Cohn

**Computing with Maple**
Francis Wright

**Dynamical Systems: Differential equations, maps, and chaotic behaviour**
D. K. Arrowsmith and C. M. Place

**Control and Optimization**
B. D. Craven

**Elements of Linear Algebra**
P. M. Cohn

**Error-Correcting Codes**
D. J. Bayliss

**Introduction to the Calculus of Variations**
U. Brechtken-Manderscheid

**Integration Theory**
W. Filter and K. Weber

**Algebraic Combinatorics**
C. D. Godsil

**An Introduction to Abstract Analysis-PB**
W. A. Light

**The Dynamic Cosmos**
M. Madsen

**Algorithms for Approximation II**
J. C. Mason and M. G. Cox

**Introduction to Combinatorics**
A. Slomson

**Elements of Algebraic Coding Theory**
L. R. Vermani

**Linear Algebra: A geometric approach**
E. Sernesi

**A Concise Introduction to Pure Mathematics**
M. W. Liebeck

**Geometry of Curves**
J. W. Rutter

**Experimental Mathematics with Maple**
Franco Vivaldi

**Solution Techniques for Elementary Partial Differential Equations**
Christian Constanda

**Basic Matrix Algebra with Algorithms and Applications**
Robert A. Liebler

**Computability Theory**
S. Barry Cooper

*Full information on the complete range of Chapman & Hall/CRC Mathematics books is available from the publishers.*

# GEOMETRY OF CURVES

## JOHN W. RUTTER

CHAPMAN & HALL/CRC

Boca Raton   London   New York   Washington, D.C.

## Library of Congress Cataloging-in-Publication Data

Rutter, John W., 1935-
    Geometry of curves / John W. Rutter.
        p. cm. — (Champman & Hall mathematics series)
    Includes index.
    ISBN 1-58488-166-6
    1. Curves, Plane. I. Title. II. Chapman and Hall mathematics series.
QA567.R78 2000
516.3′52—dc21
                                        99-088667
                                             CIP

### Visit the CRC Press Web site at www.crcpress.com

© 2000 by Chapman & Hall/CRC

No claim to original U.S. Government works
International Standard Book Number 1-58488-166-6
Library of Congress Card Number 99-088667
Printed in the United States of America      5 6 7 8 9 0
Printed on acid-free paper

TO MY MOTHER

AND FATHER

# Contents

**Introduction**  **1**
 0.1 Cartesian coordinates  1
 0.2 Polar coordinates  2
 0.3 The Argand diagram  4
 0.4 Polar equations  4
 0.5 Angles  5
 0.6 Orthogonal and parallel vectors  6
 0.7 Trigonometry  7

**1 Lines, circles, and conics**  **9**
 1.1 Lines  9
 1.2 The circle  13
 1.3 Conics  15
 1.4 The ellipse in canonical position  16
 1.5 The hyperbola in canonical position  18
 1.6 The parabola in canonical position  21
 1.7 Classical geometric constructions of conics  22
 1.8 Polar equation of a conic with a focus as pole  23
 1.9 History and applications of conics  26

**2 Conics: general position**  **31**
 2.1 Geometrical method for diagonalisation  31
 2.2 Algebra  33
 2.3 Algebraic method for diagonalisation  37
 2.4 Translating to canonical form  38
 2.5 Central conics referred to their centre  40
 2.6 Practical procedures for dealing with the general conic  45
 2.7 Rational parametrisations of conics  55

**3   Some higher curves**                                                        **59**
    3.1    The semicubical parabola: a cuspidal cubic                    61
    3.2    A crunodal cubic                                                      62
    3.3    An acnodal cubic                                                     63
    3.4    A cubic with two parts                                             64
    3.5    History and applications of algebraic curves              66
    3.6    A tachnodal quartic curve                                        66
    3.7    Limaçons                                                               68
    3.8    Equi-angular (logarithmic) spiral                             69
    3.9    Archimedean spiral                                                  71
    3.10   Application – Watt's curves                                     73

**4   Parameters, tangents, normal**                                      **77**
    4.1    Parametric curves                                                    77
    4.2    Tangents and normals at regular points                     85
    4.3    Non-singular points of algebraic curves                    88
    4.4    Parametrisation of algebraic curves                          89
    4.5    Tangents and normals at non-singular points             91
    4.6    Arc-length parametrisation                                       93
    4.7    Some results in analysis                                           96

**5   Contact, inflexions, undulations**                                 **103**
    5.1    Contact                                                                   103
    5.2    Invariance of point-contact order                              108
    5.3    Inflexions and undulations                                        112
    5.4    Geometrical interpretation of $n$–point contact         120
    5.5    An analytical interpretation of contact                      121

**6   Cusps, non-regular points**                                         **123**
    6.1    Cusps                                                                      123
    6.2    Tangents at cusps                                                    124
    6.3    Contact between a line and a curve at a cusp            126
    6.4    Higher singularities                                                  128

**7   Curvature**                                                                  **133**
    7.1    Cartesian coordinates                                              133
    7.2    Curves given by polar equation                                 142
    7.3    Curves in the Argand diagram                                  146
    7.4    An alternative formula                                             147

**8   Curvature: applications**                                             **151**
    8.1    Inflexions of parametric curves at regular points      151
    8.2    Vertices and undulations at regular points                153
    8.3    Curvature of algebraic curves                                  157

| | | |
|---|---|---|
| 8.4 | Limiting curvature of algebraic curves at cusps | 160 |
| **9** | **Circle of curvature** | **167** |
| 9.1 | Centre of curvature and circle of curvature | 167 |
| 9.2 | Contact between curves and circles | 173 |
| **10** | **Limaçons** | **177** |
| 10.1 | The equation | 177 |
| 10.2 | Curvature | 179 |
| 10.3 | Non-regular points | 179 |
| 10.4 | Inflexions | 180 |
| 10.5 | Vertices | 180 |
| 10.6 | Undulations | 181 |
| 10.7 | The five classes of limaçons | 181 |
| 10.8 | An alternative equation | 182 |
| **11** | **Evolutes** | **183** |
| 11.1 | Definition and special points | 183 |
| 11.2 | A matrix method for calculating evolutes | 186 |
| 11.3 | Evolutes of the cycloid and the cardioid | 187 |
| **12** | **Parallels, involutes** | **195** |
| 12.1 | Parallels of a curve | 195 |
| 12.2 | Involutes | 203 |
| **13** | **Roulettes** | **209** |
| 13.1 | General roulettes | 209 |
| 13.2 | Parametrisation of circles | 214 |
| 13.3 | Cycloids: rolling a circle on a line | 215 |
| 13.4 | Trochoids: rolling a circle on or in a circle | 218 |
| 13.5 | Rigid motions | 234 |
| 13.6 | Non-regular points and inflexions of roulettes | 237 |
| **14** | **Envelopes** | **243** |
| 14.1 | Evolutes as a model | 244 |
| 14.2 | Singular-set envelopes | 245 |
| 14.3 | Discriminant envelopes | 255 |
| 14.4 | Different definitions and singularities of envelopes | 259 |
| 14.5 | Limiting-position envelopes | 260 |
| 14.6 | Orthotomics and caustics | 264 |
| 14.7 | The relation between orthotomics and caustics | 266 |
| 14.8 | Orthotomics of a circle | 266 |
| 14.9 | Caustics of a circle | 268 |

**15 Singular points of algebraic curves**                                **271**
   15.1   Intersection multiplicity with a given line                 271
   15.2   Homogeneous polynomials                                     273
   15.3   Multiplicity of a point                                     274
   15.4   Singular lines at the origin                                276
   15.5   Isolated singular points                                    278
   15.6   Tangents and branches at non-isolated singular points      280
   15.7   Branches for non-repeated linear factors                    282
   15.8   Branches for repeated linear factors                        286
   15.9   Cubic curves                                                291
   15.10 Curvature at singular points                               295

**16 Projective curves**                                                  **297**
   16.1   The projective line                                         297
   16.2   The projective plane                                        298
   16.3   Projective curves                                           302
   16.4   The projective curve determined by a plane curve            303
   16.5   Affine views of a projective curve                         304
   16.6   Plane curves as views of a projective curve                305
   16.7   Tangent lines to projective curves                         307
   16.8   Boundedness of the associated affine curve                 309
   16.9   Summary of the analytic viewpoint                          311
   16.10 Asymptotes                                                 312
   16.11 Singular points and inflexions of projective curves       314
   16.12 Equivalence of curves                                      316
   16.13 Examples of asymptotic behaviour                           318
   16.14 Worked example                                             320

**17 Practical work**                                                     **329**

**18 Drawn curves**                                                       **345**
   18.1   Personalising MATLAB for metric printing                    345
   18.2   Ellipse 1 and Ellipse 2                                     346
   18.3   Ellipse 3                                                   347
   18.4   Parabola 1                                                  347
   18.5   Parabola 2                                                  348
   18.6   Parabola 3                                                  348
   18.7   Hyperbola                                                   349
   18.8   Semicubical parabola                                        350
   18.9   Polar graph paper                                           350

**19 Further reading**                                                    **353**

**Index**                                                                 **355**

# Preface

This book covers the material for a first full university course in geometry; specifically it is an introduction to the geometry of plane curves, given as parametric curves, algebraic curves, or projective curves. It is intended for students who have previously studied elementary calculus including partial differentiation, the elementary theory of complex numbers, and elementary coordinate geometry. It is developed from a one-semester geometry course I have given over a number of years to undergraduates specialising in mathematics, statistics or computing, or following degree courses involving a substantial study of mathematics.

My aim in writing the book is to make the material covered readily available in one book and in a form suitable for a modern first university course in geometry. Previously in order to cover the material in the book, it was necessary to read isolated sections of a number of other texts of varying levels of sophistication. Topics covered here have been integrated and presented in a manner suitable for a first course, and new elementary proofs have been developed where possible. Most of the material included is what I believe would be termed elementary in modern university terms, and as far as possible the proofs I have given use only elementary ideas. I have starred a small number of conceptually or technically more difficult sections and proofs; these may be left for a second reading. I have also starred a small number of sections and results which can be omitted depending on the time available. A large number of exercises of varying difficulty are included as are many worked examples. I believe that in geometry, as in most areas of mathematics, doing exercises helps the student more quickly to understand and to appreciate the subject. Solutions to exercises are included roughly on an alternate basis. One of my main aims has been to lay out a mathematical structure, understandable to modern students, which can be used for solving problems, rather than to provide a catalogue of theorems; this, I believe, is also a main aim of many modern first courses in calculus or algebra. I have included numerous figures to illustrate the ideas, proofs, and solutions, and to illustrate specific classical curves.

Many students have obtained degrees in mathematics having studied little or no geometry at university or at school, a situation which is, I believe, regrettable. The publication of this book coincides, I believe, with the rising interest in the return to the study of geometry at university level. The book will be suitable for use in many mathematics departments, including those where a complete geometry course is not currently taught and that wish to introduce one, since it provides an elementary introduction to a number of important areas. Students who successfully completed the course on which the book is based developed a heightened appreciation of geometry and many of them went on to study more advanced courses in differential geometry and/or algebraic and projective geometry.

An introductory chapter contains, for clarification and reference, basic material which will already be known by many readers. In the first chapter the basic equations of lines, circles, and conics are given; the relationship between parametric, algebraic, and polar equations is considered. The techniques for classification of conics in general position are given in the second chapter. The third chapter presents examples of some higher algebraic and transcendental curves having features such as cusps, nodes, or isolated points, which do not occur in the case of conics. In the fourth to ninth chapters, the standard properties of parametric curves are obtained, including tangents and normals, inflexions, undulations, cusps, and curvature; some of these properties are applied to give properties of algebraic curves such as tangents, normals, and curvature. In the tenth chapter, features such as cusps, inflexions, and curvature are used to classify limaçons into five classes. In the eleventh, twelfth, and thirteenth chapters, the evolute, parallel, involute, and roulette of parametric curves are considered. The fourteenth chapter gives an account of envelopes of families of parametric, algebraic, and other curves. In the fifteenth chapter tangents and branches of algebraic curves at singular points are investigated. The sixteenth chapter studies projective curves and their relationship with algebraic curves, including applications to asymptotes and boundedness. Throughout the book many classical curves are considered as examples and some are studied in more detail. I have included sections on the history and applications of several classes of curves such as conics, spirals, cubics, trochoids, and Watt's curves.

As well as giving the classification of conics in Chapter 2, a classification of cubic algebraic curves is given in Chapter 15 using the results on singular points.

I have followed the analytic method almost exclusively in the sections on algebraic and projective geometry, and have often used the calculus in proofs. Early in the twentieth century, certain purists would have objected to these methods, but with changing fashions, needs, and current school syllabuses, there are, I believe, few now who would. A university course in synthetic geometry would in any case have objectives quite different to the ones of this book.

I give the analytic description of projective space and base, for example, the proofs and techniques for asymptotes and boundedness on that description. Additionally, I have, in Chapter 16, indicated how the projective plane can be obtained by identifying opposite pairs of points on the sphere; this is perhaps the most sophisticated concept in the book, but I believe that understanding this geometrical construction will lead the reader to a fuller appreciation of projective space, the projective method and projective curves. However this geometrical construction could be omitted until a second reading. I have also included a number of ways in which projective curves can be drawn or pictured, since I believe that such representations will aid the reader to achieve a fuller understanding of these projective ideas.

The book is essentially self-contained. I have included in Chapter 2 results on and methods of orthogonal diagonalisation of quadratic forms in two variables. These are used in the classification of conics, in moving a conic to canonical position. I have also included at the end of Chapter 4 results in calculus and analysis which are used in the book.

The lecture course given was supplemented by practical classes. In the practical classes the students, collaborating in small groups, draw curves by hand using a variety of techniques, including rectangular and polar plotting, enveloping, and the methods of conchoids, cissoids, and strophoids. Although not essential to the course, practical classes are particularly popular among students and I recommend their adoption. Completion of the practical work helps and motivates students to understand the theory. The drawing of curves is one of the visual-art forms of mathematics and gives students the opportunity to achieve satisfaction in a non-theoretical part of the subject. I have included in Chapter 17 a list of practical projects suitable for students to share in groups of six, with each student in the group generally drawing a different curve. This can be modified as required by the lecturer. As an alternative to their use in practical classes, a selection of these projects could be used for take-home assignments. Plotting curves using computer packages is also popular, and a number of packages are available including Maple, MATLAB, and Mathematica. The drawing of curves by hand could be partially or wholly replaced by computer drawing in the practical classes and curve-drawing exercises. Some programs for drawing sized curves in MATLAB are given in Chapter 18.

In the practical work and some of the exercises involving curve-drawing, some standard ready-drawn curves are needed. There are many packages which can be used for drawing curves. In Chapter 18 programs are given for use with MATLAB for drawing these standard curves. A program for drawing polar graph paper is included for localities where such paper is not available.

A list of books for further reading is given in Chapter 19.

As well as being suitable for students aiming for degrees having a high content of mathematics, the book is also appropriate for students of mathematically based subjects such as engineering, who also may be required to study plane curves at some depth.

The book could also be used as a supplementary text for courses in calculus, vector calculus, linear algebra, differential geometry, singularity theory, algebraic geometry, and computer graphics.

My thanks are due to a number of people including Ian Porteous for his support in the course, Victor Flynn and several reviewers for reading some of the chapters and suggesting improvements, Peter Giblin for advice on computer graphics, Rachid Chalabi and Steve Downing for advice on the use of LaTeX and for its smooth running, and Dave Alliot of Chapman and Hall/CRC production for his detailed reading and advice. I am also grateful to students of the University of Liverpool who tried out drafts of the manuscript in class; the high satisfaction rating they expressed in student surveys and individually was an incentive.

Relevant documents and developments subsequent to publication may be available on the following linked websites.

www.crcpress.com

www.liv.ac.uk/~jwrutter/curves

# List of figures

| 0.1 | Cartesian coordinates. | 1 |
| 0.2 | Polar coordinates. | 2 |
| 0.3 | Angles. | 5 |
| | | |
| 1.1 | Equations of lines. | 10 |
| 1.2 | Polar equations of lines. | 12 |
| 1.3 | Polar equation of a circle. | 14 |
| 1.4 | Sections of a complete cone. | 16 |
| 1.5 | The ellipse. | 17 |
| 1.6 | The hyperbola. | 19 |
| 1.7 | The parabola. | 21 |
| 1.8 | Classical geometric constructions of conics. | 22 |
| 1.9 | Polar equation of a conic with a focus as pole. | 23 |
| 1.10 | Range of a projectile. | 28 |
| | | |
| 2.1 | Rotating the axes. | 32 |
| 2.2 | Hyperbola. | 47 |
| 2.3 | Ellipse. | 50 |
| 2.4 | Parabola. | 52 |
| 2.5 | Rational parametrisation of the circle. | 56 |
| | | |
| 3.1 | Some cubic curves. | 60 |
| 3.2 | A tachnodal quartic. | 67 |
| 3.3 | Equi-angular spiral. | 70 |
| 3.4 | Archimedean spiral. | 72 |
| 3.5 | Watt's linkage. | 74 |
| 3.6 | Watt's curves. | 75 |
| | | |
| 4.1 | Self-crossings of curves. | 81 |
| 4.2 | A smooth curve with a corner. | 82 |
| 4.3 | Tangents and normals of parametrised curves. | 85 |

| | | |
|---|---|---|
| 4.4 | Parametrising a circle by $x$. | 90 |
| 4.5 | Tangents and normals of algebraic curves. | 92 |
| 4.6 | Length of a curve. | 94 |
| | | |
| 5.1 | Simple inflexion. | 112 |
| 5.2 | Simple undulation. | 114 |
| 5.3 | Inflexions of a graph. | 117 |
| 5.4 | $n$–point contact of two curves. | 121 |
| 5.5 | $n$–point contact of a line and a curve. | 121 |
| | | |
| 6.1 | Non-tangent lines through a cusp. | 126 |
| 6.2 | Cusp and cuspidal tangent line. | 128 |
| | | |
| 7.1 | Curvature. | 134 |
| 7.2 | Sign of the curvature. | 135 |
| | | |
| 8.1 | Limiting curvature at cusps. | 160 |
| | | |
| 9.1 | Centre of curvature. | 168 |
| | | |
| 10.1 | The five classes of limaçons. | 178 |
| | | |
| 11.1 | The evolute of a curve. | 184 |
| 11.2 | The cycloid and its evolute. | 188 |
| 11.3 | The cardioid and its evolute. | 189 |
| | | |
| 12.1 | Parallel curves. | 196 |
| 12.2 | Parallels to an ellipse. | 200 |
| 12.3 | Involutes by wrapping or unwrapping. | 204 |
| 12.4 | Evolute and involute are orthogonal. | 207 |
| | | |
| 13.1 | Rolling curves. | 210 |
| 13.2 | The general roulette. | 212 |
| 13.3 | Unit-speed parametrisations of the circle. | 215 |
| 13.4 | Rolling a circle on a line. | 216 |
| 13.5 | Cycloids. | 218 |
| 13.6 | Rolling a circle outside a circle. | 219 |
| 13.7 | Epitrochoids: limaçons. | 221 |
| 13.8 | Epitrochoids II. | 222 |
| 13.9 | Rolling a circle 'inside' a circle. | 224 |
| 13.10 | Hypotrochoids. | 225 |
| 13.11 | Hypotrochoids: rhodonea. | 227 |
| 13.12 | Circle 1. | 239 |
| 13.13 | The inflexion circle. | 240 |

14.1 The evolute as an envelope. 244
14.2 An envelope of a family of lines. 248
14.3 A family of parabolae and their envelope. 250
14.4 A family of ellipses and their envelope. 251
14.5 Parallels as envelopes. 253
14.6 Family of cuspidal cubics with a singular envelope. 258
14.7 Family of crunodal and acnodal cubics with a singular envelope. 259
14.8 Orthotomics. 265
14.9 Reflexion caustics. 265
14.10 Caustic as evolute of the orthotomic. 266
14.11 Caustics of a circle. 269

15.1 Intersection multiplicity. 273

16.1 The point at infinity of a line. 298
16.2 The projective line as a circle. 299
16.3 The axes of the projective plane. 301
16.4 A right circular cone in 3-space. 307
16.5 Tangents at points at infinity. 312
16.6 Asymptotes of second order projective curves. 315
16.7 More examples of asymptotic behaviour. 319
16.8 Different affine views of a curve. 321

# Introduction

We have collected a number of definitions and results in elementary Plane Geometry, and listed some trigonometric formulae. The reader should absorb these carefully.

## 0.1 Cartesian coordinates

Points in the plane can be represented by coordinates $(x, y)$. We choose as origin a fixed point $O$ in the plane. The positive $x$-axis is a half-line[†] $Ox$ beginning at the origin, and the positive $y$-axis $Oy$ is the half-line also beginning at the origin obtained by rotating $Ox$ counter-clockwise through an angle $\frac{\pi}{2}$. Rarely we may consider the $y$-axis $Oy$ to be obtained by rotating the first clockwise through an angle $\frac{\pi}{2}$. The $x$-axis is the whole

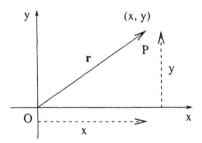

Figure 0.1 *Cartesian coordinates.*

line containing $Ox$ and the $y$-axis is the whole line containing $Oy$. Any point $P$ in the plane can be written uniquely in the form $(x, y)$ where $x$ is

---

[†] A half-line beginning at the origin consists of the origin together with those points on a line through the origin which lie on 'one side' of the origin.

the directed length (it can be negative) in the direction $Ox$ of the projection of $OP$ onto the $x$–axis, and $y$ is the directed length in the direction $Oy$ of the projection of $OP$ onto the $y$–axis (see Figure 0.1). These lengths are measured in the appropriate units or scaled units. The position vector of $P$ is $\mathbf{r} = (x, y)$. The set of ordered pairs $(x, y)$ of real numbers is denoted by $\mathbf{R}^2$. A choice of origin and (directed) axes gives an identification of the plane with $\mathbf{R}^2$. Where appropriate we write 'the plane $\mathbf{R}^2$'.

Let $x(t)$ and $y(t)$ be differentiable functions of $t$. As $t$ varies the point $\mathbf{r}(t) = (x(t), y(t))$ will in general describe a curve in the plane. The function $t \to \mathbf{r}(t)$ is a parametric equation of this curve and $t$ is the parameter.

## 0.2 Polar coordinates

Points in the plane may alternatively be described by polar coordinates. For polar coordinates we choose as pole a fixed point $O$ in the plane and an initial half-line from $O$ (see Figure 0.2). Any point $P$ in the plane, other

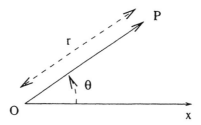

Figure 0.2 *Polar coordinates.*

than $O$, can be written in the form $(r, \theta)$, where $r > 0$ is the length $OP$ and the polar angle $\theta$ is the directed angle measured counter-clockwise from the initial half-line to $OP$. The angle $\theta$ is not uniquely determined, though it is determined up to the addition of $2n\pi$ where $n$ is an integer. The origin $O$ is specified by $r = 0$, but has no (unique) value of $\theta$ associated with it. When we consider both the Cartesian coordinates and the polar coordinates, we usually choose the pole to be at the origin and the initial half-line to be $Ox$. A point given by polar coordinates $(r, \theta)$ has Cartesian coordinates $(x, y)$ given by

$$\boxed{x = r\cos\theta, \; y = r\sin\theta.}$$

Conversely a point with Cartesian coordinates $\mathbf{r} = (x, y)$ ($\mathbf{r} \neq \mathbf{0}$) may be represented by the polar coordinates $(r, \theta)$, where

$$\boxed{r = \sqrt{x^2 + y^2} > 0, \;\; \cos\theta = \frac{x}{\sqrt{x^2 + y^2}} \;\; \text{and} \;\; \sin\theta = \frac{y}{\sqrt{x^2 + y^2}}.}$$

It is important to be aware that $\theta$ is not equal to $\tan^{-1}\frac{y}{x}$ in general, and that $\tan^{-1}\frac{y}{x}$ where defined satisfies

$$-\frac{\pi}{2} < \tan^{-1}\frac{y}{x} < \frac{\pi}{2} :$$

for example, the polar angle of $(-1,1)$ is $\frac{3\pi}{4}$ whilst $\tan^{-1}\frac{1}{(-1)}$ is $-\frac{\pi}{4}$.

> The polar angle is not $\tan^{-1}\frac{y}{x}$.

Indeed there is no differentiable (or continuous) solution $\theta$ of the equations

$$\cos\theta = \frac{x}{\sqrt{x^2 + y^2}} \quad \text{and} \quad \sin\theta = \frac{y}{\sqrt{x^2 + y^2}}$$

valid in the whole $xy$–plane, since taking a simple counter-clockwise circuit of the origin increases the value of $\theta$ by $2\pi$; a differentiable solution can be given, for example, in the $xy$–plane from which a half-line through the origin has been removed.

**Worked example 0.1.** Find the polar coordinates of the points having Cartesian coordinates $(1, -1)$ and $(-3, 4)$.

*Solution.*

i) We have $r = |(1, -1)| = \sqrt{2}$. Also $\theta$ is given by

$$\cos\theta = \frac{1}{\sqrt{2}} \quad \text{and} \quad \sin\theta = -\frac{1}{\sqrt{2}}.$$

Therefore $\theta$ is $-\frac{\pi}{4}$. The polar coordinates of $(1, -1)$ are

$$(r, \theta) = \left(\sqrt{2}, -\frac{\pi}{4}\right).$$

ii) We have $r = |(-3, 4)| = 5$. Also $\theta$ is given by

$$\cos\theta = -\tfrac{3}{5} \quad \text{and} \quad \sin\theta = \tfrac{4}{5}.$$

Therefore $\theta$ is an angle in the second quadrant. Using a calculator, we see that $\theta$ is represented by $126.87°$. The polar coordinates of $(-3, 4)$ are

$$(r, \theta) = (5, 126.87°).$$

□

In the first solution above the angle is given in radians and, in the second, in degrees.

> $2\pi$ radians is equal to 360 degrees.

## 0.3 The Argand diagram

Points in the plane may be represented by Cartesian coordinates $(x, y)$ or by polar coordinates $(r, \theta)$. A third way of representing points in the plane is by complex numbers $z = x + iy$. For a given origin and given (directed) axes, the complex number $z = x + iy$ is associated with the point $(x, y)$ in the plane; in particular the complex numbers 1 and $i$ are associated with the points $(1, 0)$ and $(0, 1)$. The Argand diagram is the plane with complex numbers associated with each of its points in this way. For the Argand diagram we insist that $Oy$ is obtained from $Ox$ by rotating $Ox$ counter-clockwise through an angle of $\dfrac{\pi}{2}$. The polar coordinates of the point $z = x + iy \neq 0$ are $(|z|, \theta)$ where $z = |z|e^{i\theta}$, $|z| = \sqrt{x^2 + y^2}$ is the modulus of $z$, and $e^{i\theta} = \cos\theta + i\sin\theta$. The argument of a complex number $z$ is the polar angle $\theta$. Complex numbers are multiplied by

$$(x + iy)(u + iv) = (xu - yv) + i(xv + yu)$$

and by $re^{i\theta} Re^{i\varphi} = rRe^{i(\theta + \varphi)}$. In particular multiplication by $i = e^{i\frac{\pi}{2}}$ rotates the vector $z = |z|e^{i\theta}$ counter-clockwise through an angle $\dfrac{\pi}{2}$. The conjugate $\bar{z}$ of a complex number $z = x + iy$ is $\bar{z} = x - iy$, in particular

$$\overline{e^{i\theta}} = e^{-i\theta}.$$

Also $|z|^2 = z\bar{z} = x^2 + y^2$.

**Worked example 0.2.** Write $-\sqrt{3} + i$ in the form $re^{i\theta}$.

*Solution.* We have $r = |-\sqrt{3} + i| = \sqrt{3 + 1} = 2$. Also $\theta$ is given by

$$\cos\theta = -\frac{\sqrt{3}}{2} \quad \text{and} \quad \sin\theta = \frac{1}{2}.$$

Therefore $\theta = \dfrac{2\pi}{3}$, and $z = 2e^{\frac{2}{3}i\pi}$.                    □

## 0.4 Polar equations

A polar equation may be given by $g(r, \theta) = 0$, where $g$ is some function; the variables $r$ and $\theta$ are traditionally used here where $r$ is a 'signed' distance (i.e., can be positive or negative) measured in the direction of the unit vector which has oriented angle $\theta$ measured counter-clockwise from the initial half-line. In this case we allow the possibility that $r$ may take negative values: in the case where $g(r, \theta) = 0$ and $r < 0$, the polar coordinates of the point given by the polar equation are not $(r, \theta)$, but are $(-r, \theta + \pi)$, so that in this case the polar angle is $\theta + \pi$ rather than $\theta$.

> For $r < 0$ the polar coordinates are $(-r, \theta + \pi)$.

When changing from a polar equation $r = h(\theta)$ to the parametric equation, we use $\mathbf{r} = (r\cos\theta, r\sin\theta)$ where $r$ is given by the polar equation and can therefore take negative values. We then replace $\theta$ by $\varphi$ since the parameter will not in general be the polar angle $\theta$. We obtain the parametrisation

$$\mathbf{r} = (h(\varphi)\cos\varphi, h(\varphi)\sin\varphi).$$

## 0.5 Angles

Angles in the plane are of different types as follows (see Figure 0.3).

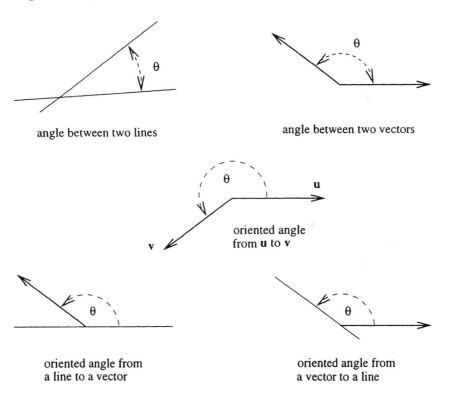

angle between two lines

angle between two vectors

oriented angle
from **u** to **v**

oriented angle from
a line to a vector

oriented angle from
a vector to a line

Figure 0.3 *Angles.*

The angle $\theta$ between two lines is usually taken to be the acute angle

$$0 \le \theta \le \frac{\pi}{2}.$$

The angle between two non-zero vectors is usually taken to be the angle

$$0 \le \theta \le \pi.$$

Each of these is a non-oriented angle.

The oriented angle from a non-zero vector **u** to a non-zero vector **v** is usually taken to be the oriented angle $0 \leq \theta < 2\pi$ measured counter-clockwise from **u** to **v**. In particular the polar angle of a non-zero vector **r** $= (x, y)$ is the oriented angle from the half-line $Ox$ to the vector **r**. The oriented angle from a vector to a line and the oriented angle from a line to a vector are measured counter-clockwise in a similar way: in these cases we can choose $0 \leq \theta < \pi$.

## 0.6 Orthogonal and parallel vectors

The scalar product or inner product of two vectors $(x, y)$ and $(u, v)$ is

$$(x, y) \cdot (u, v) = xu + yv.$$

We write $\mathbf{r}^2 = \mathbf{r} \cdot \mathbf{r} = x^2 + y^2$. The length of the vector $\mathbf{r} = (x, y)$ is

$$|\mathbf{r}| = \sqrt{\mathbf{r}^2} = \sqrt{\mathbf{r} \cdot \mathbf{r}} = \sqrt{x^2 + y^2}.$$

A unit vector is a vector whose length is 1, that is a vector **r** for which $|\mathbf{r}| = 1$. Given any non-zero vector **r**, there is a unique unit vector $\dfrac{1}{|\mathbf{r}|} \mathbf{r}$ having the same direction as **r**.

Two vectors $(x, y)$ and $(u, v)$ are orthogonal if

$$(x, y) \cdot (u, v) = xu + yv = 0.$$

Two vectors form an orthonormal set if they are orthogonal unit vectors. Given any non-zero vector $(x, y)$, there are precisely two vectors $(-y, x)$ and $(y, -x)$ which are orthogonal to $(x, y)$ and which have the same length as $(x, y)$: rotating $(x, y) = (r \cos \theta, r \sin \theta)$ counter-clockwise through an angle $\dfrac{\pi}{2}$ gives the vector (see §0.7)

$$\left( r \cos \left( \theta + \frac{\pi}{2} \right), r \sin \left( \theta + \frac{\pi}{2} \right) \right) = (-r \sin \theta, r \cos \theta) = (-y, x).$$

> Rotating $(x, y)$ counter-clockwise through $\dfrac{\pi}{2}$ gives $(-y, x)$.

Similarly rotating $(x, y)$ clockwise through $\dfrac{\pi}{2}$ gives $(y, -x)$.

Two vectors in the plane are parallel (or linearly dependent) if, and only if, one is a multiple of the other; this allows the possibility that one of the vectors is zero. Two vectors are linearly independent if they are not linearly dependent.

> **r** and **z** are parallel if one is a multiple of the other:
> $(x, y)$ and $(u, v)$ are parallel if $xv = yu$.

This second definition is precisely the condition that $(a, b)$ and $(-v, u)$ are orthogonal, that is $(a, b) \cdot (-v, u) = 0$.

**Worked example 0.3.** Show that the vectors $(1 + t, 1 - t)$ and $(1, 2)$ are parallel if, and only if, $3t = -1$.

*Solution.* $(1 + t, 1 - t)$ and $(1, 2)$ are parallel if, and only if, $2(1 + t) = 1 - t$, that is if, and only if, $3t = -1$. □

## 0.7 Trigonometry

We list some standard formulae in trigonometry which the reader may refer to.

$$\sin(\alpha + \beta) = \sin \alpha \cos \beta + \sin \beta \cos \alpha,$$
$$\sin(\alpha - \beta) = \sin \alpha \cos \beta - \sin \beta \cos \alpha,$$
$$\cos(\alpha + \beta) = \cos \alpha \cos \beta - \sin \alpha \sin \beta,$$
$$\cos(\alpha - \beta) = \cos \alpha \cos \beta + \sin \alpha \sin \beta,$$
$$\sin \alpha \cos \beta = \tfrac{1}{2}(\sin(\alpha + \beta) + \sin(\alpha - \beta)),$$
$$\cos \alpha \cos \beta = \tfrac{1}{2}(\cos(\alpha + \beta) + \cos(\alpha - \beta)),$$
$$\sin \alpha \sin \beta = \tfrac{1}{2}(-\cos(\alpha + \beta) + \cos(\alpha - \beta)),$$
$$\sin 2\alpha = 2 \sin \alpha \cos \alpha,$$
$$\cos 2\alpha = \cos^2 \alpha - \sin^2 \alpha$$
$$= 2\cos^2 \alpha - 1$$
$$= 1 - 2\sin^2 \alpha.$$

# 1

# Lines, circles, and conics

## 1.1 Lines

We first revise the standard equations of the line.

### 1.1.1 The intersection form

A line which is parallel to neither axis has an algebraic equation in intersection form

$$\frac{x}{a} + \frac{y}{b} = 1;$$

it meets the coordinate axes at $(a, 0)$ and $(0, b)$ respectively. A line which is parallel to one of the axes, but equal to neither, has an equation in intersection form either $\frac{x}{a} = 1$, if it is parallel to the $y$–axis, or $\frac{y}{b} = 1$, if it is parallel to the $x$–axis; the intersections with the appropriate axis are respectively $x = a$ or $y = b$. See Figure 1.1.

### 1.1.2 The 'tangent' form

A line which is not parallel to the $y$-axis has an algebraic equation in tangent form

$$y = mx + c$$

where $m = \tan \beta$ is the slope; $\beta$ is measured counter-clockwise from the initial half-line $\overrightarrow{Ox}$, $0 \le \beta < \pi$ and $\beta \neq \frac{\pi}{2}$. This line intersects the $y$-axis at $y = c$. Lines not covered by this general form have algebraic equation

$$x = a;$$

they are lines whose slope is infinite. See Figure 1.1.

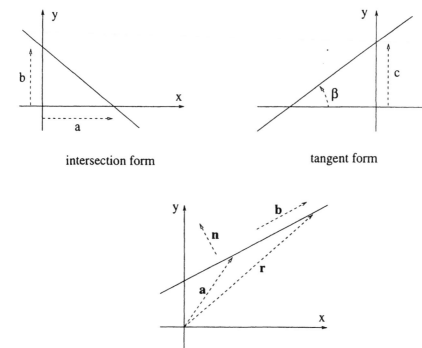

intersection form                                    tangent form

parametric form

Figure 1.1 *Equations of lines.*

### 1.1.3 The parametric and vector forms in $\mathbf{R}^2$ or $\mathbf{C}$

The parametric equation in vector form of a line in $\mathbf{R}^2$ is

$$\mathbf{r} = \mathbf{a} + t\mathbf{b} \qquad (\mathbf{b} \neq \mathbf{0})$$

for all real numbers $t$, where $\mathbf{a}$ is the position vector of a fixed point $P$ on the line and $\mathbf{b}$ is a non-zero free vector parallel to the line (e.g., $\mathbf{b}$ is determined by a directed line segment $\overrightarrow{PQ}$ ). See Figure 1.1. The parametric equations of the line in component form are

$$x = a_1 + tb_1,$$
$$y = a_2 + tb_2.$$

For all lines except ones of the form $x = c$ (a line parallel to the $y$-axis) we could use $x$ as a parameter; thus we should have

$$x = x, \ y = a_2 + \frac{b_2}{b_1}(x - a_1).$$

We can similarly consider the parametric form in the Argand diagram $\mathbf{C}$; in this case we replace $\mathbf{r}$ by $z = x + iy$, $\mathbf{a}$ by $a = a_1 + ia_2$ and $\mathbf{b}$ by $b = b_1 + ib_2 \neq 0$, and the parametric equation is

$$z = a + tb$$

for all real numbers $t$.

The non-parametric equation of the line in $\mathbf{R}^2$ in vector form is

$$(\mathbf{r} - \mathbf{a}) . \mathbf{n} = 0.$$

Here $\mathbf{n}$ is a non-zero vector which is orthogonal (perpendicular) to the line, and is therefore orthogonal to the vector $\mathbf{b}$; for example, we can take $\mathbf{n} = (n_1, n_2) = (-b_2, b_1)$. This non-parametric form can be written

$$(x - a_1, y - a_2) \cdot (n_1, n_2) = 0,$$

or

$$n_1 x + n_2 y = n_1 a_1 + n_2 a_2 = d.$$

This algebraic form of the equation can in turn generally be written in the intersection form or the tangent form.

Conversely given a line with algebraic equation

$$n_1 x + n_2 y = d,$$

it follows that the non-zero vector $\mathbf{n} = (n_1, n_2)$ is orthogonal to the line.

### 1.1.4 Polar equation

The polar equation of a general line is

$$r \cos(\theta - \alpha) = d.$$

Here $(r, \theta)$ are the polar coordinates of a general point on the line. In case the line does not pass through the origin, $(d, \alpha)$ are the polar coordinates of the point on the line which is closest to the origin. In case the line passes through the origin, $d = 0$ and $\alpha$ is the direction of a vector which is perpendicular to the line. In either case the distance of the line from the origin is $d$. The values which can be taken by $\theta$ are restricted by the equation; for $d \neq 0$, we cannot have $\cos(\theta - \alpha) = 0$, that is we cannot have $\theta = \alpha + \dfrac{\pi}{2}$ or $\theta = \alpha + \dfrac{3\pi}{2}$. In the case where $d = 0$, the equation of the line can alternatively be written

$$r = 0 \quad \text{or} \quad \theta = \alpha + \frac{\pi}{2} \quad \text{or} \quad \theta = \alpha + \frac{3\pi}{2}.$$

The equation $\theta = \beta$, for fixed $\beta$, gives a half-line and excludes the origin. See Figure 1.2.

From the polar equation $r \cos(\theta - \alpha) = d$, we obtain

$$r \cos \theta \cos \alpha + r \sin \theta \sin \alpha = d.$$

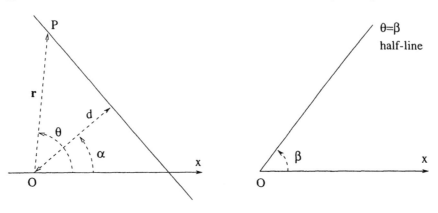

Line not through the origin                Half-line through the origin

Figure 1.2 *Polar equations of lines.*

On substituting $x = r \cos \theta$ and $y = r \sin \theta$, we obtain the algebraic equation

$$x \cos \alpha + y\theta \sin \alpha = d.$$

Conversely given an algebraic equation $ax + by = c$ of a line, we divide each side by $k = \sqrt{a^2 + b^2}$, to obtain $x \cos \alpha + y \cos \beta = d$, where $(k, \alpha)$ are the polar coordinates of the non-zero vector $(a, b)$. On substituting $x = r \cos \theta$ and $y = r \sin \theta$, we obtain the equation $r \cos(\theta - \alpha) = d$.

The reader should become adept at obtaining the parametric equation from the algebraic equation and at obtaining the algebraic equation from the parametric equation.

**Worked example 1.1.** Find the parametric equation of the line

$$2x + 3y = 3.$$

*Solution.* We want to find the general solution of the equation $2x + 3y = 3$ in terms of a parameter $t$. The general solution is $\mathbf{r} = \mathbf{a} + t\mathbf{b}$ where $\mathbf{a}$ is a particular solution and $\mathbf{b}$ is a non-zero vector parallel to the line. Thus $\mathbf{b}$ is a solution of the associated equation $2x + 3y = 0$. We choose $\mathbf{b} = (-3, 2)$; this can be considered as being given alternatively by rotating the normal vector $(2, 3)$ to the line counter-clockwise through an angle $\dfrac{\pi}{2}$ to obtain the vector $(-3, 2)$, which is parallel to the line. A particular solution is $\mathbf{a} = (0, 1)$. Therefore a parametric equation of the line is $\mathbf{r} = (0, 1) + t(-3, 2)$.    □

**Worked example 1.2.** Find the algebraic equation of the line

$$\mathbf{r} = (1, 3) + t(-1, 2).$$

*Solution.* We have $x = 1 - t$ and $y = 3 + 2t$, therefore

$$\frac{x-1}{-1} = t = \frac{y-3}{2},$$

and the algebraic equation, given by eliminating $t$, is $2(x-1)+(y-3) = 0$, that is $2x + y - 5 = 0$. □

**Worked example 1.3.** Find the algebraic equation of the line

$$\mathbf{r} = (3,1) + t(2,0).$$

*Solution.* We have $x = 3 + 2t$ and $y = 1$, therefore

$$\frac{x-3}{2} = t = \frac{y-1}{0},$$

and the algebraic equation is $y = 1$. □

## Exercises

**1.1.** Find parametric equations of the lines

a) $3x - 2y = 1$ $\qquad\qquad$ $[(1,1) + t(2,3)]$

b) $4x + y = 3$

c) $2x + y = 5$ $\qquad\qquad$ $[(2,1) + t(-1,2)]$

d) $2x + 5y = 1$

**1.2.** Find algebraic equations of the lines

a) $\mathbf{r} = (-1,3) + t(2,5)$ $\qquad$ $[5x - 2y + 11 = 0]$

b) $\mathbf{r} = (2,1) + t(-3,2)$

c) $\mathbf{r} = (1,1) + t(1,0)$ $\qquad\qquad\qquad$ $[y = 1]$

d) $\mathbf{r} = (5,7) + t(0,1)$

e) $z = 1 + it$

f) $z = (2 - i) + t(1 + 3i)$ $\qquad\qquad\qquad$ $[x = 1]$

## 1.2 The circle

Circles are special cases of conic sections. However unlike almost all other conic sections, they do not have two unique axes associated with them because of their lack of unique axes of symmetry. We assume that the circle is in general position with centre $(a, b)$ and radius $\rho > 0$.

### 1.2.1 Algebraic equation

The algebraic equation of the circle is

$$(x - a)^2 + (y - b)^2 = \rho^2.$$

For points $z = x + iy$ in the Argand diagram, the algebraic equation of the circle may be written

$$(z - (a + ib)) \cdot \overline{(z - (a + ib))} = \rho^2 \quad \text{or} \quad |z - (a + ib)| = \rho.$$

This follows since $|(x + iy) - (a + ib)|^2 = (x - a)^2 + (y - b)^2$.

### 1.2.2 Parametric equation

The standard parametric equation of the circle is

$$x = a + \rho\cos\varphi, \quad y = b + \rho\sin\varphi \quad (0 \le \varphi \le 2\pi).$$

The parameter $\varphi$ is not the polar angle $\theta$ of $(x, y)$, except in the case where the centre of the circle is at the origin.

The equation can be written in vector form as

$$\mathbf{r} = (a, b) + \rho(\cos\varphi, \sin\varphi) \quad (0 \le \varphi \le 2\pi).$$

In the Argand representation, the corresponding equation is

$$z = (a + ib) + \rho e^{i\varphi} \quad (0 \le \varphi \le 2\pi),$$

where $e^{i\varphi} = \cos\varphi + i\sin\varphi$.

### 1.2.3 Polar equation

The polar equation of a circle of radius $\rho > 0$ whose centre has polar coordinates $(d, \alpha)$ is

$$\rho^2 = r^2 + d^2 - 2rd\cos(\theta - \alpha),$$

where $(r, \theta)$ are the polar coordinates of a general point on the circle. The

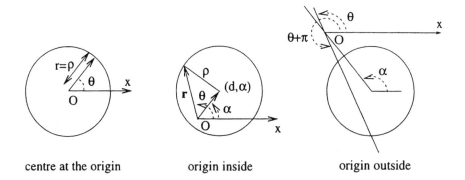

centre at the origin            origin inside            origin outside

Figure 1.3 *Polar equation of a circle.*

equation is obtained by applying the cosine formula to the triangle with vertices at the origin, the centre of the circle and the point $(r, \theta)$ on the circle (see Figure 1.3). If the pole is outside the circle, the equation imposes a restriction on the values which $\theta$ can take: in this case some lines through the pole do not intersect the circle. In the case where the pole is on the circle, we have $d = \rho$, and the equation reduces to

$$r = 2\rho\cos(\theta - \alpha).$$

In the case where the pole is at the origin, the equation is

$$r = \rho.$$

**Worked example 1.4.** Find the centre and radius of the circle

$$x^2 + y^2 - 2x + 4y + 3 = 0.$$

*Solution.* We collect all non-constant terms into the sum of two squares. We have $(x - 1)^2 + (y + 2)^2 = 1 + 4 - 3 = 2$. Therefore the centre is $(1, -2)$ and the radius is $\sqrt{2}$. □

**Exercises**

**1.3.** Find the centre and radius of each of the circles
  i) $x^2 + y^2 + 6x - 8y = 0$, and $\qquad\qquad\qquad$ [$(-3, 4), 5$]
  ii) $x^2 + y^2 - 4x + 6y = 3$.

## 1.3 Conics

Conics, or conic sections, are curves which are the intersection of a complete cone in $\mathbf{R}^3$ with a plane. In Figure 1.4 the axis of the cone lies in the plane of the page. The cone meets this plane in the two solid lines. The dotted lines indicate the intersection of this plane with three planes perpendicular to it. These three planes intersect the cone in an ellipse, a hyperbola, and a parabola as indicated. Other cases which can occur as a geometrical intersection of a plane and a cone are a pair of distinct intersecting lines, a single (iterated) line, or a single point; in each of these cases the plane passes through the 'point' of the cone.

The general quadratic equation is

$$ax^2 + 2hxy + by^2 + 2cx + 2dy + k = 0.$$

This is sometimes called the 'general equation of a conic', though not all such equations determine a geometrical conic. Ellipses, hyperbolae, parabolae, and pairs of distinct intersecting or parallel lines are all determined by such an equation as are certain degenerate cases. We can use a rotation of the plane which diagonalises the quadratic form $ax^2 + 2hxy + by^2$ to a quadratic form $\lambda X^2 + \mu Y^2$, then we can use a suitable translation (change of origin) to obtain the equation in one of the canonical forms; the proof that this can be done and the procedure for doing it are given in §2.1.

We first consider the canonical forms for the ellipse, the hyperbola, and

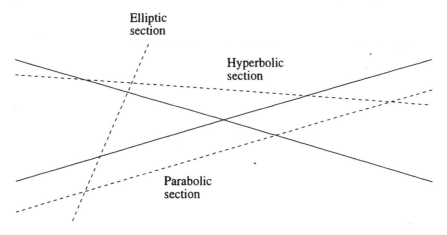

Figure 1.4 *Sections of a complete cone.*

the parabola. The word *canonical* here relates to the position of the co-ordinate axes for the given curve; in the case of the ellipse (other than a circle) and hyperbola these are the two (unique) axes of symmetry.

## 1.4 The ellipse in canonical position

In many situations the circle (see §2.1) can be regarded as a special case of an ellipse. However we consider in this section only ellipses which are not circles, that is we assume that $a > b$ (see below). The ellipse in canonical position is shown in Figure 1.5.

### 1.4.1 Algebraic equation

The algebraic equation of the ellipse in canonical position is

$$\frac{x^2}{a^2} + \frac{y^2}{b^2} = 1 \qquad (a > b > 0).$$

This curve meets the coordinate axes at $(\pm a, 0)$ and $(0, \pm b)$.

### 1.4.2 Parametric equation

The standard parametric equation of the ellipse in canonical position is

$$\mathbf{r} = (x, y) = (a \cos \varphi, b \sin \varphi) \qquad (0 \leq \varphi \leq 2\pi) \qquad (a > b > 0).$$

In this case $\varphi$ is not the polar angle $\theta$. It is the polar angle of the point $Q$ having coordinates $(x, \pm\sqrt{a^2 - x^2})$ on the circle $x^2 + y^2 = a^2$ (the auxiliary circle). The point $Q$ is vertically above/below the point $P$ having coordinates $(x, y)$ on the ellipse. A rational parametric equation can also be given as in §2.7.2.

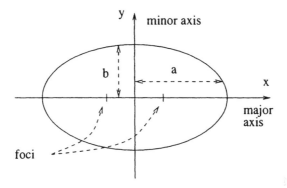

Figure 1.5 *The ellipse.*

### 1.4.3 Polar equation

Substituting $x = r \cos \theta$ and $y = r \sin \theta$ into the algebraic equation, gives the polar equation

$$\frac{1}{r^2} = \frac{\cos^2 \theta}{a^2} + \frac{\sin^2 \theta}{b^2} \qquad (0 \leq \theta \leq 2\pi) \qquad (a > b > 0).$$

In this equation the centre of the ellipse (see below) is the pole, and the positive $x$–axis is the initial half-line. The polar equation with a focus as the pole is given in §1.8.

### 1.4.4 Terminology

The centre of this ellipse is at the origin: in the case of general position, the centre of the ellipse is its centre of symmetry. There are two foci at $(\pm ae, 0)$, where

$$e = \sqrt{1 - \frac{b^2}{a^2}} \qquad (0 < e < 1)$$

is the eccentricity of the ellipse. Here $a$ and $b$ are selected by $b < a$. The singular of foci is focus. Notice that the distance of the foci from the centre is

$$ae = \sqrt{a^2 - b^2}.$$

The major axis is the chord or the line through the foci, and the minor axis is the chord or the line through the centre perpendicular to the major axis.

Ellipses in general position in space will have corresponding foci and major and minor axes. The major and minor semi-axes are the distances $a$ and $b$. The axes of an ellipse in canonical position coincide with the coordinate axes: for an ellipse in general position this will not normally be the case.

## 1.5 The hyperbola in canonical position

The hyperbola in canonical position is shown in Figure 1.6.

### 1.5.1 Algebraic equation

The algebraic equation of the hyperbola in canonical position is

$$\frac{x^2}{a^2} - \frac{y^2}{b^2} = 1 \qquad (a, b > 0).$$

This curve meets the coordinate axes at $(\pm a, 0)$. Compare this equation with the equation for the ellipse. For the ellipse, $a$ and $b$ are selected by $b < a$; for the hyperbola, $a$ and $b$ are selected by the relative signs of the terms in the equation.

### 1.5.2 Parametric equations

The hyperbola consists of two disjoint curves, so two parametrisations, one for each of these disjoint curves, are required. The parametric equations of the hyperbola in canonical position are

$$\mathbf{r} = (x, y) = (a \sec \varphi, b \tan \varphi) \qquad \left( -\frac{\pi}{2} < \varphi < \frac{\pi}{2} \right),$$

which gives the curve on the right, and

$$\mathbf{r} = (x, y) = (-a \sec \varphi, b \tan \varphi) \qquad \left( -\frac{\pi}{2} < \varphi < \frac{\pi}{2} \right),$$

which gives the curve on the left.

In this case $\varphi$ is not the polar angle $\theta$. A rational parametrisation can be obtained as in §2.7.2. An alternative parametrisation can be obtained by using hyperbolic functions: the two curves are given by

$$\mathbf{r} = (\pm a \cosh t, b \sinh t),$$

where $t$ varies over all real values.

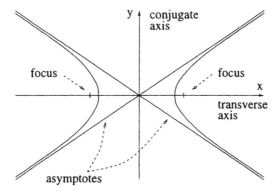

Figure 1.6 *The hyperbola.*

### 1.5.3 Polar equation

Substituting $x = r\cos\theta$ and $y = r\sin\theta$ into the algebraic equation gives the polar equation

$$\frac{1}{r^2} = \frac{\cos^2\theta}{a^2} - \frac{\sin^2\theta}{b^2} \qquad \left(|\tan\theta| < \frac{b}{a}\right).$$

In this equation the centre of the hyperbola (see below) is the pole. The polar equation with a focus as pole is given in §1.8. That the restriction on $\theta$ given here is necessary follows from

$$\frac{1}{r^2} = \frac{\cos^2\theta}{a^2} - \frac{\sin^2\theta}{b^2} = \frac{\cos^2\theta}{b^2}\left(\frac{b^2}{a^2} - \tan^2\theta\right).$$

Each side of this equation must be positive.

### 1.5.4 Terminology

The centre of this hyperbola is at the origin: in the general case the centre is the centre of symmetry. There are two foci at $(\pm ae, 0)$, where

$$e = \sqrt{1 + \frac{b^2}{a^2}} \qquad (e > 1)$$

is the eccentricity of the hyperbola. The reader should note that the formula for eccentricity is different to the corresponding one for the ellipse, and also that $a$ and $b$ are selected in different ways. Notice that the distance of the foci from the centre is

$$ae = \sqrt{a^2 + b^2}.$$

The transverse axis is the chord or the line through the foci, and the conjugate axis is the line through the centre perpendicular to the transverse axis.

The conjugate axis does not meet the hyperbola. The axes of a hyperbola in canonical position coincide with the coordinate axes: for a hyperbola in general position this will not normally be the case. Hyperbolae in general position in space will have corresponding foci, and transverse and conjugate axes. A rectangular hyperbola is a hyperbola whose asymptotes (see below) are orthogonal: the asymptotes are orthogonal if, and only if, $a = b$. The hyperbola differs from the ellipse in that it is an unbounded curve. Moreover it possess two asymptotes. Asymptotes are lines which, in a sense which will be made precise when we consider projective curves, are tangents to the curve 'at infinity'. Knowledge of asymptotes is of major importance in curve sketching and we will learn later how to determine them for more general curves.

### 1.5.5 Asymptotes

The asymptotes of a hyperbola in canonical position are

$$\frac{x}{a} = \pm\frac{y}{b}, \quad \text{or equivalently} \quad \frac{x^2}{a^2} - \frac{y^2}{b^2} = 0.$$

The axes of a hyperbola always bisect the asymptotes. A hyperbola is rectangular if, and only if, the asymptotes bisect the axes.

We now sketch an (intuitive) proof that the asymptotes to the hyperbola in canonical position are as given above. A proof using the techniques of projective geometry will be given later (see §16.10.1) following our precise definition of asymptotes.

*Proof.* The line $y = mx + c$ meets the hyperbola $\dfrac{x^2}{a^2} - \dfrac{y^2}{b^2} = 1$ where

$$\left(\frac{1}{a^2} - \frac{m^2}{b^2}\right)x^2 - \frac{2mc}{b^2}x - \frac{c^2}{b^2} = 1.$$

Both roots of the equation will be 'infinite' (and therefore 'equal') if the coefficients of $x^2$ and of $x$ are zero: this is the case if, and only if,

$$m = \pm\frac{b}{a}$$

and $c = 0$. The result follows easily.                                    □

For $m = \pm\dfrac{b}{a}$ and other values of $c$ the equation in the above proof has one infinite and one finite solution; and for other values of $m$ there are either two distinct real solutions, two equal solutions for $x$ (this case may correspond to a tangent or to two distinct points of intersection with different values for $y$), or no real solution. The reader should illustrate lines corresponding to each of these five cases.

## 1.6 The parabola in canonical position

The parabola in canonical position is shown in Figure 1.7. .

### 1.6.1 Algebraic equation

The algebraic equation of the parabola in canonical position is

$$y^2 = 4ax \qquad (a > 0).$$

This meets the coordinate axes at $(0,0)$.

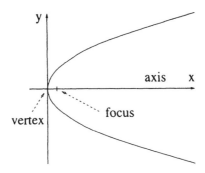

Figure 1.7 *The parabola.*

### 1.6.2 Polynomial parametric equation

The standard parametric equation of the parabola in canonical position is

$$\mathbf{r} = (x,y) = (at^2, 2at),$$

where $t$ varies over all real numbers. The coordinate functions are polynomials in $t$.

### 1.6.3 Polar equation

Substituting $x = r\cos\theta$ and $y = r\sin\theta$ into the algebraic equation gives the polar equation

$$r = 4a\cot\theta\,\mathrm{cosec}\,\theta \qquad \left(-\frac{\pi}{2} < \theta < \frac{\pi}{2}\right).$$

This gives all points on the parabola except $r = 0$. For this polar equation the vertex is the pole and the positive $x$–axis is the initial half-line. The polar equation with a focus as pole is given in §1.8.

### 1.6.4  Terminology

The parabola has no centre. There is (only) one focus at $(a, 0)$. The axis
of a parabola is its axis of symmetry; in this case of the canonical position
the axis is the $x$-axis. The vertex of a parabola is the point where it meets
its axis. The parabola is an unbounded curve but possesses no asymptote.
The eccentricity $e$ of a parabola is 1.

## 1.7  Classical geometric constructions of conics

A conic can be defined as the locus of a point $P$ for which the distance $PF$
to a fixed point $F$ (a focus) is in a fixed ratio $e$ (the eccentricity) to the
distance of $P$ from a fixed line $L$ (the directrix). See Figure 1.9. We have

$$
\begin{array}{l}
0 < e < 1 \text{ gives an ellipse,} \\
e > 1 \text{ gives a hyperbola, and} \\
e = 1 \text{ gives a parabola.}
\end{array}
$$

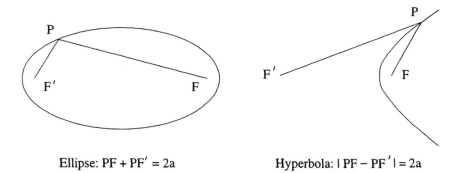

Ellipse: PF + PF$'$ = 2a                    Hyperbola: | PF − PF$'$| = 2a

Figure 1.8 *Classical geometric constructions of conics.*

An ellipse can be defined as the locus of points for which the sum of
the distances to two fixed points (the foci) is constant. A hyperbola can be
defined as the locus of points for which the absolute value of the difference
of the distances to two fixed points (the foci) is constant. See Figure 1.8

## Exercises

**1.4.** Find the equations of the directrix for

    i) the ellipse $\dfrac{x^2}{a^2} + \dfrac{y^2}{b^2} = 1$,                    $\left[ x = -\dfrac{a}{e} \right]$

    ii) the hyperbola $\dfrac{x^2}{a^2} - \dfrac{y^2}{b^2} = 1$, and

    iii) the parabola $y^2 = 4ax$.                    $[x = -a]$

## 1.8 Polar equation of a conic with a focus as pole

The polar equation of a conic with a focus as pole and initial line perpendicular to the latus rectum is

$$\frac{l}{r} = 1 - e\cos\theta \qquad \left(\text{or} \quad \frac{l}{r} = 1 + e\cos\theta\right). \qquad (1.1)$$

The latus rectum of a conic is the chord $LFL'$ through the focus $F$ which is bisected at the focus, or the line containing this chord. The semi-latus rectum $l$ is one half of the length of the chord $LFL'$. See Figure 1.9.

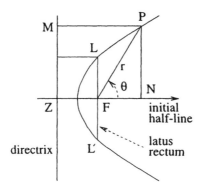

Figure 1.9 *Polar equation of a conic with a focus as pole.*

**Theorem 1.5.** *The semi-latus rectum of*

*the ellipse* $\dfrac{x^2}{a^2} + \dfrac{y^2}{b^2} = 1$ *(b < a) is* $l = \dfrac{b^2}{a}$,

*the hyperbola* $\dfrac{x^2}{a^2} - \dfrac{y^2}{b^2} = 1$ *is* $l = \dfrac{b^2}{a}$, *and*

*the parabola* $y^2 = 4ax$ *is* $l = 2a$.

*Proof\*.* For $e \neq 1$ we have successively

$$\frac{l}{r} = 1 - e\cos\theta,$$

$$l = r - er\cos\theta,$$

$$(l + ex)^2 = r^2 = x^2 + y^2,$$

$$x^2(1 - e^2) + y^2 - 2elx = l^2,$$

$$(1 - e^2)\left(x - \frac{el}{1 - e^2}\right)^2 + y^2 = \frac{l^2}{1 - e^2}, \text{ and}$$

$$(1 - e^2)X^2 + y^2 = \frac{l^2}{1 - e^2}, \quad \text{where} \quad X = x - \frac{el}{1 - e^2}.$$

In case $e < 1$, this gives

$$\frac{X^2}{a^2} + \frac{y^2}{b^2} = 1,$$

where

$$a^2 = \frac{l^2}{(1-e^2)^2} > \frac{l^2}{1-e^2} = b^2.$$

Therefore

$$\frac{b^2}{a^2} = 1 - e^2 \quad \text{and} \quad \frac{b^2}{a} = l.$$

In case $e > 1$, we have

$$\frac{X^2}{a^2} - \frac{y^2}{b^2} = 1,$$

where

$$a^2 = \frac{l^2}{(e^2-1)^2} \quad \text{and} \quad b^2 = \frac{l^2}{e^2-1}.$$

Therefore

$$\frac{b^2}{a^2} = e^2 - 1 \quad \text{and} \quad \frac{b^2}{a} = l.$$

In case $e = 1$, we have as above

$$y^2 = 2lx + l^2 = 2l\left(x + \frac{1}{2}l\right) = 2lX \quad \text{where} \quad X = x + \frac{1}{2}l,$$

which is the parabola $y^2 = 4aX$ with $l = 2a$.                                        □

The reader should note the different ways in which $b$ and $a$ are selected in these equations for the ellipse and the hyperbola. In order to calculate the semi-latus rectum for a conic in general position, we must calculate the values of $a$ and/or $b$ which would occur if the conic were moved to canonical position. We can choose either one of the polar equations in Equation 1.1 rather than the other: a change from one to the other corresponds to a rotation through an angle $\pi$ about the pole.

Values of $\theta$ which make the right hand side of the equation equal to zero correspond to 'points at infinity' of the curve: such points can only exist for $e \geq 1$, that is for parabolae and hyperbolae. In the case of the parabola there is one such value, namely $\theta = 0$. In the case of the hyperbola there are two such values, given by $\cos\theta = \dfrac{1}{e}$. These values are also the angles which the asymptotes make with the initial half-line. Also for the hyperbola, the polar equation gives negative values of $r$ in case $e\cos\theta > 1$; these correspond to points on the second component of the hyperbola.

In order to prove that a conic satisfies the above polar equation, we choose the initial half-line to begin at the focus as pole and to be perpendicular

---

* For the meaning of * see the preface.

to the directrix, and therefore also perpendicular to the latus rectum, as indicated in Figure 1.9. We thus have $e \cdot ZF = FL = l$, where $e$ is the eccentricity. For a general point $P$ on the conic we have

$$FP = e \cdot MP = e \cdot (ZF + FN) = l + er\cos\theta,$$

and the result follows.

**Worked example 1.6.** Determine the algebraic (polynomial) equation of the conic

$$\mathbf{r}(t) = (\cos\varphi + 1, 3\sin\varphi - 2).$$

Determine the type of conic, the position of the centre and the axes/vertex and the axis for a parabola, the eccentricity, the position of the foci/focus, and the points at which the conic intercepts its axes. Give a rough sketch indicating both the coordinate axes and the axes/axis of the conic, and all the above information which you have determined.

*Solution.* We have

$$(x - 1)^2 + \frac{(y + 2)^2}{9} = 1,$$

and therefore the conic is an ellipse with centre $(1, -2)$. Its axes, which are parallel to the coordinate axes, are $x = 1$ (major axis) and $y = -2$ (minor axis), and the eccentricity is $\sqrt{1 - \frac{1}{9}} = \frac{2\sqrt{2}}{3}$. The foci are at $(1, -2 \pm 2\sqrt{2})$, and the intercepts are at $(1, -2 \pm 3)$ and $(1 \pm 1, -2)$. The major axis is parallel to the $y$-axis, in this case. $\qquad\square$

## Exercises

**1.5.** Determine the algebraic (polynomial) equations for the conics which have the following parametrisations:

    i) $\mathbf{r}(t) = (\sin\varphi + \cos\varphi, \sin\varphi - \cos\varphi)$,
       [circle centre $(0,0)$, radius $\sqrt{2}$, coordinate axes $(\pm\sqrt{2}, 0)$, $(0, \pm\sqrt{2})$, eccentricity 0.]
    ii) $\mathbf{r}(t) = (2\cos\varphi, \sin\varphi - 1)$,
    iii) $\mathbf{r}(t) = (\sec\varphi + 1, 2\tan\varphi)$.
       [hyperbola, $(1,0)$, transverse $y = 0$, $x = 1$, $(0,0)$ and $(2,0)$, $2x - 2 = \pm y$, $\sqrt{5}$, $(-\pi/2, \pi/2)$, the right-hand branch]
    iv) $\mathbf{r}(t) = (1 - t, 1 + t^2)$.

In each case determine the type of conic, the position of the centre and the axes/vertex and axis, the asymptotes, the eccentricity, the position of the foci/focus, the points at which the conic intercepts its axes, the points at which the conic intercepts the coordinate axes, the points at which the conic intercepts the coordinate axes, and, for the hyperbola, its asymptotes. In each case give a rough sketch indicating the coordinate axes, the axes/axis of the conic, and all the above information which you have determined. Also write down the domain of the parameter and how much

of the conic is covered by the parametrisation. Indicate on your diagram
the direction of the given parametrisation and the initial and final points
(where appropriate).

(*Hint:* For those conics which are central, eliminate the parameter by using
trigonometric equations such as $\sin^2 \varphi + \cos^2 \varphi = 1$, and $\sec^2 \varphi = 1 + \tan^2 \varphi$.
After eliminating the parameter you should obtain an equation of the form

$$\frac{u^2}{a^2} \pm \frac{v^2}{b^2} = 1,$$

where $u$ and $v$ are functions of $x$ and $y$. The information required is immediate from this equation – do not write it in the form of the general
quadratic. The method for the parabola is somewhat similar.)

## 1.9 History and applications of conics

We now consider some history, occurrences in nature, and uses of conics.

### 1.9.1 History

Conics were studied by Greek mathematicians, with Mênaechmus (c. 350
BC) being credited with the description of conics as sections of a cone. Eu-
clid (c. 300 BC) wrote four books on conics, all now lost. A high point of
Greek mathematics was the publication by Apollonius of Perga (c. 265–170
BC) of eight books on conics in which he described all their simple proper-
ties using geometric techniques and regarding them as sections of a cone.
Pappus of Alexandria (300 AD) gave the focus-directrix property of conics.
A subsequent major advance was the algebraic description and study of
conics using Cartesian coordinates and polynomial equations of degree two
by Descartes (1596–1650), Fermat (1601–1665), and Wallis (1616–1703).
Many of the results in this chapter and the next were obtained by these
mathematicians. The arc-length of an ellipse (from $\varphi = 0$) defined a new
function

$$E(\varphi) = a \int_0^{\varphi} \sqrt{1 - e^2 \cos^2 \varphi}\, d\varphi,$$

which was called an elliptic integral.

### 1.9.2 The motion of celestial bodies

According to Newton's Law of Gravitation (Newton 1687), based on the
experimental results of Kepler (1609), large bodies are attracted to one
another by an inverse square law force

$$\kappa \frac{M_1 M_2}{d^2}$$

where $M_1$ and $M_2$ are their masses, $d$ is the distance between them and $\kappa$ is a constant. It can be deduced from this law that planets travel around the Sun in elliptical orbits, and that comets travel around the Sun in orbits which are ellipses or hyperbolae; theoretically they could also be parabolae. In each case the Sun is at a focus of the conic. This result neglects the normally small perturbations in the orbits caused by the attraction of (other) planets. The eccentricity of the elliptical orbit of a comet is large compared with that of a planet. The planets whose orbits have the greatest eccentricity are Mercury (0.2056234, about $\frac{1}{5}$) and Pluto (0.2495200, about $\frac{1}{4}$). The eccentricity of the orbit of the Earth is 0.0167322 (about $\frac{1}{60}$). The eccentricity of the orbit of Halley's comet is about 0.9675, and it passes close to the sun at intervals of approximately 76 years. A comet initially having an elliptical orbit which approaches a planet closely can be thrown by the planet's gravitational field into a hyperbolic orbit and thus escape from the solar system. The Moon and satellites similarly travel in an elliptical orbit around the Earth. The orbit of a communications satellite, which is positioned above a fixed spot on the Earth, will be almost circular, and therefore will have eccentricity almost zero.

### 1.9.3 Optical, radio wave, and sound reflectors

Rays of light arriving from stars and distant objects can be regarded as being parallel. Such rays travelling parallel to the axis of a parabolic mirror are all reflected to pass through the focus of the parabola. The same applies to radio and sound signals from distant objects, where in this case the reflector of the radio or sound signals is parabolic. A parabolic reflector in three-dimensions is given by rotating the parabola about its axis. Because of this optical property parabolic mirrors are normally used for astronomical and some terrestrial telescopes. They are also used for searchlights and automobile headlights. Parabolic reflectors are often used when transmitting and receiving radio or microwave signals between two fixed distant points, and for radar antennae. Parabolic sound reflectors can be used in espionage for listening in to private conversations some distance away. Where light, radio waves or sound is received and/or transmitted between two fixed points close together, it would be more accurate to use elliptical reflectors since rays from a focus of an ellipse are reflected by the ellipse to pass through the other focus; moreover, the wave emitted by the source at one focus at a given instant reaches the other focus after a fixed time which is independent of its path, since the length of any such path is equal to the length of the major axis. Astronomical reflecting telescopes sometimes have a secondary mirror which is elliptical (Gregory's design 1663) or hyperbolic (Cassegraine's design 1672). This can give a more comfortable or suitable viewing position for the observer.

### 1.9.4 Ballistics

It was shown by Galileo (1564–1642) that the trajectory of a projectile fired from a gun is parabolic. Similarly a bomb falling freely from an airplane travels along a parabola.

**Worked example 1.7.** A projectile is fired with velocity $\kappa$ and inclination $\alpha$. Find the range.

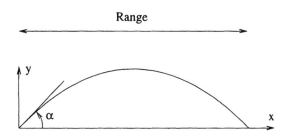

Figure 1.10 *Range of a projectile.*

*Solution.* The horizontal and vertical components of the velocity are initially

$$u = \kappa \cos \alpha \quad \text{and} \quad v = \kappa \sin \alpha.$$

Using the laws of motion, the position of the projectile at time $t$ (see Figure 1.10) is

$$x = (\kappa \cos \alpha)t$$
$$y = -\tfrac{1}{2}gt^2 + (\kappa \sin \alpha)t.$$

Eliminating $t$ we obtain

$$y - \frac{\kappa^2 \sin^2 \alpha}{2g} = -\frac{g}{2\kappa^2 \cos^2 \alpha}\left(x - \frac{\kappa^2 \sin(2\alpha)}{2g}\right)^2,$$

which is the equation of a parabola. Solving for $y = 0$, we obtain

$$\left(x - \frac{\kappa^2 \sin(2\alpha)}{2g}\right)^2 = \left(\frac{\kappa^2 \sin(2\alpha)}{2g}\right)^2,$$

and therefore

$$x = 0 \quad \text{or} \quad x = \frac{\kappa^2 \sin(2\alpha)}{g}.$$

The range of the projectile is therefore

$$R = \frac{\kappa^2 \sin(2\alpha)}{g}.$$

□

## 1.9.5 Global positioning

If two receiving stations receive a radio or sound signal from a ship with a time difference $\tau$ the distances $r$ and $r'$ of the stations from the ship will satisfy

$$r - r' = 2a,$$

where $2a$ is the distance the signal will travel in the time $\tau$. The set of points satisfying this equation lie on a well determined branch of the hyperbola whose foci are at the two stations (see Figure 1.8). If a third station also receives the signal, the ship's position can be determined as the point of intersection of specific branches of two appropriate hyperbolae. Similarly with three satellites, fixed in position above the Earth, emitting a signal simultaneously, a hand-held computer can determine its own position by considering the differences in the times at which the three signals are received. A display screen on the computer can then display a local map and the position of the observer on this map.

# 2

# Conics: general position

We now classify the curves given by a general polynomial equation of the second degree. We prove that such a general quadratic equation can be reduced to a canonical form of equation. At the same time we show how this canonical form can be obtained. The method of obtaining the canonical form involves rotations and translations of the coordinate axes, or equivalently of the plane, and enables us to determine whether the given quadratic equation is the equation of an ellipse, a hyperbola, or a parabola, and to find information about its axes, semi-axes, foci, asymptotes, etc. We give two methods for diagonalising the quadratic terms, namely the geometric method and the algebraic method. For those familiar with the diagonalisation of simple quadratic forms the algebraic method can give the results more rapidly and is recommended. However the examples and exercises considered in this chapter can also be solved using the geometric method. A resumé of the theory of matrices and determinants used in the algebraic method of diagonalising quadratic forms is included in §2.2 for reference.

## 2.1 Geometrical method for diagonalisation

The general quadratic equation is

$$ax^2 + 2hxy + by^2 + 2cx + 2dy + k = 0.$$

We assume that $a$, $h$, $b$ are not all zero. It is important to note that the symbols $a$ and $b$, traditionally used for the coefficients here, do not coincide, in meaning, with the same symbols used for the canonical forms in §1.4, §1.5 and §1.6. We give a simple geometrical method for diagonalising the quadratic form

$$ax^2 + 2hxy + by^2$$

in the general quadratic equation in order to determine the canonical form of the associated conic.

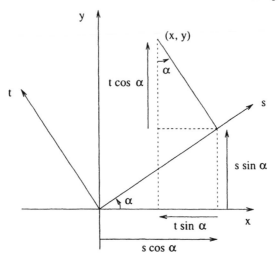

Figure 2.1 *Rotating the axes.*

**Definition 2.1.** A *diagonal* quadratic form is a form

$$ax^2 + by^2,$$

that is the coefficient of $xy$ is zero.

In order to diagonalise a quadratic form

$$ax^2 + 2hxy + by^2,$$

we rotate the axes from the $xy$–position through an angle $\alpha$ counterclockwise to the $st$–axes, so that after the rotation the quadratic form becomes a diagonal form

$$\lambda s^2 + \mu t^2.$$

We choose a new set of rectangular axes, an $s$–axis and a $t$–axis through the origin of the $xy$–plane. The $xy$–coordinates are obtained from the $st$–coordinates (see Figure 2.1) by

$$x = s \cos \alpha - t \sin \alpha \qquad\qquad (2.1)$$
$$y = s \sin \alpha + t \cos \alpha.$$

On solving these equations for $s$ and $t$, we obtain

$$s = x \cos \alpha + y \sin \alpha$$
$$t = x \sin \alpha - y \cos \alpha.$$

Upon substituting for $x$ and $y$, we have

$$ax^2 + 2hxy + vy^2 = a's^2 + 2h'st + b't^2.$$

First assume that $h \neq 0$, that is, the quadratic form is not already diagonal. The coefficient of the $st$ term on the right-hand side is

$$2h' = -2a \sin \alpha \cos \alpha + 2h(\cos^2 \alpha - \sin^2 \alpha) + 2b \sin \alpha \cos \alpha$$

$$= 2h \cos 2\alpha + (b - a) \sin 2\alpha$$

$$= 2h \left( \cos 2\alpha + \frac{b - a}{2h} \sin 2\alpha \right)$$

Thus $h'$ is zero if, and only if,

$$\boxed{\cot 2\alpha = \frac{a - b}{2h}.}$$

We have obtained this equation from the previous one by dividing by $h \sin(2\alpha)$. Note that, since we are assuming $h \neq 0$, $\sin(2\alpha) = 0$ would give $h' = \pm h \neq 0$. There is precisely one angle $\alpha$ satisfying this equation and $0 < 2\alpha < \pi$, and therefore satisfying

$$0 < \alpha < \frac{\pi}{2}.$$

Allowing now the possibility that $h = 0$ already, that is, the original form was diagonal, we have proved the following theorem.

**Theorem 2.2.** *There is precisely one angle $\alpha$ of rotation in the range*

$$0 \leq \alpha < \frac{\pi}{2}$$

*which diagonalises the quadratic form $ax^2 + 2hxy + by^2$.*

For this choice of $\alpha$, after making the substitution given by Equations 2.1 into the general equation,

$$ax^2 + 2hxy + by^2 + 2cx + 2dy + k = 0,$$

we obtain the equation

$$\lambda s^2 + \mu t^2 + 2es + 2fv + g = 0,$$

referred to the $st$–axes.

We must next move the coordinate origin to the centre (in case of an ellipse or a hyperbola) or the vertex (in the case of a parabola), and make a further rotation through an angle $\frac{\pi}{2}$, $\pi$, or $\frac{3\pi}{2}$, if necessary, in order to obtain the canonical form of equation. We show how this can be done in §2.4.

## 2.2 Algebra*

We give for reference a summary of the notation, results, and methods needed for diagonalising quadratic forms.

---

* For the meaning of * see the preface.

### 2.2.1 Vectors

A *column vector* or *column matrix* $X$ is an $n \times 1$ array, that is it has $n$ rows and 1 column. The number $n$ will depend on the context. In the case where $n = 2$,

$$X = \begin{pmatrix} x_1 \\ x_2 \end{pmatrix},$$

and in the case where $n = 3$,

$$X = \begin{pmatrix} x_1 \\ x_2 \\ x_3 \end{pmatrix}.$$

In these cases we often replace $x_1$, $x_2$, and $x_3$ by $x$, $y$ and $z$ in the applications. The transpose $X^T$ of a column vector $X$ is the row vector

$$X^T = (x_1, \dots, x_n) = \mathbf{x}.$$

The *inner product* or *scalar product* of two vectors $X$ and $Y$ is given by matrix multiplication, that is

$$X^T Y = \sum_{i=1}^{n} x_i y_i.$$

The *length* of the vector $X$ is $\| X \| = |\mathbf{x}|$ where

$$\| X \|^2 = X^T X = \sum_{i=1}^{n} x_i^2.$$

### 2.2.2 Matrices

An $n \times n$ *matrix* $A$ is an $n \times n$ array $A = (a_{ij})$, which has $n$–rows and $n$–columns. The element in the $i$-th row and $j$-th column is $a_{ij}$. We are especially interested in the case where $n = 2$, in which case

$$A = \begin{pmatrix} a_{11} & a_{12} \\ a_{21} & a_{22} \end{pmatrix},$$

and the case where $n = 3$, in which case

$$A = \begin{pmatrix} a_{11} & a_{12} & a_{13} \\ a_{21} & a_{22} & a_{23} \\ a_{31} & a_{32} & a_{33} \end{pmatrix}.$$

The *identity matrix* is the matrix $I$ all of whose entries are zero except that, for each $i$, the entry lying in the $i$-th row and $i$-th column is equal to 1. In the case where $n = 2$,

$$I = \begin{pmatrix} 1 & 0 \\ 0 & 1 \end{pmatrix},$$

and in the case where $n = 3$,

$$I = \begin{pmatrix} 1 & 0 & 0 \\ 0 & 1 & 0 \\ 0 & 0 & 1 \end{pmatrix}.$$

The *product* of two $n \times n$ matrices $A$ and $B$, is the matrix $C = AB$ where

$$c_{ij} = \sum_{k=1}^{n} a_{ik} b_{kj}.$$

Thus the element $c_{ij}$ is given by multiplying componentwise the $i$-th row of $A$ by the $j$-th column of $B$. For example

$$\begin{pmatrix} 1 & 2 \\ -3 & 1 \end{pmatrix} \begin{pmatrix} 2 & -1 \\ 1 & 5 \end{pmatrix} = \begin{pmatrix} 4 & 9 \\ -5 & 8 \end{pmatrix}.$$

In general $AB \neq BA$. However $AI = IA = A$ for each $A$.

The *transpose* of a matrix $A = (a_{ij})$ is $A^T = (b_{ij})$, where $b_{ij} = a_{ji}$. The transpose $A^T$ is given by interchanging the rows and columns of $A$, thus the $j$-th column of $A^T$ is the $j$-th row of $A$. For example

$$\begin{pmatrix} 1 & 2 & 3 \\ 4 & 5 & 6 \\ 7 & 8 & 9 \end{pmatrix}^T = \begin{pmatrix} 1 & 4 & 7 \\ 2 & 5 & 8 \\ 3 & 6 & 9 \end{pmatrix}.$$

In general the order of a product of matrices is reversed on taking the transpose, that is

$$(AB)^T = B^T A^T.$$

A *symmetric matrix* is a matrix $A$ for which $A^T = A$. For example

$$\begin{pmatrix} 1 & 2 & 3 \\ 2 & 4 & 5 \\ 3 & 5 & 6 \end{pmatrix}$$

is symmetric.

The *determinant* of a matrix $A$ is $|A| = |a_{ij}|$. Thus

$$|A| = \begin{vmatrix} a_{11} & a_{12} \\ a_{21} & a_{22} \end{vmatrix} = a_{11}a_{22} - a_{21}a_{12}$$

in the case $n = 2$ for example. Also $|I| = 1$ for each $n$, and $|AB| = |A||B|$. A *non-singular matrix* is a matrix $A$ for which $|A| \neq 0$.

A matrix $A$ is *invertible* if there is a matrix $B$ such that $AB = I$ or equivalently $BA = I$. The *inverse* of $A$ is the (unique) matrix $B$ which satisfies these equations. The inverse of $A$ is denoted $A^{-1}$, and satisfies

$$AA^{-1} = A^{-1}A = I.$$

The inverse exists if, and only if, $|A| \neq 0$. The inverse of a $2 \times 2$ matrix $A$

is

$$A^{-1} = \frac{1}{|A|} \begin{pmatrix} a_{22} & -a_{12} \\ -a_{21} & a_{11} \end{pmatrix}.$$

In general the order of a product of two invertible matrices is reversed on taking the inverse, that is

$$(AB)^{-1} = B^{-1}A^{-1}.$$

An *orthogonal matrix* is a (real) matrix $U$ for which $U^T U = I$: this is equivalent to $UU^T = I$, and is equivalent to $U^{-1} = U^T$.

**Mnemonic**

> The inverse of an orthogonal matrix $U$ is $U^T$.

For an orthogonal matrix $U$, we have $|U| = \pm 1$. A *rotation matrix* is an orthogonal matrix $U$ for which $|U| = +1$.

A *diagonal matrix* is a matrix $\Lambda = \mathrm{dg}(\lambda_1, \ldots, \lambda_n)$ all of whose entries are zero except that, for each $i$ the entry lying in the $i$-th row and $i$-th column is equal to $\lambda_i$. In case $n = 2$, a diagonal matrix has the form

$$\Lambda = \begin{pmatrix} \lambda_1 & 0 \\ 0 & \lambda_2 \end{pmatrix},$$

and in case $n = 3$, a diagonal matrix has the form

$$\Lambda = \begin{pmatrix} \lambda_1 & 0 & 0 \\ 0 & \lambda_2 & 0 \\ 0 & 0 & \lambda_3 \end{pmatrix}.$$

The *characteristic polynomial* of an $n \times n$ matrix $A$ is the polynomial

$$|\lambda I - A| = \lambda^n + \cdots + (-1)^n |A|$$
$$= (\lambda - \lambda_1)(\lambda - \lambda_2) \ldots (\lambda - \lambda_n),$$

and the *characteristic equation* is $|\lambda I - A| = 0$. An *eigenvalue* of $A$ is a solution $\lambda_i$ of the characteristic equation. In general the solutions of a real polynomial equation are either real or occur in conjugate pairs.

**Mnemonic**

> The eigenvalues of a real symmetric matrix are all real.

An *eigenvector* for the eigenvalue $\lambda_i$ is a non-zero column vector $X$ which satisfies

$$(\lambda_i I - A)X = 0.$$

Since $|\lambda_i I - A| = 0$, such a non-zero vector exists.

**Theorem 2.3.** *Let $A$ be a real symmetric matrix, then there is a real orthogonal matrix $U$ such that*

$$U^T A U = U^{-1} A U = \Lambda = \mathrm{dg}(\lambda_1, \ldots, \lambda_n)$$

is a diagonal matrix where the $\lambda_i$ are the eigenvalues of $A$. The $i$-th column $U_i$ of $U$ is a unit eigenvector for the eigenvalue $\lambda_i$.

Conversely let $U$ be a real orthogonal matrix whose $i$-th column $U_i$ is a unit eigenvector for the eigenvalue $\lambda_i$, then $U^T A U = dg(\lambda_1, \ldots, \lambda_n)$.

We demonstrate how to construct $U$ in the case $n = 2$.

1. Find the eigenvalues $\lambda_1$ and $\lambda_2$. If these are equal, $A$ is already the diagonal matrix $dg(\lambda_1, \lambda_2)$.

2. Given that the eigenvalues are distinct, find, for each eigenvalue $\lambda_i$, a unit eigenvector $X_i$, that is a unit vector which is a solution of

$$(\lambda_i I - A)X = 0.$$

A non-zero eigenvector always exists: $X_i$ is given by dividing any non-zero eigenvector by its length. There are two possible unit eigenvectors $X_i$ and $-X_i$. Define

$$U = (X_1, X_2),$$

the matrix whose first column is $X_1$ and whose second column is $X_2$. Then $U$ is an orthogonal matrix and $U^T A U = dg(\lambda_1, \lambda_2)$. Notice that the eigenvector $X_1$ which is the first column of $U$ corresponds to the eigenvalue which is the first entry of the matrix $dg(\lambda_1, \lambda_2)$. The matrix $U$ can be chosen to be a rotation matrix; if $|U| = -1$ for the unit eigenvectors chosen, then multiplying the first column of $U$ by $-1$ will give a new matrix for which $|U| = +1$.

**Remark* 2.4.** For $n > 2$, the orthogonal matrix $U$ is constructed in the same way if all the eigenvalues are distinct. Where not all the eigenvalues of a real symmetric matrix are distinct, it is necessary, for each distinct eigenvalue $\mu$, to find an orthonormal basis for the space of vectors satisfying

$$(\mu I - A)X = 0.$$

The matrix $U$ has as its columns the vectors in these orthonormal bases. Further details in this case $n > 2$ can be found in books on linear algebra.

## 2.3 Algebraic method for diagonalisation

The general quadratic equation is

$$ax^2 + 2hxy + by^2 + 2cx + 2dy + k = 0.$$

We assume that $a$, $h$, $b$ are not all zero. We now give an algebraic method for diagonalising the quadratic form.

We use a rotation, with real orthogonal matrix $P$ of determinant $+1$, to

---

* For the meaning of * see the preface.

diagonalise the quadratic form

$$ax^2 + 2hxy + by^2 = \begin{pmatrix} x & y \end{pmatrix} A \begin{pmatrix} x \\ y \end{pmatrix}.$$

Thus we make the substitution

$$\begin{pmatrix} x \\ y \end{pmatrix} = P \begin{pmatrix} s \\ t \end{pmatrix}$$

to obtain

$$\begin{pmatrix} x & y \end{pmatrix} A \begin{pmatrix} x \\ y \end{pmatrix} = \begin{pmatrix} s & t \end{pmatrix} P^T A P \begin{pmatrix} s \\ t \end{pmatrix}$$

$$= \begin{pmatrix} s & t \end{pmatrix} \begin{pmatrix} \lambda & 0 \\ 0 & \mu \end{pmatrix} \begin{pmatrix} s \\ t \end{pmatrix}$$

$$= \lambda s^2 + \mu t^2.$$

Here $\lambda$ and $\mu$ are the eigenvalues of the real symmetric matrix

$$A = \begin{pmatrix} a & h \\ h & b \end{pmatrix} \qquad (A \neq 0).$$

The first and second columns of $P$ are mutually orthogonal unit eigenvectors corresponding to $\lambda$ and $\mu$ respectively. These eigenvectors are chosen so that the determinant of $P$ is $+1$. This can always done by multiplying one of the eigenvectors by $-1$ if necessary.

Making this same substitution into the general quadratic equation, we obtain the equation

$$\lambda s^2 + \mu t^2 + 2es + 2ft + g = 0,$$

in which the coefficient of $st$ is zero.

We give some worked examples illustrating the use of this method in §2.6.

## 2.4 Translating to canonical form

We now show how a general quadratic equation

$$\lambda s^2 + \mu t^2 + 2es + 2ft + g = 0,$$

where the quadratic terms are in diagonal form, can be reduced to a canonical form of the equation of the conic.

There are two cases to consider.

Suppose first that both $\lambda$ and $\mu$ (the eigenvalues) are non-zero. Then the equation can be written

$$\lambda \left( s + \frac{e}{\lambda} \right)^2 + \mu \left( t + \frac{f}{\mu} \right)^2 + k' = 0,$$

and using the translation (shift of origin)

$$u = s + \frac{e}{\lambda}, \quad v = t + \frac{f}{\mu},$$

we obtain the canonical form

$$\lambda u^2 + \mu v^2 + k' = 0. \tag{2.2}$$

The second case is where either $\lambda$ or $\mu$ is zero. By interchanging the coordinates, if necessary, we can assume that $\lambda \neq 0$. A counter-clockwise rotation through $\frac{\pi}{2}$ interchanges the coordinates. Such a rotation is given by the matrix

$$\begin{pmatrix} 0 & -1 \\ 1 & 0 \end{pmatrix}.$$

Then the equation can be written

$$\lambda \left(s + \frac{e}{\lambda}\right)^2 = -(2ft + k'),$$

and a shift of origin gives

$$\lambda u^2 = -2fv \quad (f \neq 0) \quad \text{where} \quad u = s + \frac{e}{\lambda}, \quad v = t + \frac{k'}{2f}, \tag{2.3}$$

or

$$\lambda u^2 = -k' \quad (f = 0) \quad \text{where} \quad u = s + \frac{e}{\lambda}, \quad v = t.$$

It now follows, in each of the cases given by Equation 2.2 and Equations 2.3, that, after a further rotation if necessary through $\pi$ given by the matrix

$$\begin{pmatrix} -1 & 0 \\ 0 & -1 \end{pmatrix}$$

or through $\pm\frac{\pi}{2}$, the resulting equation can be written in one of the following forms:

i) an ellipse

$$\frac{u^2}{p^2} + \frac{v^2}{q^2} = 1 \quad (p > q > 0),$$

ii) a hyperbola,

$$\frac{u^2}{p^2} - \frac{v^2}{q^2} = 1 \quad (p, q > 0),$$

iii) a parabola

$$v^2 = 4pu \quad (p > 0),$$

iv) a pair of distinct non-parallel lines

$$p^2 u^2 - q^2 v^2 = 0 \quad (p, q > 0),$$

v) a pair of distinct parallel lines

$$u^2 = p^2 \quad (p > 0),$$

vi) an iterated line (two copies of the line $u = 0$)

$$u^2 = 0, \text{ or}$$

vii) some other equation having no real solution or no interesting solution, such as

$$u^2 + v^2 = -1, \text{ or } u^2 = -1, \ u^2 + v^2 = 0.$$

In particular the equations of the ellipse, hyperbola, and parabola are now reduced to canonical form. Notice that case v) does not occur as a geometrical section of a cone.

Given that the equation

$$ax^2 + 2hxy + by^2 + 2cx + 2dy + k = 0$$

represents a conic, we can obtain some information about the conic by studying the quadratic form $ax^2 + 2hxy + by^2$. The eigenvalues of the matrix $A$, that is the numbers $\lambda$ and $\mu$ in a diagonal form for $ax^2 + 2hxy + by^2$, are unchanged by a rotation from the $xy$–axes, and therefore we have the following criterion.

**Criterion 2.5.** *Let the equation*

$$ax^2 + 2hxy + by^2 + 2cx + 2dy + k = 0$$

*determine a conic. Then we have the following cases.*

*i) $h^2 > ab$ (equivalently $\lambda$ and $\mu$ are non-zero and have opposite signs). In this case the conic is a hyperbola if $k' \neq 0$, and a pair of distinct intersecting lines if $k' = 0$.*

*ii) $h^2 = ab$ (equivalently $\lambda \neq 0$ and $\mu = 0$ for example). In this case the conic is a parabola if $f \neq 0$, and an iterated line or a pair of parallel lines if $f = 0$.*

*iii) $h^2 < ab$ (equivalently $\lambda$ and $\mu$ are non-zero and have the same sign). In this case the conic is an ellipse if $k'$ is not zero and has the opposite sign to $\lambda$ and $\mu$, or consists of one point if $k' = 0$.*

## 2.5 Central conics referred to their centre

A *central* conic is a conic with a *centre* (of symmetry). A centre is a point such that all chords through the centre are bisected at the centre. Thus a general conic with equation

$$ax^2 + 2hxy + by^2 + 2cx + 2dy + k = 0,$$

has a centre at the origin if $(-x, -y)$ lies on the conic whenever $(x, y)$ lies on the conic: this happens if, and only if, $c = d = 0$.

The parabola is not central, but ellipses, hyperbolae, and pairs of distinct intersecting lines have unique centres. A pair of parallel or iterated lines will have a line of centres.

### 2.5.1  To find the centre of a central conic

Given that a conic has a centre, we prove below that such a centre is a solution of the pair of linear equations

$$ax + hy + c = 0,$$
$$hx + by + d = 0. \tag{2.4}$$

As a mnemonic for these equations notice that the left-hand sides are, up to a constant multiple, the partial derivatives of the function

$$\varphi(x,y) = ax^2 + 2hxy + by^2 + 2cx + 2dy + k$$

used to define the conic.

**Mnemonic**

> The centre of a central conic $\varphi(x,y) = 0$ is given by
> $$\frac{\partial \varphi}{\partial x} = \frac{\partial \varphi}{\partial y} = 0.$$

The following cases arise.

i) The Equations 2.4 have a unique solution: in this case the conic has an unique centre, or $\varphi(x,y) = 0$ has a unique solution or has no solution (e.g., $x^2 + y^2 = 0$ or $x^2 + y^2 + 1 = 0$ respectively).

ii) The Equations 2.4 have many solutions: in this case we have a pair of parallel or iterated lines, or $\varphi(x,y) = 0$ has no solution (e.g., $\varphi(x,y) = x^2 + 1 = 0$).

iii) The Equations 2.4 have no solution: in this case the conic is a parabola.

**Mnemonic**

> Unique solution – central conic.
> Many solutions – parallel lines.
> No solution – parabola.

*Proof that a centre satisfies the Equations 2.4.* The equation of the general conic is

$$ax^2 + 2hxy + by^2 + 2cx + 2dy + k = 0.$$

Given any point $(\alpha, \beta)$, we can rewrite this as

$$a(x - \alpha)^2 + 2h(x - \alpha)(y - \beta) + b(y - \beta)^2 +$$
$$(2a\alpha + 2h\beta + 2c)x + (2h\alpha + 2b\beta + 2d)y + k' = 0, \tag{2.5}$$

where $k'$ is a constant. On using the substitution $x = s + \alpha$ and $y = t + \beta$, corresponding to the translation (or change of origin), the equation takes the form

$$as^2 + 2hst + bt^2 + k' = 0$$

if, and only if,

$$2a\alpha + 2h\beta + 2c = 0 \quad \text{and} \quad 2h\alpha + 2b\beta + 2d = 0,$$

that is if, and only if, $(\alpha, \beta)$ satisfies the Equations 2.4. The condition that $(\alpha, \beta)$ is the centre is that the Equation 2.5 is unchanged when $(s, t)$ is replaced by $(-s, -t)$; and this is true if, and only if, the coefficients of $x$ and $y$ in Equation 2.5 are zero. The result follows easily.  □

After using the translation to the centre to obtain

$$as^2 + 2hst + bt^2 + k' = 0,$$

we can now perform a rotation to obtain one of the canonical forms. This gives an alternative method to that of §2.1 or §2.3 and §2.4 for determining the canonical form of the equation of a general central conic.

The method which we used in §2.1 or §2.3 and §2.4 was to perform the rotation first, followed by a translation. In the method described here we perform a translation first, followed by a rotation.

### 2.5.2 The axes of a central conic

We now describe how the axes of a central conic can be identified where the algebraic method is used.

The axes of a central conic are lines through the centre of the conic which are parallel to the (mutually perpendicular) eigenvectors of the matrix $A$. In the case of a circle the eigenvalues are equal and there are no (unique) axes. In the case of a hyperbola the axes are also the lines through the centre which bisect the angles between the asymptotes. This can readily be seen by considering the canonical form of the equation of the hyperbola. To distinguish the major and minor axes of an ellipse, we note that the major axis is parallel to the eigenvector corresponding to the eigenvalue whose absolute value is least. To distinguish the transverse and conjugate axes of a hyperbola, we note similarly that the transverse axis corresponds to the positive eigenvalue in the case where $k' < 0$ and to the negative eigenvalue in the case where $k' > 0$.

In practice, to determine the axes of a specific conic, we find its centre and then write down the parametric equations of the lines through its centre which are parallel to the eigenvectors.

### Mnemonic

| |
|---|
| Ellipse: major axis corresponds to smallest absolute eigenvalue. |
| Hyperbola: tranverse axis corresponds to positive eigenvalue for $k' < 0$. |

### 2.5.3 The asymptotes of a hyperbola

The asymptotes of a hyperbola are lines through the centre of the hyperbola which are parallel to the lines

$$ax^2 + 2hxy + by^2 = 0.$$

The asymptotes are not orthogonal in general. A rectangular hyperbola is a hyperbola whose asymptotes are orthogonal to one another.

In practice, to determine the asymptotes of a specific hyperbola, we find its centre and then write down the parametric equations of the lines through its centre which are parallel to the lines $ax^2 + 2hxy + by^2 = 0$.

### 2.5.4 The discriminant*

In the general case of a central conic, the constant $k'$ and specific features of the conic can be determined using the discriminant.

**Definition 2.6.** The *discriminant* of the quadratic polynomial

$$ax^2 + 2hxy + by^2 + 2cx + 2dy + k$$

is

$$\Delta = \begin{vmatrix} a & h & c \\ h & b & d \\ c & d & k \end{vmatrix} = abk + 2cdh - ac^2 - bd^2 - kh^2.$$

When the centre $(\alpha, \beta)$ of a central conic

$$ax^2 + 2hxy + by^2 + 2cx + 2dy + k = 0$$

is moved to the origin, we have shown that the conic becomes

$$as^2 + 2hst + bt^2 + k' = 0,$$

where $k'$ is determined by the fact that $(\alpha, \beta)$ satisfies Equations 2.4. By elementary algebra Equations 2.4 have a unique solution if, and only if, $ab - h^2 \neq 0$.

**Theorem 2.7.**

$$k' = \frac{\Delta}{ab - h^2}.$$

*Proof.* Multiplying the equations

$$a\alpha + h\beta + c = 0 \quad \text{and}$$
$$h\alpha + b\beta + d = 0$$

by $\alpha$ and $\beta$ respectively and subtracting each of them from the equation

$$k' = a\alpha^2 + 2h\alpha\beta + b\beta^2 + 2c\alpha + 2d\beta + k,$$

---

* For the meaning of * see the preface.

given by Equations 2.5, we obtain

$$k' = c\alpha + d\beta + k.$$

Since $(\alpha, \beta)$ is the centre of the conic, the following system of homogeneous linear equations have the non-trivial solution $(\alpha, \beta, 1)$.

$$a\alpha + h\beta + c\gamma = 0$$
$$h\alpha + b\beta + d\gamma = 0$$
$$c\alpha + d\beta + (k - k')\gamma = 0.$$

Therefore we have

$$0 = \begin{vmatrix} a & h & c \\ h & b & d \\ c & d & k - k' \end{vmatrix} = \begin{vmatrix} a & h & c \\ h & b & d \\ c & d & k \end{vmatrix} - k'(ab - h^2) = \Delta - k'(ab - h^2),$$

and the result follows.                                                                       □

In practice, when determining the canonical form of a specific quadratic equation, we do not use Theorem 2.7, but use the method of §2.5.1.

*The condition that a conic is a pair of non-parallel lines*

From Theorem 2.7, the condition for this is that $k' = 0$ and that

$$as^2 + 2hst + bt^2$$

is a product of distinct real linear factors. Thus the required condition is that

$$\Delta = 0 \quad \text{and} \quad h^2 > ab.$$

*To find the asymptotes of a hyperbola*

Suppose that

$$ax^2 + 2hxy + by^2 + 2cx + 2dy + k = 0$$

is the equation of a hyperbola. A necessary and sufficient condition for this is that $A$ has one strictly positive and one strictly negative eigenvalue or equivalently that $h^2 > ab$ (see Criterion 2.5), and that $k' \neq 0$.

**Theorem 2.8.** *The algebraic equation of the asymptotes of a hyperbola*

$$ax^2 + 2hxy + by^2 + 2cx + 2dy + k = 0$$

*is*

$$ax^2 + 2hxy + by^2 + 2cx + 2dy + k - k' = 0,$$

*where*

$$k' = \frac{\Delta}{ab - h^2}.$$

*Proof.* The equation of the asymptotes of a hyperbola

$$as^2 + 2hst + bt^2 + k' = 0$$

is given by subtracting $k'$ from the left-hand side (see §1.5.5). Therefore the equation of the asymptotes, of a hyperbola

$$ax^2 + 2hxy + by^2 + 2cx + 2dy + k = 0,$$

is also given by subtracting $k'$ from the left-hand side. □

Notice that the asymptotes are parallel to the pair of lines

$$ax^2 + 2hxy + by^2 = 0$$

through the origin.

## 2.6 Practical procedures for dealing with the general conic

Diagonalising the quadratic terms of the general quadratic equation first by a rotation, as in §2.1 or §2.3, can involve introducing surds at an early stage.

Unless the conic is known to be a parabola, or there is some reason to diagonalise the quadratic form first, the following procedure is recommended.

i) Find the general solution of Equations 2.4 to determine a centre[†].

ii) In case the Equations 2.4 have a solution, the conic is central and the solution is a centre. To find the canonical form of the equation of the conic in this case, use the method given in §2.5.1, that is translate the centre of the conic to the origin and then rotate to obtain the canonical diagonal form.

iii) If the Equations 2.4 have no solution, the conic is a parabola. To find the canonical form of the equation of the conic in this case, use the method given in §2.1 or §2.3 and §2.4, that is first rotate to obtain a suitable diagonal form for the quadratic terms, then translate the vertex to the origin to obtain the canonical form. Since one of the $\lambda$ or $\mu$ is zero, surds may not be involved in the formula for the rotation.

iv) In the case of a hyperbola, if the asymptotes only are required, they can be found as in §2.5.3 above. The axes can then be obtained by bisecting the asymptotes. This gives a method for finding the axes of a hyperbola in cases where the eigenvectors do not need to be determined.

v) In the case of a central conic, if the axes only are required, they can be found by determining the lines through the centre given by rotating the $xy$–axes directions through the angle $\alpha$ determined in §2.1. Equivalently they are parallel to two mutually orthogonal eigenvectors of the matrix $A$.

---

[†] In case the conic is a pair of parallel lines or an iterated line, the centres of the conic will form a line. An ellipse, a hyperbola, a parabola, and a pair of lines intersecting only at one point have a unique centre.

vi) In the case of a central conic, if only the semi-axes (see §1.4 and §1.5) are required, they can be determined from the eigenvalues of the matrix $A$ and the number $k'$. Indeed it follows from

$$as^2 + 2hst + bt^2 + k' = 0$$

that the semi-axes are, in some order, the two numbers

$$\sqrt{\left|\frac{k'}{\lambda}\right|} \quad \text{and} \quad \sqrt{\left|\frac{k'}{\mu}\right|}.$$

We now give three worked examples in which the canonical forms of equation of a hyperbola, an ellipse and a parabola are obtained using the algebraic method.

**Worked example 2.9.** For the hyperbola

$$3x^2 - 10xy + 3y^2 + 16x - 16y + 8 = 0$$

find i) its centre, ii) its axes, iii) its asymptotes, iv) the matrix of a rotation which diagonalises the quadratic form, v) the canonical form of its equation, vi) the length of the semi-transverse axis and the equation of the transverse axis, and vii) the eccentricity and the coordinates of the foci.

*Solution.* i) The centre is obtained by solving the equations obtained from the partial derivatives of the defining polynomial. These are

$$6x - 10y + 16 = 0,$$
$$-10x + 6y - 16 = 0.$$

Using the row-equivalence method we obtain the unique centre $(-1, 1)$.
    Making the substitution

$$(x, y) = (-1, 1) + (s, t),$$

that is $x = s - 1$ and $y = t + 1$, moves the centre of the conic to the origin (mnemonic: $s = 0$ and $t = 0$ gives $x = -1$ and $y = 1$), and we obtain the equation

$$3s^2 - 10st + 3t^2 = 8.$$

ii) The characteristic equation of the matrix of the quadratic form is

$$0 = |\lambda I - A| = \begin{vmatrix} \lambda - 3 & 5 \\ 5 & \lambda - 3 \end{vmatrix} = (\lambda - 8)(\lambda + 2),$$

and therefore the eigenvalues are $\lambda = 8$ and $\lambda = -2$. Eigenvectors for these, obtained by solving the equations

$$(\lambda I - A) \begin{pmatrix} x \\ y \end{pmatrix} = 0,$$

are $(1, -1)$ and $(1, 1)$ respectively.

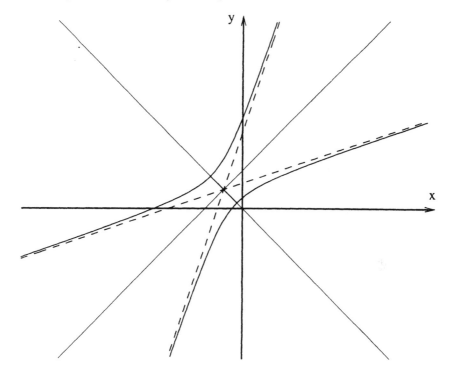

Figure 2.2 *Hyperbola.*

The axes of the conic are the lines through the centre which are parallel to the eigenvectors. In parametric form these are

$$\mathbf{r} = (-1, 1) + \varphi(1, -1) \quad \text{and} \quad \mathbf{r} = (-1, 1) + \varphi(1, 1)$$

respectively. Converting these equations to algebraic form we obtain

$$\frac{x+1}{1} = \frac{y-1}{-1} \quad \text{and} \quad \frac{x+1}{1} = \frac{y-1}{1}$$

respectively.

iii) The asymptotes are the lines through the centre which are parallel to the line pair

$$3x^2 - 10xy + 3y^2 = 0,$$

that is to

$$(3x - y)(x - 3y) = 0.$$

The lines $3x - y = 0$ and $x - 3y = 0$ are parallel to the vectors $(1, 3)$ and

$(3, 1)$ respectively. Thus the asymptotes are

$$\mathbf{r} = (-1, 1) + \varphi(1, 3) \quad \text{and} \quad \mathbf{r} = (-1, 1) + \varphi(3, 1).$$

Converting these equations to algebraic form we obtain

$$\frac{x+1}{1} = \frac{y-1}{-2} \quad \text{and} \quad \frac{x+1}{2} = \frac{y-1}{-1}.$$

Notice that the hyperbola is not rectangular since its asymptotes are not orthogonal. Therefore, in this case, the asymptotes are not the bisectors of the axes. In general, the asymptotes bisect the axes if, and only if, the hyperbola is rectangular.

iv) Unit eigenvectors for the eigenvalues $\lambda = 8$ and $\lambda = -2$ are

$$\frac{1}{\sqrt{2}}(1, -1) \quad \text{and} \quad \frac{1}{\sqrt{2}}(1, 1)$$

respectively, and the matrix

$$P = \begin{pmatrix} \frac{1}{\sqrt{2}} & \frac{1}{\sqrt{2}} \\ \frac{-1}{\sqrt{2}} & \frac{1}{\sqrt{2}} \end{pmatrix},$$

obtained by taking these as its columns, has determinant $+1$, and therefore is a rotation matrix, which, of course, satisfies

$$P^T A P = P^{-1} A P = \begin{pmatrix} 8 & 0 \\ 0 & -2 \end{pmatrix},$$

and therefore diagonalises the quadratic form.

v) The rotation corresponding to the substitution

$$\begin{pmatrix} s \\ t \end{pmatrix} = P \begin{pmatrix} u \\ v \end{pmatrix} = \begin{pmatrix} \frac{1}{\sqrt{2}} & \frac{1}{\sqrt{2}} \\ \frac{-1}{\sqrt{2}} & \frac{1}{\sqrt{2}} \end{pmatrix} \begin{pmatrix} u \\ v \end{pmatrix} = \begin{pmatrix} \frac{1}{\sqrt{2}}u + \frac{1}{\sqrt{2}}v \\ \frac{-1}{\sqrt{2}}u + \frac{1}{\sqrt{2}}v \end{pmatrix},$$

with inverse

$$\begin{pmatrix} u \\ v \end{pmatrix} = P^T \begin{pmatrix} s \\ t \end{pmatrix} = \begin{pmatrix} \frac{1}{\sqrt{2}} & \frac{-1}{\sqrt{2}} \\ \frac{1}{\sqrt{2}} & \frac{1}{\sqrt{2}} \end{pmatrix} \begin{pmatrix} s \\ t \end{pmatrix} = \begin{pmatrix} \frac{1}{\sqrt{2}}s + \frac{-1}{\sqrt{2}}t \\ \frac{1}{\sqrt{2}}s + \frac{1}{\sqrt{2}}t \end{pmatrix},$$

gives (immediately) the canonical form

$$u^2 - \frac{v^2}{2^2} = 1,$$

and therefore the conic is a hyperbola.

vi) The transverse axis is the axis corresponding to the eigenvalue 8, thus it is the line $x + y = 0$. From the canonical form, the (length of the) semi-transverse axis is 1.

vii) The eccentricity is

$$\sqrt{1 + \frac{b^2}{a^2}} = \sqrt{1 + \frac{2^2}{1^2}} = \sqrt{5},$$

and the foci are at

$$(-1, 1) \pm \sqrt{5} \left( \frac{1}{\sqrt{2}}, \frac{-1}{\sqrt{2}} \right).$$

The hyperbola is drawn in Figure 2.2.

We have determined the axes above. However we illustrate also the alternative method for doing this in the case of a hyperbola. The axes of the hyperbola are parallel to the bisectors of the asymptotes. We first choose unit normal vectors parallel to the asymptotes,

$$\mathbf{p} = \frac{1}{\sqrt{10}}(1, 3) \quad \text{and} \quad \mathbf{q} = \frac{1}{\sqrt{10}}(3, 1).$$

The required bisectors are parallel to the vectors $\mathbf{p} \pm \mathbf{q}$, and therefore to $(1, 1)$ and $(1, -1)$ respectively. The determination of the equations of the axes then proceeds as above. □

**Worked example 2.10.** For the ellipse

$$3x^2 + 2xy + 3y^2 - 6x + 14y - 101 = 0$$

find i) its centre, ii) its axes, iii) the matrix of a rotation which diagonalises the quadratic form, iv) the canonical form of its equation, v) the lengths of the semi-major and semi-minor axes, and the equation of the major axis, and vi) the eccentricity and the coordinates of the foci.

*Solution.* i) The centre is obtained by solving the equations obtained from the partial derivatives of the defining polynomial. These are

$$6x + 2y - 6 = 0,$$
$$2x + 6y + 14 = 0.$$

Using the row-equivalence method we obtain the unique centre $(2, -3)$.

Making the substitution $x = s + 2$ and $y = t - 3$ moves the centre of the conic to the origin, and we obtain the equation

$$3s^2 + 2st + 3t^2 = 128.$$

ii) The characteristic equation of the matrix of the quadratic form is

$$0 = |\lambda I - A| = \begin{vmatrix} \lambda - 3 & -1 \\ -1 & \lambda - 3 \end{vmatrix} = (\lambda - 4)(\lambda - 2),$$

and therefore the eigenvalues are $\lambda = 2$ and $\lambda = 4$. Eigenvectors for these, obtained by solving the equations

$$(\lambda I - A) \begin{pmatrix} x \\ y \end{pmatrix} = 0,$$

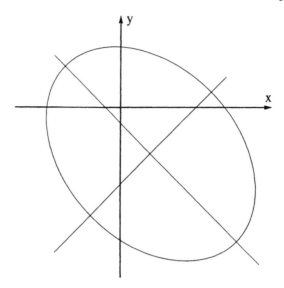

Figure 2.3 *Ellipse.*

are $(1, -1)$ and $(1, 1)$ respectively.

The axes of the conic are the lines through the centre which are parallel to the eigenvectors. In parametric form these are

$$\mathbf{r} = (2, -3) + \theta(1, -1), \quad \text{and} \quad \mathbf{r} = (2, -3) + \theta(1, 1).$$

Converting these equations to algebraic form we obtain

$$\frac{x - 2}{1} = \frac{y + 3}{-1} \quad \text{and} \quad \frac{x - 2}{1} = \frac{y + 3}{1},$$

respectively.

iii) Unit eigenvectors for the eigenvalues $\lambda = 2$ and $\lambda = 4$ are $\frac{1}{\sqrt{2}}(1, -1)$ and $\frac{1}{\sqrt{2}}(1, 1)$ respectively, and the matrix

$$P = \begin{pmatrix} \frac{1}{\sqrt{2}} & \frac{1}{\sqrt{2}} \\ \frac{-1}{\sqrt{2}} & \frac{1}{\sqrt{2}} \end{pmatrix},$$

obtained by taking these, up to multiplication by $-1$, as its columns, has determinant $+1$, and therefore is a rotation matrix, which, of course, satisfies

$$P^T A P = P^{-1} A P = \begin{pmatrix} 2 & 0 \\ 0 & 4 \end{pmatrix}.$$

The rotation corresponding to the substitution

$$\begin{pmatrix} s \\ t \end{pmatrix} = P \begin{pmatrix} u \\ v \end{pmatrix} = \begin{pmatrix} \frac{1}{\sqrt{2}} & \frac{1}{\sqrt{2}} \\ \frac{-1}{\sqrt{2}} & \frac{1}{\sqrt{2}} \end{pmatrix} \begin{pmatrix} u \\ v \end{pmatrix} = \begin{pmatrix} \frac{1}{\sqrt{2}} u + \frac{1}{\sqrt{2}} v \\ \frac{-1}{\sqrt{2}} u + \frac{1}{\sqrt{2}} v \end{pmatrix},$$

with inverse

$$\begin{pmatrix} u \\ v \end{pmatrix} = P^T \begin{pmatrix} s \\ t \end{pmatrix} = \begin{pmatrix} \frac{1}{\sqrt{2}} & \frac{-1}{\sqrt{2}} \\ \frac{1}{\sqrt{2}} & \frac{1}{\sqrt{2}} \end{pmatrix} \begin{pmatrix} s \\ t \end{pmatrix} = \begin{pmatrix} \frac{1}{\sqrt{2}} s + \frac{-1}{\sqrt{2}} t \\ \frac{1}{\sqrt{2}} s + \frac{1}{\sqrt{2}} t \end{pmatrix},$$

gives (immediately) the canonical form

$$\frac{s^2}{64} + \frac{t^2}{32} = 1,$$

and therefore the conic is an ellipse.

    v) The major axis is

$$(x - 2) = -(y + 3)$$

(for an ellipse the smallest eigenvalue corresponds to the major axis). The semi-major axis is 8 and the semi-minor axis is $4\sqrt{2}$.

    vi) The eccentricity is

$$\sqrt{1 - \frac{b^2}{a^2}} = \sqrt{1 - \frac{32}{64}} = \frac{1}{\sqrt{2}}.$$

The distance of the centre from the foci is $ae = 8\sqrt{2}$, and the foci are at the points

$$(2, -3) \pm 8\frac{1}{\sqrt{2}}(\frac{1}{\sqrt{2}}, \frac{-1}{\sqrt{2}}) = (2, -3) \pm (4, -4).$$

The ellipse is indicated in Figure 2.3.     □

**Worked example 2.11.** Show that the conic

$$x^2 + 4xy + 4y^2 - 4\sqrt{5}x - 3\sqrt{5}y = 0$$

has no centre. Show that it is a parabola. Find i) its axis, ii) the matrix of a rotation which diagonalises the quadratic form, iii) the canonical form of its equation, and iv) the coordinates of the focus.

*Solution.* The centre is obtained by solving the equations obtained from the partial derivatives of the defining polynomial. These are

$$2x + 4y - 4\sqrt{5} = 0,$$
$$4x + 8y - 3\sqrt{5} = 0.$$

They have no solution, and therefore the conic is a parabola.

    The characteristic equation of the matrix of the quadratic form is

$$0 = |\lambda I - A| = \begin{vmatrix} \lambda - 1 & -2 \\ -2 & \lambda - 4 \end{vmatrix} = \lambda(\lambda - 5),$$

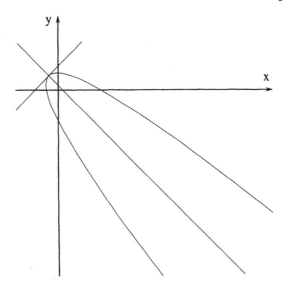

Figure 2.4 *Parabola.*

and therefore the eigenvalues are $\lambda = 0$ and $\lambda = 5$. Eigenvectors for these, obtained by solving the equations

$$(\lambda I - A) \begin{pmatrix} x \\ y \end{pmatrix} = 0,$$

are $(2, -1)$ and $(1, 2)$ respectively.

Unit eigenvectors for the eigenvalues $\lambda = 0$ and $\lambda = 5$ are

$$\frac{1}{\sqrt{5}}(2, -1) \quad \text{and} \quad \frac{1}{\sqrt{5}}(1, 2)$$

respectively, and the matrix

$$P = \begin{pmatrix} \frac{2}{\sqrt{5}} & \frac{1}{\sqrt{5}} \\ \frac{-1}{\sqrt{5}} & \frac{2}{\sqrt{5}} \end{pmatrix},$$

obtained by taking these as its columns, has determinant $+1$, and therefore is a rotation matrix, which, of course, satisfies

$$P^T A P = P^{-1} A P = \begin{pmatrix} 0 & 0 \\ 0 & 5 \end{pmatrix}.$$

The rotation corresponding to the substitution

$$\cdot \begin{pmatrix} x \\ y \end{pmatrix} = P \begin{pmatrix} u \\ v \end{pmatrix} = \begin{pmatrix} \frac{2}{\sqrt{5}} & \frac{1}{\sqrt{5}} \\ \frac{-1}{\sqrt{5}} & \frac{2}{\sqrt{5}} \end{pmatrix} \begin{pmatrix} u \\ v \end{pmatrix} = \begin{pmatrix} \frac{2}{\sqrt{5}}u + \frac{1}{\sqrt{5}}v \\ \frac{-1}{\sqrt{5}}u + \frac{2}{\sqrt{5}}v \end{pmatrix},$$

with inverse

$$\begin{pmatrix} u \\ v \end{pmatrix} = P^T \begin{pmatrix} x \\ y \end{pmatrix} = \begin{pmatrix} \frac{2}{\sqrt{5}} & \frac{-1}{\sqrt{5}} \\ \frac{1}{\sqrt{5}} & \frac{2}{\sqrt{5}} \end{pmatrix} \begin{pmatrix} x \\ y \end{pmatrix} = \begin{pmatrix} \frac{2}{\sqrt{5}}x + \frac{-1}{\sqrt{5}}y \\ \frac{1}{\sqrt{5}}x + \frac{2}{\sqrt{5}}y \end{pmatrix}$$

(recall that $P^{-1} = P^T$), gives (immediately) the parabola

$$(v - 1)^2 = (u + 1),$$

in the $(u, v)$-plane.

In the $(u, v)$-plane, the vertex is at $(u, v) = (-1, +1)$, the axis is $v = 1$ and the focus is at

$$(u, v) = (-1, 1) + \frac{1}{4}(1, 0) = (-\frac{3}{4}, 1).$$

In the $(x, y)$-plane, the vertex is at

$$\begin{pmatrix} x \\ y \end{pmatrix} = P \begin{pmatrix} -1 \\ 1 \end{pmatrix} = \begin{pmatrix} -\frac{1}{\sqrt{5}} \\ \frac{3}{\sqrt{5}} \end{pmatrix},$$

the axis is

$$v = \frac{x}{\sqrt{5}} + \frac{2y}{\sqrt{5}} = 1, \quad \text{that is} \quad x + 2y = \sqrt{5},$$

and similarly the focus is at

$$(x, y) = \frac{1}{4\sqrt{5}}(-2, 11).$$

The eccentricity is of course equal to 1.

The parabola is drawn in Figure 2.4. □

## NOTES

1. Once the rotation matrix $P$ is determined we can immediately write down $P^T A P = \mathrm{dg}(\lambda, \mu)$. It is superfluous formally to multiply the matrices $P^T$, $A$ and $P$.

2. When marking on a diagram the intersects of the conic with its axes and when marking the foci, we measure the distances along the appropriate axis from the centre or vertex. This method is quicker than calculating the coordinates.

3. For an ellipse, the smallest eigenvalue corresponds to the major axis. For a hyperbola, the transverse axis is determined by the sign of the constant term and the signs of the eigenvalues. The positive eigenvalue does

not necessarily correspond to the transverse axis; this will depend on the sign of the constant term after the centre is shifted to the origin.

4. When drawing the sketch of a parabola, it is necessary to determine which side of the line through the vertex and perpendicular to the axis the parabola lies on. To do this we choose a vector having the direction from the vertex to the focus in the canonical form, and then determine the corresponding vector in the original position of the parabola in the $xy$–plane before the rotations were applied.

**Exercises**

**2.1.** Prove that the conics having equations

    i)    $3x^2 + 10xy + 3y^2 + 46x + 34y + 93 = 0$,

    ii)   $x^2 - 6xy - 7y^2 - 16x - 48y - 88 = 0$,

    iii)   $4x^2 - 10xy + 4y^2 + 6x - 12y - 9 = 0$, and

    iv)   $3x^2 - 10xy + 3y^2 + 8x - 24y - 8 = 0$,

are hyperbolae. In each case find the centre, and the equations of the axes and the asymptotes. Determine the canonical form of the equation and the coordinate changes which determine the canonical form, including the rotation matrix which rotates the semi-transverse axis to a line parallel to the $x$–axis. Write down the length of the semi-transverse axis, and the equation of the transverse axis. Determine the eccentricity and the coordinates of the foci.

Make a rough sketch of the hyperbola on graph paper after drawing precisely its axes and its asymptotes and marking precisely the foci and the points of intersection of the hyperbola with its transverse axis.

(Take 1 centimetre as unit, and choose the centre of the hyperbola near the centre of the graph paper and the coordinate axes parallel to the edges of the paper.)

   i)  $\left[ (-1, -4), x + y + 5 = 0, \dfrac{1}{\sqrt{2}} \begin{pmatrix} 1 & 1 \\ -1 & 1 \end{pmatrix}, X^2 - 4Y^2 = 1 \right]$

  iii)  $\left[ (-2, -1), x - y + 1 = 0, \dfrac{1}{\sqrt{2}} \begin{pmatrix} 1 & 1 \\ -1 & 1 \end{pmatrix}, X^2 - \dfrac{1}{9}Y^2 = 1 \right]$

**2.2.** Prove that the conics having equations

    i)    $6x^2 - 4xy + 3y^2 + 20x - 16y - 198 = 0$,

    ii)   $5x^2 - 4xy + 8y^2 - 18x + 36y - 279 = 0$, and

    iii)   $3x^2 + 2xy + 3y^2 + 14x + 20y - 183 = 0$,

are ellipses. In each case find the centre, and the equations of the axes. Determine the canonical form of the equation and the coordinate changes which determine the canonical form, including the rotation matrix which rotates the semi-major axis to a line parallel to the $x$–axis. Write down the length of the semi-major axis, and the equation of the major axis. Determine the eccentricity and the coordinates of the foci.

Make a rough sketch of the ellipse on graph paper after drawing precisely its axes and its asymptotes and marking precisely the foci and the points of intersection of the hyperbola with its major axis.

(Take 1 centimetre as unit, and choose the centre of the ellipse near the centre of the graph paper and the coordinate axes parallel to the edges of the paper.)

i) $\left[(-1,2), 2x - y + \sqrt{5} = 0, \dfrac{1}{\sqrt{5}}\begin{pmatrix} 1 & -2 \\ 2 & 1 \end{pmatrix}, \dfrac{X^2}{112} + \dfrac{Y^2}{32} = 1\right]$

iii) $\left[(-2,1), 2x - y + \sqrt{5} = 0, \dfrac{1}{\sqrt{2}}\begin{pmatrix} 1 & 1 \\ -1 & 1 \end{pmatrix}, \dfrac{X^2}{100} + \dfrac{Y^2}{50} = 1\right]$

**2.3.** Prove that the conics having equations
i)  $4x^2 - 4xy + y^2 + 8\sqrt{5}x + 6\sqrt{5}y - 15 = 0$, and
ii) $x^2 - 2xy + y^2 - 6\sqrt{2}x - 2\sqrt{2}y - 6 = 0$,
have no centres. In each case find the vertex, and the equation of the axis. Determine the canonical form of the equation and the coordinate changes which determine the canonical form, including the rotation matrix which rotates the axis of the parabola to a line parallel to the $x$–axis. Write down the distance of the focus from the centre. Determine the coordinates of the foci.

Make a rough sketch of the ellipse on graph paper after drawing precisely its axes and its asymptotes and marking precisely the foci and the points of intersection of the hyperbola with its major axis.

(Take 1 centimetre as unit, and choose the centre of the ellipse near the centre of the graph paper and the coordinate axes parallel to the edges of the paper.)

i) $\left[\frac{1}{\sqrt{5}}(-1,3), 2x - y + \sqrt{5} = 0, \begin{pmatrix} 2 & 1 \\ -1 & 2 \end{pmatrix}, X^2 + 4Y = 0\right]$

## 2.7 Rational parametrisations of conics*

Geometers may want to find parametrisations for algebraic curves such as conics using rational or polynomial functions. We give here methods for obtaining rational or polynomial parametrisations of conics.

### 2.7.1 Use of trigonometric parametrisations

If the parametrisation of a curve is given by rational functions of $\sin\theta$ and $\cos\theta$, the substitutions

$$\sin\theta = \frac{2t}{1+t^2} \quad \text{and} \quad \cos\theta = \frac{1-t^2}{1+t^2}$$

---

* For the meaning of * see the preface.

will give a rational parametrisation of the curve in terms of

$$t = \tan \frac{\theta}{2}.$$

Notice however that such a change in parameter will in general result in points of the curve not being covered by the new parametrisation (see 2.7.2).

### 2.7.2 Geometrical methods for rational parametrisations

We now show how conics can be given rational parametrisations without the intermediate step of a trigonometric parametrisation.

### The circle

The line through $(-k, 0)$ of slope $t = \tan \theta$, that is the line $y = t(x + k)$, cuts the circle $x^2 + y^2 = k^2$ $(k > 0)$ again at

$$\mathbf{r}(t) = \left( \frac{k(1 - t^2)}{1 + t^2}, \frac{2tk}{1 + t^2} \right)$$

for all real numbers $t$. The function $t \mapsto \mathbf{r}(t)$ is a rational parametrisation of the circle. The point $(-k, 0)$ corresponds in a sense to $t = \pm\infty$ (see Figure 2.5), so that this parametrisation of the circle does not cover the point $(-k, 0)$.

The circle of radius $k > 0$ with centre at $(a, b)$ is similarly parametrised by

$$\mathbf{r}(t) = \left( a + \frac{k(1 - t^2)}{1 + t^2}, \ b + \frac{2tk}{1 + t^2} \right).$$

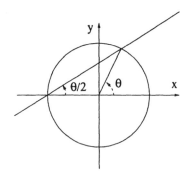

Figure 2.5 *Rational parametrisation of the circle.*

*The general conic*

Given that the quadratic equation

$$ax^2 + 2hxy + by^2 + 2cx + 2dy + k = 0$$

determines a conic, we have two cases as follows.

    Case i) The degree 2 homogeneous polynomial

$$ax^2 + 2hxy + by^2 = (\alpha_1 x + \beta_1 y)(\alpha_2 x + \beta_2 y)$$

is a product of two real linear factors. A necessary and sufficient condition for this is $h^2 \geq ab$. These factors are essentially different for a hyperbola ($h^2 > ab$), but equal, up to proportionality, for a parabola ($h^2 = ab$). We can then perform a rotation of the form

$$\begin{pmatrix} x \\ y \end{pmatrix} = \begin{pmatrix} p & q \\ r & s \end{pmatrix} \begin{pmatrix} X \\ Y \end{pmatrix}$$

to bring the line $\alpha_1 x + \beta_1 y = 0$ to the line $Y = 0$. The quadratic equation then becomes

$$2HXY + BY^2 + 2CX + 2DY + K = 0,$$

which can then be solved to give $X$ as a rational function of $Y$. The variables $x$ and $y$ are polynomial functions of $X$ and $Y$ and therefore are also rational functions of $Y$. The rotation which is used brings an asymptote, in the case of the hyperbola, or the axis, in the case of the parabola, to a line

$$Y = \text{ constant.}$$

    Case ii) The degree 2 homogeneous polynomial

$$ax^2 + 2hxy + by^2$$

is not a product of two real linear factors. A necessary and sufficient condition for this is $h^2 < ab$. In this case we have an ellipse. We rotate the ellipse to diagonalise its quadratic terms, and move its centre to the origin. Its equation then becomes

$$\frac{X^2}{A^2} + \frac{Y^2}{B^2} = 1.$$

We parametrise this by considering lines through the vertex $(-A, 0)$ of slope $t$, as with the circle. The line cuts the curve a second time where at

$$(X, Y) = \left( \frac{A(B^2 - t^2 A^2)}{B^2 + t^2 A^2}, \frac{2tAB^2}{B^2 + t^2 A^2} \right),$$

which gives a rational parametrisation of the ellipse in the $XY$–plane. Since $x$ and $y$ are polynomial functions of $X$ and $Y$, we obtain a rational parametrisation by $t$ of the original ellipse in the $xy$–plane. The vertex corresponding to $(X, Y) = (-A, 0)$ is not covered by this parametrisation.

This method also gives parametrisations for hyperbolae and for parabolae, using lines through a vertex as for the ellipse.

### 2.7.3 Parametrisation by polynomials

The parabola can be parametrised by polynomials: this follows from §2.7.2 above. We have $H = 0$ and therefore $X$ and $Y$ are polynomials in $Y$. Since

$$\begin{pmatrix} x \\ y \end{pmatrix} = \begin{pmatrix} p & q \\ r & s \end{pmatrix} \begin{pmatrix} X \\ Y \end{pmatrix}$$

we deduce that $x$ and $y$ are polynomials in $Y$. The hyperbola and the ellipse cannot be parametrised by polynomials.

# 3

# Some higher curves

An algebraic curve is a curve which is determined by some polynomial equation

$$f(x, y) = \sum a_{ij} x^i y^j = 0$$

in $x$ and $y$. The degree of the curve is the degree of the polynomial used to define it. In Chapter 2 we gave a complete classification, up to rotation and translation of those curves which are determined by a polynomial equation of degree two. For equations of higher degree there is no such simple classification, though in Chapter 15 we give a classification of cubic curves by their singular points. We give in this chapter some examples of curves determined by polynomials of degree three or more, and of other curves which are not determined by a polynomial equation. Like the canonical forms given in Chapter 1, their positions in space are chosen so that one or more of the axes or the origin play a important role, with respect to symmetry or some other factor. Some curves considered here exhibit features such as branches, cusps, double points, and nodes, which we shall be studying later in more detail. Such features do not occur in algebraic curves given by a polynomial of degree two, except for the case of the double point of intersection of a pair of non-parallel lines. We consider the parametric, polar, Cartesian, and algebraic equations, and illustrate the techniques which can be used to determine one type of equation from another. The reader should become familiar with these techniques. The reader should take care to check whether points are being introduced or deleted when changing the type of equation. Squaring each side of an equation, or cancelling a factor which can take the value 0 can result in an incorrect answer being obtained. For example squaring each side of $y = x^2$ results in the curve $(y - x^2)(y + x^2) = 0$. When changing to a parametrisation, the domain of the parametrisation and any points of the curve not described by the parametrisation should be specified carefully. Most of the curves introduced here are studied further in later chapters.

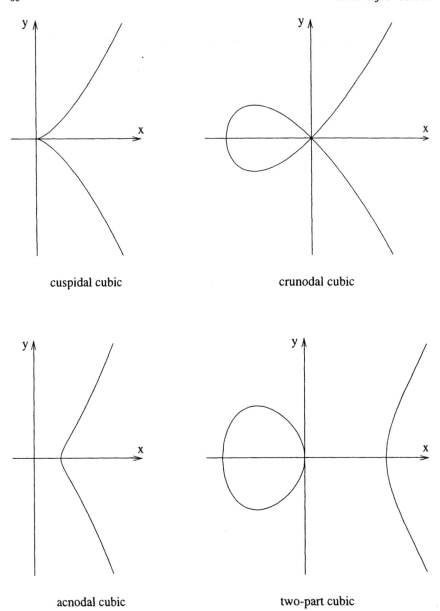

cuspidal cubic

crunodal cubic

acnodal cubic

two-part cubic

Figure 3.1 *Some cubic curves.*

## 3.1 The semicubical parabola: a cuspidal cubic

A cubic curve is a curve given by a polynomial of degree three. The semi-cubical parabola is a cubic curve which has two branches at the origin, but the tangents there to the two branches coincide. The origin is a cusp of the curve, and the common limiting tangent to the two branches at the cusp is called the cuspidal tangent. See Figure 3.1.

### 3.1.1 Algebraic equation

The algebraic equation of the semi-cubical parabola is

$$y^2 = x^3.$$

### 3.1.2 Parametric equation

The parametric equation is

$$(x, y) = (t^2, t^3)$$

where $t$ varies over all real values. This is a polynomial parametrisation, that is $x(t)$ and $y(t)$ are polynomials in $t$. To determine this from the algebraic equation, we consider the intersection of the curve with the line $y = tx$ for each fixed value of $t$. Substituting $y = tx$ into $y^2 - x^3 = 0$ gives $x^2(t^2 - x) = 0$. Solving this gives two copies of the point $(x, y) = \mathbf{0}$, together with $x = t^2$, $y = tx = t^3$. Therefore all points satisfying the algebraic equation are given by the parametric equation $(x, y) = (t^2, t^3)$.

From the parametric equation we have $y^2 = t^6 = x^3$ which gives the algebraic equation once more.

### 3.1.3 Polar equation

The polar equation is

$$r = \sin^2 \theta \sec^3 \theta \qquad \left( -\frac{\pi}{2} < \theta < \frac{\pi}{2} \right).$$

This equation is obtained by substituting $x = r \cos \theta$ and $y = r \sin \theta$ into the algebraic equation. We obtain $r^2(\sin^2 \theta - r \cos^3 \theta) = 0$, which gives $r^2 = 0$ and $\sin^2 \theta - r \cos^3 \theta = 0$. For the given range of values of $\theta$, we have $r \geq 0$, and therefore $\theta$ is the polar angle. Since $r$ may be zero, the factor $r$ must not simply be cancelled with no consideration. The iterated point $r^2 = 0$ occurs in obtaining the polar equation from the algebraic equation. However the polar equation also allows $r = 0$ as a special case; indeed, we have $r = 0$ if, and only if, $\sin \theta = 0$. As $\theta$ varies from 0 to $2\pi$ $\left( \theta \neq \frac{\pi}{2}, \frac{3\pi}{2} \right)$ the curve is covered twice, with negative values of $r$ for part of this range. Where $r$ is negative, $\theta$ will differ from the polar angle by $\pi$.

From the polar equation we have

$$\mathbf{r} = (r\cos\theta, r\sin\theta) = (\tan^2\theta, \tan^3\theta),$$

and the substitution $t = \tan\theta$ gives the parametric equation.

### 3.1.4 History and properties

The semicubical parabola is an isochronous curve; a particle descending the curve under gravity falls equal vertical distances in an equal time (Huygens 1687). It was the first algebraic curve whose arc-length was calculated (Neile 1659). Before this the arc-length had been found only for certain transcendental curves such as the cycloid and the equi-angular spiral.

## 3.2 A crunodal cubic

Two branches of the crunodal cubic curve pass through the origin, but in this case the tangents at the origin to the two branches are distinct. The origin is a node. It is a double point at which the two branches have distinct tangents. For small $x$ and $y$, the $x^3$ term is very small and the curve is close to the pair of lines $y^2 = x^2$; indeed, this is the equation of its tangent lines, and, for large $x$ and $y$, the curve resembles the semicubical parabola $y^2 = x^3$. See Figure 3.1.

### 3.2.1 Algebraic equation

The algebraic equation is

$$y^2 = x^3 + x^2 = x^2(x+1).$$

### 3.2.2 Parametric equation

The parametric equation is

$$x = t^2 - 1, \quad y = t(t^2 - 1),$$

where $t$ varies over all real values. This is a polynomial parametrisation. To determine this from the algebraic equation, we consider the intersection of the curve with the line $y = tx$ for each fixed value of $t$. Substituting $y = tx$ into $x^3 + x^2 - y^2 = 0$ gives $x^2(x+1-t^2) = 0$. Solving this gives two copies of the point $(x, y) = \mathbf{0}$, together with $x = t^2 - 1$, $y = tx = t(t^2 - 1)$. Therefore all points satisfying the algebraic equation are given by the parametric equation $(x, y) = (t^2 - 1, t(t^2 - 1))$.

From the parametric equation we have $y^2 = t^2x^2 = x^2(1+x)$ which gives the algebraic equation once more.

### 3.2.3 Polar equation

The polar equation is

$$r = \frac{\tan^2 \theta - 1}{\cos \theta} \qquad \left(-\frac{\pi}{2} < \theta < \frac{\pi}{2}\right).$$

This equation is obtained by substituting $x = r \cos \theta$ and $y = r \sin \theta$ into the algebraic equation. We obtain $r^2 (\cos^2 \theta - \sin^2 \theta) + r^3 \cos^3 \theta = 0$, which gives $r^2 = 0$ and $\sin^2 \theta - \cos^2 \theta - r \cos^3 \theta = 0$. The polar equation allows the possibility that $r$ may be negative, in which case $\theta$ will differ from the polar angle by $\pi$. The iterated point $r^2 = 0$ occurs in obtaining the polar equation from the algebraic equation. However the polar equation also allows $r = 0$ as a special case; indeed, we have $r = 0$ in the polar equation if, and only if, $\tan \theta = \pm 1$. This result is interpreted as meaning that the two tangents at the origin have slopes $\pm 1$ (compare the tachnodal quartic below). As $\theta$ varies from 0 to $2\pi$ $(\theta \neq \frac{\pi}{2}, \frac{3\pi}{2})$ the curve is covered twice. Negative values of $r$ already occur in the range $-\frac{\pi}{4} < \theta < \frac{\pi}{4}$.

From the polar equation we have

$$\mathbf{r} = (r \cos \theta, r \sin \theta) = (\tan^2 \theta - 1, \tan \theta (\tan^2 \theta - 1)),$$

and the substitution $t = \tan \theta$ gives the parametric equation.

## 3.3 An acnodal cubic

Although the algebraic equation of the acnodal cubic curve is similar to that of the crunodal cubic, its shape is quite different. The curve has an acnode, or isolated point, at the origin. Again the curve resembles the semicubical parabola $y^2 = x^3$ for large $x$ and $y$. See Figure 3.1.

### 3.3.1 Algebraic equation

The algebraic equation is

$$y^2 = x^3 - x^2 = x^2(x - 1).$$

### 3.3.2 Parametric equation

The parametric equation is

$$x = 1 + t^2 \text{ and } y = t(1 + t^2),$$

where $t$ varies over all real values. This is a polynomial parametrisation. To determine this from the algebraic equation, we consider the intersection of the curve with the line $y = tx$ for each fixed value of $t$. Substituting $y = tx$ into $x^3 - x^2 - y^2 = 0$ gives $x^2(x - 1 - t^2) = 0$. Solving this gives two copies of the point $(x, y) = \mathbf{0}$, together with $x = 1 + t^2$, $y = tx = t(1 + t^2)$. Therefore all points satisfying the algebraic equation, other than the isolated point at the origin, are given by the parametric equation $(x, y) = (1 + t^2, t(1 + t^2))$. The algebraic equation gives a point, the isolated point, which is not covered by the parametric equation.

From the parametric equation we have $y^2 = t^2 x^2 = x^2(x - 1)$ which gives the algebraic equation once more.

### 3.3.3 Polar equation

The polar equation is

$$r = 0 \quad \text{or} \quad r = \sec^3 \theta \qquad (-\pi/2 < \theta < \pi/2).$$

This equation is obtained by substituting $x = r \cos \theta$ and $y = r \sin \theta$ into the algebraic equation. We obtain $r^2(\cos^2 \theta + \sin^2 \theta) - r^3 \cos^3 \theta$, which gives $r^2 = 0$ and $1 - r \cos^3 \theta = 0$. For the given range of values of $\theta$, we have $r \geq 0$, and therefore $\theta$ is the polar angle. The iterated point $r^2 = 0$ occurs in obtaining the polar equation from the algebraic equation. The polar equation $r = \sec^3 \theta$ does not allow $r = 0$ as a special case, since $|\sec \theta| \geq 1$. As $\theta$ varies from 0 to $2\pi$ $\left(\theta \neq \dfrac{\pi}{2}, \dfrac{3\pi}{2}\right)$ the curve is covered twice, with negative values of $r$ for part of this range. It is important not to cancel a term, in this case $r$, which could be zero. Since $\sec \theta$ is never zero, ignoring $r = 0$ would result in one of the points given by the algebraic equation being missed. Where $r$ is negative in the extended range, $\theta$ will differ from the polar angle by $\pi$.

From the polar equation we have

$$\begin{aligned} \mathbf{r} &= (r \cos \theta, r \sin \theta) \\ &= (\sec^2 \theta, \sec^2 \theta \tan \theta) \\ &= (1 + \tan^2 \theta, (1 + \tan^2 \theta) \tan \theta), \end{aligned}$$

and the parametric equation follows on substituting $t = \tan \theta$.

## 3.4 A cubic with two parts

Some plane curves appear to have several parts. The simplest example of this is the hyperbola, which appears to have two parts. The two parts of the hyperbola are both unbounded, and do not meet in the plane[†]. We give now an example of a cubic curve which appears to have two parts one of which is bounded and the other of which is unbounded. See Figure 3.1.

[†] They do meet in projective space as we shall see later.

### 3.4.1 Algebraic equation

The algebraic equation is

$$y^2 - x^3 + x = 0.$$

### 3.4.2 Parametric equation

A parametric equation is

$$x = t, \quad y = \pm\sqrt{t(t^2 - 1)} \quad (-1 \le t \le 0 \text{ or } t \ge 1).$$

This parametrisation by $x$ is given by solving the algebraic equation to give $y$ as a function of $x$. We determine where the line $x = t$ meets the curve for each fixed value of $t$. Substituting $x = t$ into $y^2 - x^3 + x = 0$ gives $y^2 - t^3 + t = 0$. Solving this gives $x = t$, $y = \pm\sqrt{t(t^2 - 1)}$. This gives parametrisations of four parts of the curve, one in each of the four quadrants $\{x \ge 0, y \ge 0\}$, $\{x \le 0, y \ge 0\}$, $\{x \le 0, y \le 0\}$, and $\{x \ge 0, y \le 0\}$. These parametric equations are not differentiable at $t = 0, \pm 1$.

An algebraic curve can be parametrised in many different ways. An alternative parametric equation is given by determining where the line $y = tx$ meets the curve for each fixed value of $t$. Substituting $y = tx$ into $y^2 - x^3 + x = 0$ gives $x(t^2 x - x^2 + 1) = 0$. Solving this gives the origin $(x, y) = 0$, the parametrisation

$$\mathbf{r}(t) = \left( \tfrac{1}{2}\left( t^2 - \sqrt{t^4 + 4} \right), \tfrac{1}{2}t\left( t^2 - \sqrt{t^4 + 4} \right) \right)$$

of the bounded part of the curve and the parametrisation

$$\mathbf{r}(t) = \left( \tfrac{1}{2}\left( t^2 + \sqrt{t^4 + 4} \right), \tfrac{1}{2}t\left( t^2 + \sqrt{t^4 + 4} \right) \right)$$

of the unbounded part. The origin is not covered by the parametrisation of the bounded part, because it corresponds to the value $\infty$ of the parameter, that is to the vertical line $x = 0$ . These two parametrisations are differentiable for all values of $t$. In general we are interested only in differentiable parametrisations.

### 3.4.3 Polar equation

The polar equation is

$$-r^2 \cos^3 \theta + r \sin^2 \theta + \cos \theta = 0.$$

This is obtained by substituting $x = r \cos \theta$ and $y = r \sin \theta$ into the algebraic equation. There are two values of $r$ corresponding to each value of $\theta$.

Solving the polar equation for $r$ and substituting into $\mathbf{r} = (r \cos \theta, r \sin \theta)$, then substituting $t = \tan \theta$ gives the alternative parametrisations considered above.

## 3.5 History and applications of algebraic curves

When Cartesian coordinates were introduced by Descartes, many curves which had been known from the time of the Ancient Greeks were shown to be algebraic. However it was now possible to define a general cubic curve for example by means of an equation. There were known to be three types of curves defined by an equation of degree two, namely ellipses, hyperbolae, and parabolae. For curves of higher degree, the results are less simple. Newton classified cubic curves into 72 different types: six further types he had omitted were found subsequently. Plücker (1835) using different classification criteria obtained 219 types. Later in the book we classify cubics into four broad classes determined by the type of singular points, if any, which they possess. Another problem we have already considered is to parametrise algebraic curves using rational parametrisations. We shall see later that any non-degenerate cubic curve with a singular point can be parametrised in this way, as can any quartic curve with a triple point. However the Fermat curves

$$x^n + y^n = 1 \quad (n \geq 3)$$

have no rational parametrisation.

## 3.6 A tachnodal quartic curve

A quartic curve is a curve given by a polynomial of degree four. A tachnode is a double point with only one tangent, but two branches (see Figure 3.2). It occurs where two intersecting branches of the curve both extend each side of the intersection point and have the same tangent line at the intersection point. In this case the two branches near the origin approximate the parabolas $x = \pm y^2$; indeed, the algebraic curve

$$x^2 - y^4 = (x - y^2)(x + y^2) = 0$$

given by these two parabolas also has a tachnode at the origin. Compare the tachnode with the cusp in the semi-cubical parabola and the node in the crunodal cubic. In the case of the node there are two distinct tangent lines, and in the case of the cusp the two branches do not extend each side of the point of intersection.

### 3.6.1 Algebraic equation

The algebraic equation is

$$x^3 + x^2 - y^4 = 0.$$

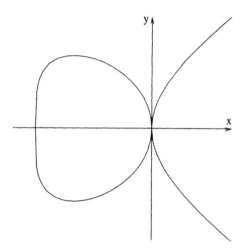

Figure 3.2 *A tachnodal quartic.*

### 3.6.2 *Parametric equations*

A parametric equation is

$$x = t, \quad y = \pm \sqrt[4]{t^2(1+t)} \qquad (t \geq -1),$$

where $\sqrt[4]{t^2(1+t)}$ is the unique positive fourth root of $t^2(1+t)$. This parametrisation by $x$ is given by solving the algebraic equation to give $y$ as a function of $x$. We determine where the line $x = t$ meets the curve for each fixed value of $t$. Substituting $x = t$ into $x^3 + x^2 - y^4 = 0$ gives $t^3 + t^2 - y^4 = 0$. Solving this gives $x = t$, $y = \pm \sqrt[4]{t^2(1+t)}$. This gives parametrisations of four parts of the curve, one in each of the four quadrants $\{x \geq 0, y \geq 0\}$, $\{x \leq 0, y \geq 0\}$, $\{x \leq 0, y \leq 0\}$, and $\{x \geq 0, y \leq 0\}$. Although the parametric equation chosen here involves surds, it can readily be used to plot the curve with the help of a calculator. These parametrisations are not differentiable at the points given by $t = 0$ and $t = -1$.

### 3.6.3 *Polar equation*

The polar equation is

$$r^2 \sin^4 \theta - r \cos^3 \theta - \cos^2 \theta = 0.$$

This equation is obtained by substituting $x = r \cos \theta$ and $y = r \sin \theta$ into the algebraic equation. We obtain $r^2(r^2 \sin^4 \theta - r \cos^3 \theta - \cos^2 \theta) = 0$, which gives $r^2 = 0$ and $r^2 \sin^4 \theta - r \cos^3 \theta - \cos^2 \theta = 0$. Solving this quadratic equation for $r$ in terms of $\theta$ will give two polar equations, each of which gives $r$ as a function of $\theta$. We allow the possibility that $r$ may be negative, in which case $\theta$ will differ from the polar angle by $\pi$. Since $r$ may be zero, the factor $r$ must not simply be cancelled with no consideration. The iterated point $r^2 = 0$ occurs in obtaining the polar equation from the algebraic equation. The polar equation also allows $r = 0$; indeed, we have $r = 0$ if, and only if, $\cos \theta = 0$.

## 3.7 Limaçons

Limaçons (see Figure 10.1) are usually given by their polar equation. We include the Cartesian equation to show that the curves are indeed algebraic. If we apply the function $z \mapsto z^2$ to any circle in the Argand diagram we obtain a limaçon as the image of the circle. An analysis of the various types of limaçon is given in Chapter 10. History and applications are given in Chapter 13.

### 3.7.1 Polar equation

The polar equation is

$$r = a + b \cos \theta \quad (a, b \geq 0) \quad (0 \leq \theta \leq 2\pi).$$

The parameter $\theta$, in the polar equation, is not always the polar angle for this curve in the case where $b > a > 0$; for example, in the case of the trisectrix ($b = 2a$), $r$, given by this equation, can take negative values.

### 3.7.2 Parametric equation

The parametric equation is

$$x = a \cos \varphi + b \cos^2 \varphi, \quad y = a \sin \varphi + b \sin \varphi \cos \varphi \quad (0 \leq \varphi \leq 2\pi).$$

The substitution $x = r \cos \theta$ and $y = r \sin \theta$, into the polar equations, gives $x = a \cos \theta + b \cos^2 \theta$ and $y = a \sin \theta + b \sin \theta \cos \theta$. In the case where $b > a$, the polar equation will give negative values for $r$ for some values of $\theta$, and there will be two points having a given polar angle; therefore we use $\phi$ as a parameter rather than the polar angle $\theta$.

### 3.7.3 Complex parametric equation

The complex parametric equation is

$$z = x + iy = e^{i\varphi}(a + b \cos \varphi) \quad (0 \leq \varphi \leq 2\pi).$$

This can be derived from the parametric equation since

$$x + iy = (a + b \cos \varphi)(\cos \varphi + i \sin \varphi).$$

Alternatively, using $\theta$ instead of $\varphi$, it is the polar equation in the Argand diagram.

### 3.7.4 Algebraic equation

The algebraic equation is

$$(x^2 + y^2 - bx)^2 = a^2(x^2 + y^2).$$

Thus the curve is a quartic (fourth degree) curve. To obtain the algebraic equation from the polar equation we have $r = a + b\cos\theta \Rightarrow r - b\cos\theta = a \Rightarrow r^2 - br\cos\theta = ar$ (introducing a point at the origin) $\Rightarrow (r^2 - br\cos\theta)^2 = a^2 r^2$ (on squaring) $\Rightarrow (x^2 + y^2 - bx)^2 = a^2(x^2 + y^2)$. The equation $(r^2 - br\cos\theta)^2 = a^2 r^2$, which is obtained during the above proof, represents the two equations $r = a + b\cos\theta$ and $r = -a + b\cos\theta$, the second of which gives the same curve as the first (replace $\theta$ by $\theta + \pi$ and $r$ by $-r$). At the beginning of this proof the angle $\theta$ is not necessarily the polar angle, but either $\theta$ or $\theta + \pi$ is the polar angle (see §0.4). After squaring has taken place, $\theta$ can be regarded as the polar angle.

### 3.7.5 Special cases

The cardioid is a limaçon for which $b = a$; this has a cusp at the origin. In the case where $b > a$, the limaçon has two loops and a node, a double point, at the origin. In the case where $b < a$, it has one loop, and the origin is an acnode. The trisectrix is a limaçon for which $b = 2a$; it can be used for trisecting angles. These special cases are shown in Figure 10.1.

## 3.8 Equi-angular (logarithmic) spiral

The equi-angular spiral (see Figure 3.3) is an example of a curve which is not algebraic, that is, it is not given by any polynomial equation in $x$ and $y$. It is an unending curve of finite length[†]. It can be seen in the growth of snail and seashells. This curve cuts radius vectors at a constant angle, namely $\alpha$. The curve spirals outwards as $\theta$ increases in the case where $a > 0$ and $0 < \alpha < \pi$. The Equi-angular spiral is usually given by its polar equation; but we include the Cartesian equation, which is not algebraic. There are many other types of spiral curves.

### 3.8.1 Polar equation

The polar equation is

$$r = e^{a\theta} \qquad (a = \cot\alpha),$$

[†] For $a > 0$, spiralling inwards the length of the curve from $\theta = 0$ is

$$\int_{-\infty}^{0} \left|\frac{dr}{dt}\right| d\theta = \sqrt{1 + a^2} \int_{-\infty}^{0} e^{a\theta}\, d\theta = \frac{\sqrt{1 + a^2}}{a} \left[e^{a\theta}\right]_{-\infty}^{0} = \frac{\sqrt{1 + a^2}}{a}.$$

See §4.6.

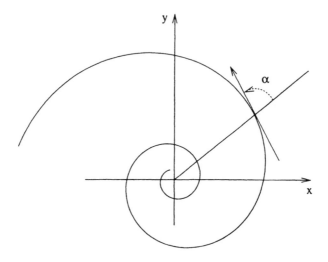

Figure 3.3 *Equi-angular spiral.*

where $\theta$ varies over the real numbers.

### 3.8.2 Cartesian equation

The Cartesian equation is

$$\tan\left(\frac{1}{a}\log\sqrt{x^2+y^2}\right) = \frac{y}{x}.$$

This is obtained from the polar equation. We have $\sqrt{x^2+y^2} = e^{a\theta}$ since $e^{a\theta} > 0$. Taking the log of each side gives $\log\sqrt{x^2+y^2} = a\theta$, and the equation follows since $\tan\theta = \frac{y}{x}$. The reader should beware of using the function $\tan^{-1}$, since

$$-\frac{\pi}{2} < \tan^{-1} u < \frac{\pi}{2}$$

for each $u$. The Cartesian equation is not given by

$$\log\sqrt{x^2+y^2} = a\tan^{-1}\frac{y}{x}.$$

The Cartesian equation is not algebraic, that is, it is not a polynomial equation.

### 3.8.3 Parametric equation

The parametric equation is

$$x = e^{a\varphi}\cos\varphi, \quad y = e^{a\varphi}\sin\varphi,$$

where $\varphi$ varies over all real values. This follows easily using the formulae $x = r\cos\theta$ and $y = r\sin\theta$ of polar coordinates. Since $r = e^{a\theta} > 0$, the parameter $\varphi$ is the polar angle $\theta$.

### 3.8.4 Complex parametric equation

The complex parametric equation is

$$z = x + iy = e^{(a+i)\varphi},$$

where $\varphi$ varies over all real values.

### 3.8.5 History and properties

The equi-angular spiral was first studied intensively by Descartes (1638). It is called equi-angular because it cuts radius vectors from the origin at a constant angle $\alpha$. It can be found in the growth of spiral snail or seashells, most famously in the shell of the Pearly Nautilus, which lives in the south-west Pacific and which grows to about 25 cm. in diameter. It is an unending curve of finite length – in a way in terms of infinite series that is comparable to the paradox of Achilles and the tortoise. Insects are said to approach a candle along this curve, thinking perhaps they are flying in a straight line at a constant angle to the rays of the sun. The boundary between two colonies of bacteria growing at different rates is an equi-angular spiral. Jakob Bernoulli (d. 1705) willed that the curve be inscribed on his tomb-stone: 'Eadem mutata resurgo' (though changed I rise unchanged). He had proved that, when many of the geometrical constructions which give 'new curves from old' were applied, this 'wonderful spiral' was essentially left unchanged.

## 3.9 Archimedean spiral

The groove in a gramophone record is an Archimedean spiral. The distance between two consecutive intersections of the curve with any line through the origin is constant. See Figure 3.4. The Archimedean spiral is usually given by its polar equation; but we include the Cartesian equation, which is not algebraic.

### 3.9.1 Polar equation

The polar equation is

$$r = a\theta \qquad (\theta \geq 0) \qquad (a \neq 0).$$

Many authors allow $\theta$ to vary over all real numbers. This results in the curve considered here together with its reflexion in the origin. The groove

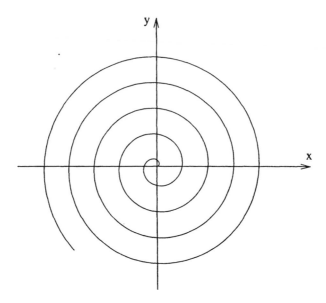

Figure 3.4 *Archimedean spiral.*

in a gramophone record corresponds to the spiral $r = a\theta$, in the case where $a < 0$.

### 3.9.2 Cartesian equation

For $a > 0$, the Cartesian equation is

$$\tan\left(\frac{\sqrt{x^2 + y^2}}{a}\right) = \frac{y}{x}.$$

We have $\sqrt{x^2 + y^2} = a\theta$ from the polar equation since we are assuming that $\theta \geq 0$. The Cartesian equation follows since

$$\tan\theta = \frac{y}{x}.$$

For the reflected curve given by $\theta < 0$ the corresponding equation is $\sqrt{x^2 + y^2} = -a\theta$, so the resulting Cartesian equation is different. As with the equi-angular spiral, $y$ is not given as a simple function of $x$. Drawing these spirals using $x$ or $y$ as parameter would involve some hard calculations. The Archimedean spiral is not an algebraic curve.

## 3.9.3 Parametric equation

The parametric equation is

$$x = a\varphi \cos\varphi, \quad y = a\varphi \sin\varphi \qquad (\varphi \geq 0).$$

This same equation also gives the reflected curve for $\varphi < 0$. For $a\varphi > 0$, $\varphi$ is the polar angle $\theta$.

## 3.9.4 Complex parametric equation

The complex parametric equation is

$$z = a\varphi e^{i\varphi} \qquad (\varphi \geq 0).$$

## 3.9.5 History and applications

The Archimedean spiral was discussed by Archimedes (225 BC). It is the spiral of a groove in a gramophone record. The constant angular velocity of the spiral rotated about its centre is transformed to the constant linear velocity of a point lying on a fixed line through the origin. A 'heart-shaped' cam comprising two pieces of this curve is used in machinery for transforming a constant angular velocity to an oscillating constant linear velocity, for example for coiling cotton onto the spool of a sewing machine.

## 3.10 Application – Watt's curves

James Watt (1784) discovered a family of algebraic curves given mechanically by a four bar linkage. The curves are given by the locus of the midpoint of a rod whose ends are constrained to move around two circles of equal radius. Linkages are important in engineering, for example in steam engines and robotics. The approximate linear motion given by a suitable linkage is important in the design of linkages connected to the driving pistons of a steam engine. Watt was investigating linkages for use in steam engines. In some cases parts of Watt's curve are very close to being a straight line. Watt is said to have been more proud of his discovery of the approximate straight line motion given by the linkage, than his development work on steam engines. Watt's curves are algebraic curves given by a polynomial equation of degree six. Conversely, it was shown by Kempe (1876) that each finite piece of any algebraic curve can be described by a suitable linkage.

We now describe Watt's curves. They are given using a four bar linkage ABCD, one of whose bars AD is fixed – see Figure 3.5. The points B and C describe equal-radii circles centre A and D respectively. The centre P of the moving bar BC describes Watt's curve[†]. Let $AB = CD = a$, $AD = 2c$

---

[†] Some authors choose $c^2 = a^2 + b^2$. On the other hand, some authors allow the two circles to have different radii.

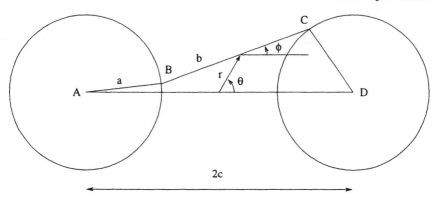

Figure 3.5 *Watt's linkage.*

and $BC = 2b$. The origin $O$ is the centre of $AD$, and the $x$–axis is parallel to $\overrightarrow{AD}$. In Figure 3.6 we give some examples of the curves.

### 3.10.1 Polar equation

The polar equation is

$$r^2 = a^2 - \left[ c \sin \theta \pm \left( b^2 - c^2 \cos^2 \theta \right)^{\frac{1}{2}} \right]^2.$$

In case $b = c$, the curve includes a circle which has the same radius as the two given circles.

### 3.10.2 Polar parametric equation

The polar coordinates can be parametrised by $\phi$ to give

$$r^2 = a^2 - b^2 - c^2 + 2bc \cos \phi, \text{ and}$$

$$\tan \theta = \frac{c - b \cos \phi}{b \sin \phi}.$$

Again in the special case $b = c$ the solution $\sin \phi = 0$ gives the two circles

$$r^2 = a^2 - (b \pm c)^2.$$

### 3.10.3 Algebraic equation

Watt's curve is a sextic algebraic curve with equation

$$\left( x^2 + y^2 \right)^3 + 2 \left( x^2 + y^2 \right)^2 \left( b^2 - a^2 \right) - 2c^2 \left( x^4 - y^4 \right) +$$
$$\left( x^2 + y^2 \right) \left( a^4 + b^4 + c^4 - 2b^2 c^2 - 2b^2 a^2 \right) + 2a^2 c^2 \left( x^2 - y^2 \right) = 0.$$

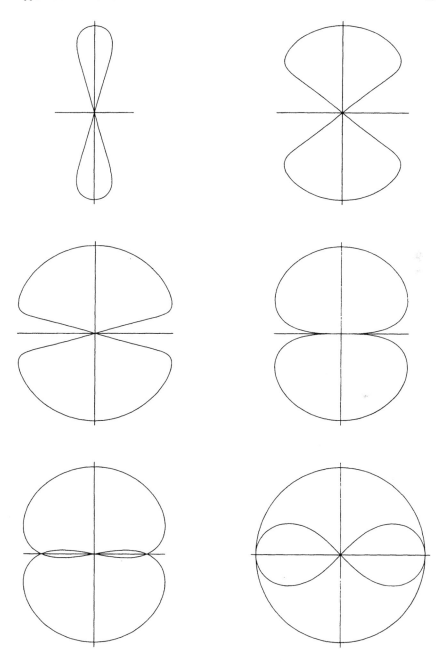

Figure 3.6 *Watt's curves.*

*Sketch of proof\**. We have

$$A = -c,$$
$$B = re^{i\theta} - be^{i\phi},$$
$$C = re^{i\theta} + be^{i\phi}, \text{ and}$$
$$D = c.$$

Since the lengths of $AB$ and $CD$ are both $a$, we have

$$|re^{i\theta} + be^{i\phi} - c| = a, \text{ and}$$
$$|re^{i\theta} - be^{i\phi} + c| = a;$$

that is

$$r^2 + b^2 + c^2 - 2cr\cos\theta - 2bc\cos\phi + 2br\cos(\theta - \phi) = a^2, \text{and}$$
$$r^2 + b^2 + c^2 + 2cr\cos\theta - 2bc\cos\phi - 2br\cos(\theta - \phi) = a^2.$$

Adding and subtracting these two equations gives

$$r^2 = a^2 - b^2 - c^2 + 2bc\cos\phi, \text{ and}$$
$$c\cos\theta = b\cos(\theta - \phi),$$

that is;

$$c\cos\theta = b\cos\theta\cos\phi + b\sin\theta\sin\phi.$$

The polar parametric equations are now immediate. The polar equation and the algebraic equation are obtained by eliminating $\phi$ from these two equations using the equation $\cos^2\phi + \sin^2\phi = 1$, and, for the algebraic equation, using also $x = r\cos\theta$, $y = r\sin\theta$, and $r^2 = x^2 + y^2$. □

In Figure 3.6, clockwise from top left, the values of $a{:}b{:}c$ are $2.2{:}0.6{:}2.1$, $1{:}1{:}1.25$, $1.2{:}2{:}1.6$, $\sqrt{2}{:}1{:}1$, $1.5{:}2{:}1.6$, and $1.1{:}3{:}3.1$. The inner curve of the bottom right is Bernoulli's lemniscate.

---

\* For the meaning of \* see the preface.

# 4

# Parameters, tangents, normal

---

In Chapters 1 and 3, we gave parametrisations of some standard curves. We consider in this chapter the formal definitions of parametrised curve and differentiability of a curve. We also give some examples of parametrisations having properties which the reader may not expect or may even find bizarre. Examples of non-differentiable curves include space-filling curves which can fill up a whole square in the plane. In the rest of the book we generally consider only curves which are differentiable and whose derivatives have certain 'nice' properties. We show how the tangent and normal at a point of a differentiable parametric curve may be defined by differentiating with respect to the parameter. Next we show how algebraic curves can be parametrised locally. The theoretical existence of such a local parametrisation enables us to determine the tangent and normals of an algebraic curve simply using the partial derivatives of the defining polynomial of the curve. We do not need to determine a local parametrisation in order to do this. Next we show how the length of a parametric curve can be determined. In §4.7 we include a number of results in calculus and analysis which we shall refer to in this chapter and elsewhere.

## 4.1 Parametric curves

Many of the curves we study in this book are parametric curves, and others, such as algebraic curves, are curves which can be parametrised, at least locally. We now formally define parametric curves.

### 4.1.1 Definitions

A parametric curve is generally defined on some interval $J$. An *interval* is a set of real numbers $J$ which satisfy the property that if $c$ and $d$ belong to $J$ and $c < d$, then $t$ also belongs to $J$ whenever $c < t < d$. Examples of intervals are

$$[\alpha, \beta] = \text{the set of real numbers } t \text{ which satisfy } \alpha \le t \le \beta,$$
$$(\alpha, \beta) = \text{the set of real numbers } t \text{ which satisfy } \alpha < t < \beta,$$
$$(\alpha, \beta] = \text{the set of real numbers } t \text{ which satisfy } \alpha < t \le \beta,$$
$$[\alpha, \infty) = \text{the set of real numbers } t \text{ which satisfy } \alpha \le t,$$
$$(-\infty, \beta) = \text{the set of real numbers } t \text{ which satisfy } t < \beta, \text{ and}$$
$$\mathbf{R} = \text{the set of all real numbers.}$$

The *end points* of the interval $[\alpha, \beta]$ are $\alpha$ and $\beta$. The interval $(\alpha, \beta)$ has no end points, since the points $\alpha$ and $\beta$ do not belong to this interval. An *interior point* of an interval $J$ is a point of $J$ which is not an end point.

**Definitions 4.1.** A *parametric plane curve* is a continuous function

$$\mathbf{r} : J \to \mathbf{R}^2$$

where $J$ is some interval. A *parametric curve from* **c** *to* **d** is a curve

$$\mathbf{r} : [\alpha, \beta] \to \mathbf{R}^2$$

where $\mathbf{r}(\alpha) = \mathbf{c}$ and $\mathbf{r}(\beta) = \mathbf{d}$; the *initial point* is $\mathbf{r}(\alpha)$ and the *final point* is $\mathbf{r}(\beta)$. A *closed parametric curve* is a curve $\mathbf{r} : [\alpha, \beta] \to \mathbf{R}^2$ for which $\mathbf{r}(\alpha) = \mathbf{r}(\beta)$. The *geometric curve* determined by the parametric curve is the image set $\mathbf{r}(J) = \{\mathbf{r}(t) \text{ for } t \text{ in } J\}$. The *domain* of $\mathbf{r} : J \to \mathbf{R}^2$ is the interval $J$.

The curves we consider are generally differentiable one or more times.

**Definitions 4.2.** A curve $t \mapsto \mathbf{r}(t)$ is *differentiable* at $t = t_0$, if

$$\mathbf{r}'(t_0) = (x'(t_0), y'(t_0))$$

exists: the *derivative* of $\mathbf{r}(t)$ at $t = t_0$ is the vector $\mathbf{r}'(t_0)$. The curve is *smooth at* $t = t_0$ if derivatives of all orders of $x$ and $y$ exist near $t_0$. The curve is *smooth* if it is smooth at each point of $J$. At the left (respectively right) end point of an interval, the above 'derivatives' mean the usual right (respectively left) derivatives. A *piecewise smooth curve* is a (continuous) curve where $J$ is the union of a finite number of subintervals $\{J_k\}$ such that $\mathbf{r}$ is smooth on each of these subintervals (the curve may, for example, give a polygon, and each $\{J_k\}$ may give a side of the polygon). We denote the higher derivatives by $\mathbf{r}''(t) = (x''(t), y''(t)), \mathbf{r}'''(t) = (x'''(t), y'''(t))$, and in general the $n$-th derivative by

$$\mathbf{r}^{(n)}(t) = \frac{d^n \mathbf{r}}{dt^n} = \left( \frac{d^n x}{dt^n}, \frac{d^n y}{dt^n} \right).$$

Sometimes we shall use as parameter $s$, $\theta$, or $\varphi$, for example, rather than $t$.

**Worked example 4.3.** Show that the cuspidal cubic $r(t) = (t^2, t^3)$, is a smooth curve. Find the parameter values for which $r'(t) = 0$, and show that $r''(t)$ is never zero.

*Solution.* We have $r'(t) = (2t, 3t^2)$, $r''(t) = (2, 6t)$, $r'''(t) = (0, 6)$, and $r^{(n)} = 0$ for $n \geq 4$. Thus the curve is a smooth curve defined on $\mathbf{R}$. Also $r'(t) = 0$ if, and only if, $t = 0$; and $r''(t)$ is never $0$. $\square$

## Exercises

**4.1.** Show that the crunodal cubic $r(t) = (t^2 - 1, t(t^2 - 1))$ is a smooth curve. Find $r'$, $r''$ and $r'''$ at $t = 0$ and $t = 1$.
$$[(0, -1), (2, 0), (0, 6); (2, 2), (2, 6), (0, 6)]$$

**4.2.** Show that the circle $r(\theta) = (\cos \theta, \sin \theta)$ is a smooth curve. Find $r'$, $r''$ and $r'''$ at $\theta = 0$ and $\theta = \dfrac{\pi}{2}$.

**4.3.** For the equi-angular spiral $r(\varphi) = (e^{a\varphi} \cos \varphi, e^{a\varphi} \sin \varphi)$, find $r'$ and $r''$ at $\varphi = 0$ and $\varphi = \dfrac{\pi}{2}$.
$$[(a, 1), (a^2 - 1, 2a); e^{\frac{a\pi}{2}}(-1, a), e^{\frac{a\pi}{2}}(-2a, a^2 - 1)]$$

Most of the parametric curves we study are smooth. The domain $J$ of the parametrisation of a curve should normally always be specified or noted.

When we use the terms *differentiable, respectively smooth*, for a closed curve, we assume that the curve is differentiable, respectively smooth, at the 'end points' in the following sense. We can extend the parametrisation of a closed curve $r : [\alpha, \beta] \to \mathbf{R}^2$ to a periodic function $r : \mathbf{R} \to \mathbf{R}^2$ of period $\rho = \beta - \alpha$ in the usual way, that is using the formula $r(t) = r(t + n\rho)$ for each integer $n$ and for each $t$ in $\mathbf{R}$. We say that the closed curve is differentiable, respectively smooth, if this periodic function $r : \mathbf{R} \to \mathbf{R}^2$ is differentiable, respectively smooth. In many of the parametrisations of closed curves which occur in practice the formulae defining them are periodic, involving functions such as sin, cot, etc., and the periodic extension uses the same formulae.

### 4.1.2 Differentiability class*

Many of the theorems and other results which follow in this book are valid more generally if 'smooth curve' in the hypotheses of the theorem, for example, is replaced by 'curve having continuous $n$-th derivative' where $n$ is some suitable integer. We have omitted to state results for the more general $n$-times continuously differentiable curves in order to simplify the statements of the theorems and other results. In case the more advanced reader may wish to consider more general forms of the statements of results

* For the meaning of * see the preface.

and theorems, we note here the standard terminology which is used for these more general classes of curves.

The *differentiability class* of a parametric curve is the largest number $n$ for which $\mathbf{r}(t)$ has continuous $n$–th derivative.

**Definitions 4.4.** A curve $t \mapsto \mathbf{r}(t)$ is a $C^n$*-curve at* $t = t_0$, or is of *class* $C^n$ $(n \geq 1)$ *at* $t = t_0$, if the $n$–th derivative $\mathbf{r}^{(n)}(t)$ exists and is continuous near $t_0$. This implies that $\mathbf{r}^{(s)}(t)$ exists and is continuous near $t_0$ for each $s$ in the range $0 \leq s \leq n$. The curve is a $C^\infty$*-curve*, or of *class* $C^\infty$, at $t = t_0$, if it is smooth at $t = t_0$. The curve is a $C^n$*-curve*, or is of *class* $C^n$, if it is of class $C^n$ at each point of its domain $J$; similarly the curve is a $C^\infty$*-curve*, or is of *class* $C^\infty$, if it is of class $C^\infty$ at each point of its domain $J$. At the left (respectively right) end point of an interval, the above 'derivatives' mean the usual right (respectively left) derivatives. A *piecewise* $C^n$*-curve* is a (continuous) curve where $J$ is the union of a finite number of subintervals $\{J_k\}$ such that $\mathbf{r}$ is $C^n$ on each of these subintervals.

Clearly a smooth curve is of class $C^n$ for each $n \geq 0$, and similarly a $C^{n+1}$–curve is of class $C^n$. The converse is not true: a $C^n$–curve need not be of class $C^{n+1}$ (see Example 4.5).

### 4.1.3 Multiple points

The parametrisations $\mathbf{r} : J \to \mathbf{R}^2$ we consider will normally be injective with the exception of two types of multiple points, that is points where different values $t_1$ and $t_2$ of the parameter give the same point $\mathbf{r}(t_1) = \mathbf{r}(t_2)$ on the geometric curve. We discuss now these two types of multiple points.

**Type 1 multiple points. Closed curves: initial point = final point.** Closed curves are parametrised on an interval $[\alpha, \beta]$ where $\alpha < \beta$. For example in the parametrisation of the circle $t \mapsto \mathbf{r}(t) = (\cos 2\pi t, \sin 2\pi t)$, where $0 \leq t \leq 1$, we have $\mathbf{r}(0) = (1, 0)$, the initial point, is equal to $\mathbf{r}(1) = (1, 0)$, the final point.

**Type 2 multiple points. Self-crossings** (or **self-intersections**). Parametrisations of curves such as those given in Figure 4.1 will have self-crossings. A point of *self-crossing* is a point $\mathbf{r}(t_1)$ for which there are only finitely many distinct values $t_1, \ldots t_n$ $(n \geq 2)$ satisfying $\mathbf{r}(t_1) = \mathbf{r}(t_k)$ $(n \geq k \geq 1)$. In general we allow only curves for which self-crossing points are *isolated*, that is if $\mathbf{r}(t_0)$ is a self-crossing point, then $\mathbf{r}(t)$ is not a self-crossing point for $0 < |\mathbf{r}(t) - \mathbf{r}(t_0)| < \epsilon$ for some $\epsilon > 0$, that is there is a suitably small circle centre $\mathbf{r}(t_0)$ inside which the only self-crossing point is $\mathbf{r}(t_0)$.

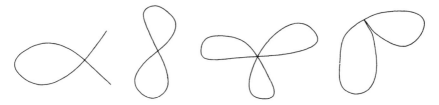

Figure 4.1 *Self-crossings of curves.*

### 4.1.4 Odd things that can happen

Parametrised curves do not always behave in the manner that one might intuitively expect. In order to help avoid the pitfalls which can occur as a result of making inadequate hypotheses or as a result of drawing invalid conclusions, we now give some examples of behaviour which may not be expected.

We first note that an $n$–times continuously differentiable curve need not be a smooth curve nor even $(n + 1)$–times continuously differentiable.

**Example 4.5.** An $n$–times continuously differentiable curve need not be $(n + 1)$–times continuously differentiable. For example the parametrisation

$$\mathbf{r}(t) = \begin{cases} (-t^3, 0) & (-1 \leq t \leq 0), \\ (t^3, 0) & (0 \leq t \leq 1) \end{cases}$$

is twice continuously differentiable with $\mathbf{r}'(0) = \mathbf{0}$ and $\mathbf{r}''(0) = \mathbf{0}$, though it is not three times continuously differentiable at $t = 1$ since

$$\mathbf{r}'''(t) = \begin{cases} (-6, 0) & (-1 < t < 0), \\ (6, 0) & (0 < t < 1). \end{cases}$$

Indeed the third derivative $\mathbf{r}'''$ does not exist at $t = 0$, that is

$$\frac{\mathbf{r}''(t) - \mathbf{r}''(0)}{t}$$

does not tend to a limit as $t$ tends to 0.

The second example illustrates a problem which can arise if no restrictions are placed on the type of non-injectivity which can occur.

**Example 4.6.** There are *space-filling curves*

$$\mathbf{r} : [0, 1] \to S = \{(x, y) : 0 \leq x, y \leq 1\}$$

which fill the whole square, that is each point of the square is of the form $\mathbf{r}(t)$ for at least one value of $t$; here $t \mapsto \mathbf{r}(t)$ is continuous but is generally many to one. Such space filling curves are usually called *Peano curves*. Examples of Peano curves can be found in texts on analysis or on topology.

The third example is of a curve all of whose derivatives vanish at one point. In general we avoid parametrisations having this property.

**Example 4.7.** Differentiable curves can have corners; for example,

$$x(t) = \begin{cases} e^{-1/t^2}, & \text{if } t < 0; \\ 0, & \text{if } t = 0; \\ e^{-1/t^2}, & \text{if } t > 0. \end{cases} \qquad y(t) = \begin{cases} 0, & \text{if } t \leq 0; \\ e^{-1/t^2}, & \text{if } t > 0. \end{cases}$$

is a smooth parametrisation $\mathbf{r} : \mathbf{R} \to \mathbf{R}^2$ of the union of the two half lines $\{y = x : x \geq 0\}$ and $\{y = 0 : x \geq 0\}$ (see Figure 4.2). This curve has a corner at the origin. It can formally be shown that all the derivatives of $\mathbf{r}$ exist and are equal to $\mathbf{0}$ at $t = 0$. Note that $\mathbf{r}(0) = \mathbf{0}$ and $\mathbf{r}'(0) = \mathbf{0}$, and thus $t = 0$ gives a non-regular point (see §4.1.6).

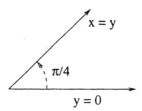

Figure 4.2 *A smooth curve with a corner.*

*4.1.5 Jordan curve theorem**

Closed curves with which we are familiar such as the circle and the ellipse divide the plane into two regions, a bounded region 'inside' the curve and an unbounded region 'outside' the curve. For closed curves with self-intersections this does not happen. We consider curves which do divide the plane into two parts.

**Definition 4.8.** A *simple closed curve* is a closed curve which has no self-crossings.

A simple closed curve is alternatively called a *Jordan curve*. We now state without proof the Jordan curve theorem which gives sufficient conditions on a closed curve for this result to hold.

**Theorem 4.9.** *Let* $\mathbf{r} : [\alpha, \beta] \to \mathbf{R}^2$ *be a simple closed curve. Then the geometric curve*[†] $K = \mathbf{r}([\alpha, \beta])$ *in the plane* $\mathbf{R}^2$ *divides the remainder of the plane into two parts; one part consists of the points inside* $K$ *and the other part consists of the points outside* $K$.

---

\* For the meaning of * see the preface.
[†] By $\mathbf{r}[\alpha, \beta]$ we mean the set of points $\mathbf{r}(t)$ for $\alpha \leq t \leq \beta$. See Definitions 4.1.

The theorem was stated by Jordan in 1892, but the first correct proof was given by Veblen in 1905. Although the reader may believe that this result is intuitively obvious, the proof is quite lengthy and uses ideas and results in topology beyond the scope of this book. Proofs of the theorem can be found in texts on analytic topology or general topology.

### 4.1.6 Velocity, speed, regularity

Many of the properties of parametric curves depend on the derivatives of $t \mapsto \mathbf{r}(t)$ and especially on the first derivative. Indeed, for the general curves we consider, the first derivative is only exceptionally zero; we single out for special consideration the points where it is zero.

**Definitions 4.10.** The *velocity vector at* $\mathbf{r}(t)$ is $\mathbf{r}'(t)$. The *speed at* $\mathbf{r}(t)$ is the length $|\mathbf{r}'(t)|$ of $\mathbf{r}'(t)$. A *unit speed parametrisation* is a parametrisation for which $|\mathbf{r}'(t)| = 1$ for all values of $t$.

**Definitions 4.11.** The parametric curve is *regular at* $t$ if $\mathbf{r}'(t) \neq 0$: it is *non-regular* (or *not regular*) at $t$ if $\mathbf{r}'(t) = 0$. A *regular parametric curve* is a parametric curve which is regular for each value of $t$ in its domain $J$.

The velocity and speed depend on the parametrisation. Given a different parametrisation of the same geometric curve, the velocity and speed will in general be different. Non-regularity at a point may be a property of the parametrisation, and need not correspond to any special feature of the geometric curve. For a different parametrisation the corresponding point of the geometric curve may correspond to a regular value of the parametrisation.

From now on we assume in general that the parametrisation is injective with the possible exception of multiple points of types 1 and 2 above. It follows that a unique *orientation* (direction) along the curve is given by increasing $t$.

**Example 4.12.** Parametrise the $x$-axis by $t \mapsto (t^3, 0)$ for all real $t$. Then the parametrisation is not regular at $t = 0$, although the arc-length parametrisation $t \mapsto (t, 0)$ is a regular parametrisation of the same geometric curve.

**Worked example 4.13.** Find the velocity and speed, as functions of the parameter, of the parametrised curve, $\mathbf{r}(t) = (a\theta \cos\theta, a\theta \sin\theta)$, and find any non-regular points. (This is an Archimedean spiral.)

*Solution.* The velocity vector is $\mathbf{r}'(\theta) = a(\cos\theta - \theta\sin\theta, \sin\theta + \theta\cos\theta)$, and the speed is (on simplifying)

$$|\mathbf{r}'(\theta)| = |a|\sqrt{(\cos\theta - \theta\sin\theta)^2 + (\sin\theta + \theta\cos\theta)^2} = a\sqrt{1 + \theta^2}.$$

Therefore the parametrisation is regular. □

**Exercises**

**4.4.** Find the velocity and speed, as functions of the parameter, of each of the following parametrised curves, and find any non-regular points.

  a) $\mathbf{r}(t) = (1 - t^2, t^2(1 + t))$,           [ $|\mathbf{r}'(t)| = |t|\sqrt{(8 + 12t + 9t^2)}$ ; 0]
  b) $\mathbf{r}(\varphi) = (4\cos\varphi, \cos 2\varphi + 1)$,
  c) $\mathbf{r}(t) = (1 + t^2, t(1 + t^2))$,        [ $|\mathbf{r}'(t)| = \sqrt{4t^2 + (1 + 3t^2)^2}$ ; none ]
  d) $\mathbf{r}(\varphi) = (\cos 2\varphi, 4\sin\varphi)$.

## *4.1.7 The Argand diagram*

Vectors in the plane $\mathbf{R}^2$ correspond to vectors in the Argand diagram $\mathbf{C}$ under the bijection given by $\mathbf{r} = (x, y) \mapsto z = x + iy$. A vector in the Argand diagram is simply a complex number. A curve, in $\mathbf{C}$, is a function $z : J \to \mathbf{C}$, where $J$ is an interval and $z(t) = x(t) + iy(t)$.

**Definitions 4.14.** The *velocity vector* at $z(t)$ is $z'(t)$. The *speed* at $z(t)$ is the length (the modulus) $|z'(t)|$ of $z'(t)$. A *unit speed* parametrisation is one for which $|z'(t)| = 1$ for all values of $t$.

**Definitions 4.15.** The parametric curve $t \mapsto z(t)$ is *regular* at $t$ if

$$z'(t) \neq 0 :$$

it is *non-regular* (or *not regular*) at $t$ if $z'(t) = 0$. A *regular* parametric curve is a parametric curve which is regular for each value of $t$ in its domain.

The use of complex number techniques can considerably shorten calculations for some types of curves. The reader should be prepared to perform calculations using $z(t) = x(t) + iy(t)$ rather than $\mathbf{r}(t) = (x(t), y(t))$ whenever it is appropriate to do so. Notice, for example, that replacing $(\cos\varphi, \sin\varphi)$ by $e^{i\varphi}$ when doing this can ease subsequent calculations.

**Worked example 4.16.** Find $z'$, $z''$, and $z'''$, for the Archimedean spiral $z = a\varphi e^{i\varphi}$.

*Solution.*    $z' = ae^{i\varphi}(1 + i\varphi)$, $z'' = ae^{i\varphi}(2i - \varphi)$, and $z''' = ae^{i\varphi}(-3 - i\varphi)$.

□

**Exercises**

**4.5.** Find the velocity and speed, as functions of the parameter, of each of the following parametrised curves, and find any non-regular points.

  a) $\mathbf{r}(t) = (e^{t^2}\cos t, e^{t^2}\sin t)$,            [ $|z'(t)| = \sqrt{4t^2 + 1}\ e^{t^2}$; none]
  b) The equi-angular spiral   $\mathbf{r}(t) = (e^{at}\cos t, e^{at}\sin t)$,
              [ $|z'(t)| = \sqrt{a^2 + 1}\ e^{at}$; none]

  c) The cardioid   $z(t) = 2e^{it} + e^{2it}$,        $\left[\ |z'(t)| = 4\left|\cos\dfrac{t}{2}\right| ; (2n + 1)\pi\right]$

  (*Hint:* $1 + e^{it} = e^{\frac{it}{2}}\left(e^{\frac{it}{2}} + e^{\frac{it}{2}}\right) = 2\cos\dfrac{t}{2}\ e^{\frac{it}{2}}$.)

  d) The limaçon   $z(t) = e^{it}(a + b\cos t)$    $(a, b > 0)$.

## 4.2 Tangents and normals at regular points

We now consider, at a regular point, the tangent and normal vectors determined by the parametrisation of a parametric curve, and the equations of the tangent line and the normal line. We defer consideration of tangent lines and normal lines at non-regular points until §6.2.

### 4.2.1 Plane curves

At a regular point $\mathbf{r}(t)$ of a parametric curve we have a (non-zero) *tangent vector* $\mathbf{r}'(t) = (x'(t), y'(t))$. The *direction of the tangent vector* and the *direction of the parametrised curve* at the point $\mathbf{r}(t)$ is the direction of the vector $\mathbf{r}'(t) \neq \mathbf{0}$. We adopt the convention that the *normal vector to the parametrised curve at* $\mathbf{r}(t)$ is given by rotating this tangent vector *counterclockwise* through an angle $\dfrac{\pi}{2}$. The resulting *normal vector* is $(-y'(t), x'(t))$. Thus as we move along the curve in the direction of increasing $t$, the normal vector points to the left.

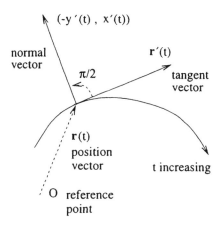

Figure 4.3 *Tangents and normals of parametrised curves.*

## Mnemonic

> Rotating $(a, b)$ counter-clockwise through $\dfrac{\pi}{2}$ gives $(-b, a)$.

If the curve $t \mapsto \mathbf{r}(t)$ is described in the opposite direction by $t \mapsto \mathbf{r}(-t)$ for example, then the tangent and normal directions determined by the new parametrisation become the opposites of the previous ones.

For a parametric curve we have a tangent line and a normal line at each fixed point $\mathbf{r}(t)$ for which $\mathbf{r}'(t) \neq \mathbf{0}$ (see Figure 4.3). The *tangent line to the curve at* $\mathbf{r}(t)$ passes through $\mathbf{r}(t)$ and is parallel to $\mathbf{r}'(t) \neq \mathbf{0}$: its equation is

$$s \mapsto \mathbf{w}(s) = \mathbf{r}(t) + s\mathbf{r}'(t) \qquad \text{(parametric equation), or}$$
$$(\mathbf{w} - \mathbf{r}(t)) \cdot (-y'(t), x'(t)) = 0 \quad \text{(algebraic equation)}.$$

Here $\mathbf{w} = (x, y)$, and $\cdot$ denotes the scalar product given by

$$(p, q) \cdot (x, y) = px + qy.$$

The *normal line at* $\mathbf{r}(t)$ passes through $\mathbf{r}(t)$ and is parallel to

$$(-y'(t), x'(t)).$$

The corresponding forms of its equation are

$$s \mapsto \mathbf{w}(s) = \mathbf{r}(t) + s(-y'(t), x'(t)), \text{ and}$$
$$(\mathbf{w} - \mathbf{r}(t)) \cdot \mathbf{r}'(t) = 0.$$

Notice that in these four equations $t$ is fixed and determines the fixed point $\mathbf{r}(t)$ on the curve, and $s$ is the (variable) parameter for the line.

**Worked example 4.17.** Find the tangent and normal lines of the crunodal cubic

$$\mathbf{r}(t) = (t^2 - 1, t(t^2 - 1))$$

at the points $t = \pm 1, 0$.

*Solution.* We have $\mathbf{r}'(t) = (2t, 3t^2 - 1)$, $\mathbf{r}'(1) = (2, 2)$, $\mathbf{r}'(-1) = (-2, 2)$, and $\mathbf{r}'(0) = (0, -1)$. Also $\mathbf{r}(\pm 1) = \mathbf{0}$ and the two parts of the curve with parameter values near $t = \pm 1$ are the two branches of the curve at the double point $\mathbf{r} = \mathbf{0}$. The tangent lines at this double point are respectively $\mathbf{w}(s) = s(1, 1)$, that is $y = x$, and $\mathbf{w}(s) = s(-1, 1)$, that is $x + y = 0$. The normal lines at the double point are respectively $\mathbf{w}(s) = s(-1, 1)$, that is $x + y = 0$, and $\mathbf{w}(s) = s(-1, -1)$, that is $y = x$. At $t = 0$, we have $\mathbf{r}'(0) = (0, -1)$, and the tangent line at $\mathbf{r}(0)$ is $\mathbf{w}(s) = (-1, 0) + s(0, -1)$, that is $\dfrac{x + 1}{0} = \dfrac{y}{-1}$, that is $x = -1$. The normal line at $\mathbf{r}(0)$ is $\mathbf{w}(s) = (-1, 0) + s(1, 0)$, that is $\dfrac{x + 1}{1} = \dfrac{y}{0}$, that is $y = 0$. $\qquad\square$

**Exercises**

**4.6.** The ellipse $\dfrac{x^2}{a^2} + \dfrac{y^2}{b^2} = 1$ is parametrised by $\varphi \mapsto \mathbf{r}(\varphi) = (p, q) = (a \cos \varphi, b \sin \varphi)$. Write down equations giving $\cos \varphi$ and $\sin \varphi$ in terms of $p$ and $q$. Calculate $\mathbf{r}'$, and hence write down, in terms of $p$ and $q$, a tangent vector, at the point $(p, q)$ of the curve, having the same direction as the parametrisation. Similarly write an outward normal vector at $(p, q)$. Hence prove that the equation of the tangent line at $(p, q)$ is $\dfrac{px}{a^2} + \dfrac{qy}{b^2} = 1$ and the equation of the normal line at $(p, q)$ is $a^2 qx - b^2 py = pq \left(a^2 - b^2\right)$.

$$\left[ \left( -\frac{aq}{b}, \frac{bp}{a} \right), \ \left( \frac{bp}{a}, \frac{aq}{b} \right) \right]$$

**4.7.** Find the equations of the tangent and normal lines to the cuspidal cubic $\mathbf{r}(t) = (t^2, t^3)$ at the point $(p, q)$.

**4.8.** Find the equations of the tangent and normal lines to the acnodal cubic $\mathbf{r}(t) = (1 + t^2, t(1 + t^2))$ at the point $(p, q)$.

$$[(2p - 3p^2)x + 2qy = -p^3, \ 2qx + p(3p - 2)y = 3q(p + q)]$$

### 4.2.2 The Argand diagram

At a regular point of a curve $t \mapsto z(t)$, the *tangent vector* is $z'(t) = \dfrac{dz}{dt} = x'(t) + iy'(t)$, and the *normal vector* is $iz'(t) = -y'(t) + ix'(t)$. We follow the same convention as above, that is that the direction of the normal vector is given by rotating the given tangent vector counter-clockwise through $\dfrac{\pi}{2}$.

**Mnemonic**

Rotating $a + ib$ counter-clockwise through $\dfrac{\pi}{2}$ gives $i(a + ib) = -b + ia$.

The *tangent line at* $z(t)$ passes through $z(t)$ and is parallel to $z'(t)$. The parametric equation of the tangent line is

$$s \mapsto w(s) = z(t) + sz'(t).$$

The *normal line at* $z(t)$ passes through $z(t)$ and is parallel to $iz'(t)$. The parametric equation of the normal line is

$$s \mapsto w(s) = z(t) + siz'(t).$$

Recall that, in the Argand diagram, if $c \neq 0$ is a real number, then $c(a + ib)$ is parallel to $a + ib$. Conversely if two vectors (complex numbers) are parallel, then one is a real multiple of the other.

**Worked example 4.18.** Find vectors parallel to the tangent and normal of the cardioid $z(\varphi) = 2e^{i\varphi} + e^{2i\varphi}$ at $z(\varphi)$.

*Solution.* $z' = 2i(e^{i\varphi} + e^{2i\varphi}) = 4ie^{\frac{3}{2}i\varphi} \cos \dfrac{\varphi}{2}$. Thus the tangent line is parallel to $ie^{\frac{3}{2}i\varphi}$ and the normal line is parallel to $e^{\frac{3}{2}i\varphi}$. □

**Exercises**

**4.9.** Find vectors parallel to the tangent and normal of the following curves.
  a) The equi-angular spiral $z(t) = e^{a+it}$ at $z(t)$.
  $$[(a + i)e^{it}, (1 - ia)e^{it}]$$
  b) The Archimedean spiral $z(\varphi) = a\varphi e^{i\varphi}$ at $z(\varphi)$.

## 4.3 Non-singular points of algebraic curves

We now wish analogously to determine normal and tangent vectors at non-singular points of algebraic curves. A non-regular point of a parametrisation is a 'singularity' of the parametrisation, though, as noted above, it need not correspond to a 'singularity' of the corresponding geometric curve. A singular point of an algebraic curve is a 'singularity' of the defining poynomial equation. It is defined as follows.

**Definition 4.19.** A *singular point* of the algebraic curve $f(x,y) = 0$ is a point $(a,b)$, on the curve, at which $\dfrac{\partial f}{\partial x} = \dfrac{\partial f}{\partial y} = 0$, that is at which $\mathbf{grad} f = \mathbf{0}$. A *non-singular point* is a point which is not a singular point, that is a point, on the curve, for which $\left( \dfrac{\partial f}{\partial x}, \dfrac{\partial f}{\partial y} \right) = \mathbf{grad} f \neq \mathbf{0}$.

To determine singular points, it is not sufficient to solve $\mathbf{grad} f = \mathbf{0}$. We must find simultaneous solutions of $\mathbf{grad} f = \mathbf{0}$ and $f = 0$; that is, we must find those 'singularities of the polynomial' which also lie on the curve.

**Worked example 4.20.** Find the singular points of the following curves.
    a) The circle $f(x,y) = x^2 + y^2 - 1$.
    b) The cuspidal cubic $f(x,y) = y^2 - x^3 = 0$.
    c) The acnodal cubic $f(x,y) = y^2 + x^2 - x^3 = 0$.
    d) The crunodal cubic $f(x,y) = y^2 - x^2 - x^3 = 0$.

*Solution.*
    a) $\mathbf{grad} f = (2x, 2y) = \mathbf{0}$ if, and only if, $x = y = 0$. Since $\mathbf{0}$ does not lie on the circle, the circle has no singular points.
    b) $\mathbf{grad} f = (-3x^2, 2y) = \mathbf{0}$ if, and only if, $(x,y) = (0,0)$. The only singular point is $(0,0)$.
    c) $\mathbf{grad} f = (2x - 3x^2, 2y) = \mathbf{0}$ if, and only if, $(x,y) = (0,0)$, or $(\frac{2}{3}, 0)$. The second point does not lie in the curve; therefore, the only singular point is $(0,0)$.
    d) $\mathbf{grad} f = (-2x - 3x^2, 2y) = \mathbf{0}$ if, and only if, $(x,y) = (0,0)$, or $(-\frac{2}{3}, 0)$. Again the second point does not lie in the curve; therefore, the only singular point is $(0,0)$.     □

In the case of the cuspidal cubic the point $(0,0)$ is a cusp, in the case of the crunodal cubic the point $(0,0)$ is a double point, and in the case of the acnodal cubic the point $(0,0)$ is an isolated point of the curve: these are three of the types of behaviour which can occur at singular points. (These curves are illustrated in Chapter 3 and a partial classification of the behaviour at singular points is given in Chapter 15.)

## Exercises

**4.10.** Find all the singular points of the following curves.

    a) The tachnodal quartic $f(x, y) = x^3 + x^2 - y^4 = 0$.      [ (0,0) ]

    b) The curves $f(x, y) = y^2(1 - x) - x^2(m + x) = 0$ $(m \neq 0, -1)$.

    c) Gutschoven's curve $f(x, y) = (x^2 + y^2)y^2 - x^2 = 0$.      [ (0,0) ]

    d) The cissoid of Diocles $f(x, y) = x^3 + xy^2 - 2y^2 = 0$.

    e) The curve $f(x, y) = x^4 + y^4 - y^2 = 0$.      [ (0,0) ]

    f) The nephroid $f(x, y) = (x^2 + y^2 - 1)^3 - \frac{27}{4}y^2 = 0$.

    g) The astroid $f(x, y) = (x^2 + y^2 - 1)^3 + 27x^2y^2 = 0$.

         [ $(1, 0), (-1, 0), (0, 1), (0, -1)$ ]

**4.11.** Show that the bicorn (cocked hat)

$$f(x, y) = (x^2 + 2y - 1)^2 - y^2(1 - x^2) = 0$$

has singular points at $(1, 0)$ and $(-1, 0)$.

## 4.4 Parametrisation of algebraic curves

Let an algebraic curve $\Gamma$ be given by $f(x, y) = 0$ where $f$ is a polynomial of degree $d \geq 1$. One question is 'Is it possible to parametrise the curve?' We now show that algebraic curves can be parametrised locally near non-singular points. In general, algebraic curves, or parts of them, can be parametrised either by $x$ or by $y$ or by both. Another important parametrisation, which we shall consider later, of suitable curves, is by their arc-length.

**Definition 4.21.** A *local parametrisation* of an algebraic curve near a point $(a, b)$ on the curve is a parametrisation $J \to \mathbf{R}^2$ of a piece of the curve including the point $(a, b)$. Usually we assume that the point $(a, b)$ in $\mathbf{R}^2$ corresponds to an interior point of the interval $J$ of parametrisation.

To obtain the local parametrisations at suitable points with respect to $x$ or $y$ as parameters, we use the following special case of the implicit function theorem. The proof of the implicit function theorem can be found in texts on analysis.

**Theorem 4.22.** *(Local parametrisation theorem). Let $f$ be a polynomial of degree $d \geq 1$ satisfying $f(a, b) = 0$ and $\dfrac{\partial f}{\partial y}(a, b) \neq 0$. Then there is an interval $J = \{x : a - \epsilon < x < a + \epsilon\}$ where $\epsilon > 0$ (equivalently an interval $J$ having the point $a$ as an interior point), and a unique smooth function $\phi : J \to \mathbf{R}$ such that $\phi(a) = b$ and $f(x, \phi(x)) = 0$ for $x$ in $J$. This solution gives a unique regular smooth local parametrisation of the curve near to the point $(a, b)$ by the $x$ coordinate, that is $x \mapsto \mathbf{r}(x) = (x, \phi(x))$ for $a - \epsilon < x < a + \epsilon$: near $(a, b)$ the curve $f(x, y) = 0$ is the graph of the function $y = \phi(x)$. Also*

$$\frac{d\phi}{dx} = -\frac{f_x}{f_y}$$

*for $x$ in $J$. Similarly, in the case where $\dfrac{\partial f}{\partial x}(a,b) \neq 0$, there exists a unique regular smooth local parametrisation $y \mapsto (\psi(y), y)$ defined near $b$ and satisfying $\psi(b) = a$, $f(\psi(y), y) = 0$, and*

$$\frac{d\psi}{dy} = -\frac{f_y}{f_x}$$

*for $y$ near $b$. Near $(a,b)$ the curve $f(x,y) = 0$ is the graph of the function $x = \psi(y)$.*

These smooth local parametrisations of algebraic curves by the $x$ or $y$ coordinates are regular; for example the derivative of $x \mapsto (x, \phi(x))$ is $x \mapsto (1, \phi'(x)) \neq 0$. These local parametrisations are not, in general, rational. In simple cases they may involve surds, local solutions of cubics, and quartics, etc. The local parametrisations by $x$ and by $y$ may give different directions to the curve locally. In the case where $(a,b)$ is a cusp, node, acnode, tachnode, or other singular point, the theorem does not apply; we consider local parametrisations near singular points in Chapter 15.

If the maximum power of $y$ in the defining polynomial of the algebraic curve is $n \geq 1$ and the curve does not contain a line parallel to the $y$–axis, then, for a given value $a$ of $x$, there will be $\leq n$ real values of $b$ satisfying $f(a,b) = 0$, since the polynomial $f(a,y)$ has degree $\leq n$. Thus the line $x = a$ intersects the algebraic curve in finitely many, indeed $\leq n$, distinct points $(a,b)$. For each of these points which satisfy the conditions of the theorem, we shall have a local parametrisation near that point.

Summarising the above result, we have that, close to a non-singular point, an algebraic curve may be parametrised locally by either the $x$ coordinate or the $y$ coordinate, or possibly by both, and the resulting parametrisation is regular and smooth.

**Worked example 4.23.** Consider the circle given by $f(x,y) = x^2 + y^2 - 1$ (see Figure 4.4). We have $f_y = 2y \neq 0$ for $y \neq 0$. Thus the theorem tells us

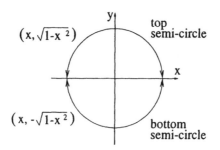

Figure 4.4 *Parametrising a circle by $x$.*

there is a local smooth parametrisation by $x$ near $(a,b)$ where $a^2 + b^2 = 1$

and $b \neq 0$. Indeed the parametrisation is given by

$$x \mapsto \left(x, \sqrt{1-x^2}\right) \quad \text{or by} \quad \left(x, -\sqrt{1-x^2}\right) \quad (-1 < x < 1)$$

depending on whether $b > 0$ or $b < 0$. In this case, for fixed $a$ $(-1 < a < 1)$, there are two values for $b$, and local parametrisations as given. Similarly, for $a \neq 0$, there is a local parametrisation by $y$ near $(a, b)$. Again this is given by

$$y \mapsto \left(\sqrt{1-y^2}, y\right) \quad \text{or by} \quad \left(-\sqrt{1-y^2}, y\right) \quad (-1 < y < 1)$$

depending on whether $a > 0$ or $a < 0$.

## Exercises

**4.12.** For each of the following curves, use Theorem 4.22 to determine the points near which a smooth local parametrisation by $x$ does not exist. Find formulae for smooth local parametrisations by $x$ at points where they exist.

a) The cuspidal cubic $f(x, y) = y^2 - x^3 = 0$,
$$[(0,0); (x, \pm x^{\frac{3}{2}}), (x > 0): \text{two domains}]$$
b) The acnodal cubic $f(x, y) = y^2 + x^2 - x^3 = 0$,
c) The crunodal cubic $f(x, y) = y^2 - x^2 - x^3 = 0$,
$$[(-1,0),(0,0); (x, \pm x\sqrt{x+1}); -1 < x < 0, x > 0: \text{four domains}]$$
d) The tachnodal quartic $f(x, y) = x^3 + x^2 - y^4 = 0$.

## 4.5 Tangents and normals at non-singular points

Let $(a, b)$ be a non-singular point of an algebraic curve $\Gamma$ given by

$$f(x, y) = 0,$$

and let $t \mapsto (x(t), y(t))$ be a regular local continuously differentiable parametrisation near $(a, b) = (x(t_0), y(t_0))$. Such a regular local parametrisation always exists; indeed by the local parametrisation theorem, we have a regular smooth local parametrisation for which the parameter is either $x$ or $y$. Differentiating $f(x(t), y(t)) = 0$, we have, by the chain rule,

$$f_x x' + f_y y' = \mathbf{r}' \cdot \mathbf{grad} f = 0,$$

where $\mathbf{r}' \neq 0$ and $\mathbf{grad} f \neq 0$. The tangent to a parametrised curve at a regular point $(a, b) = (x(t_0), y(t_0))$ is the line through $(a, b)$ which is parallel to the non-zero vector $\mathbf{r}'(t_0) = (x'(t_0), y'(t_0))$, and hence the normal to the algebraic curve at $(a, b)$ is the line through $(a, b)$ which is parallel to the non-zero vector $\mathbf{grad} f = (f_x, f_y)$ (see Figure 4.5). Rotating this vector clockwise through an angle of $\dfrac{\pi}{2}$ we obtain the (non-zero) tangent vector $(f_y, -f_x)$. Collecting these results together, we have the following theorem.

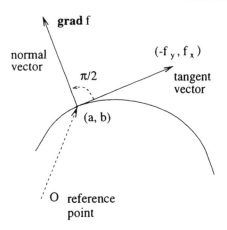

Figure 4.5 *Tangents and normals of algebraic curves.*

**Theorem 4.24.** *Let $(a, b)$ be a non-singular point of an algebraic curve $\Gamma$ given by $f(x, y) = 0$. Then the normal line and the tangent line at $(a, b)$ are parallel to the vectors $\mathbf{grad}\, f$ and $(f_y, -f_x)$ respectively.*

The relation given in this proposition between the tangent and normal vectors follows the convention we adopted above in §4.2.1 in the case of parametrised curves. Thus the normal vector is obtained from the given tangent vector by rotating it counter-clockwise through an angle $\frac{\pi}{2}$. Notice that the direction of the normal vector $\mathbf{grad}\, f$ is changed if we replace $f$ by $-f$ as the defining polynomial.

The equation of the tangent line at $(a, b)$ is

$$(\mathbf{r} - (a, b)) \cdot \mathbf{grad}\, f = 0 \quad \text{(non-parametric equation), or}$$
$$\mathbf{r}(s) = (a, b) + s(f_y, -f_x) \quad \text{(parametric equation).}$$

The first of these can be written

$$(x - a, y - b) \cdot (f_x, f_y)_{(a,b)} = (x - a)f_x + (y - b)f_y = 0.$$

The corresponding equations of the normal line are

$$(\mathbf{r} - (a, b)) \cdot (f_y, -f_x) = 0, \quad \text{and}$$
$$\mathbf{r}(s) = (a, b) + s(f_x, f_y).$$

For a regular parametrised curve the direction along the curve determined by increasing $t$ gives a direction along the tangent line, and a direction for the tangent vector. For an algebraic curve the parametrisation and the associated direction along the curve and the tangent line depend on a choice.

At singular points of an algebraic curve or non-regular points of a differentiable parametrisation (that is points for which $\mathbf{r}'(t) = 0$), determining the tangent and normal lines, where they exist, is more complicated. (See Chapter 15, and §6.2 respectively.)

## Exercises

**4.13.** Write down an outward normal vector (away from the region including the focus) at the point $(p, q)$ of the hyperbola $\dfrac{x^2}{a^2} - \dfrac{y^2}{b^2} = 1$, and also write down the tangent vector obtained by rotating this normal vector clockwise through an angle of $\dfrac{\pi}{2}$. Hence obtain the equations of the tangent and normal lines at the point $(p, q)$ of the curve.

$$\left[ \left( -\frac{p}{a^2}, \frac{q}{b^2} \right) \left( \frac{q}{b^2}, \frac{p}{a^2} \right); \frac{px}{a^2} - \frac{qy}{b^2} = 1, a^2 qx + b^2 py = (a^2 + b^2)pq \right]$$

(*Hint:* Given a non-zero vector $(x, y)$ in $\mathbf{R}^2$, the two vectors orthogonal to it and of the same length are $\pm(y, -x)$. Given a vector $x + iy$ in the Argand diagram, multiplying it by $i$ rotates it counter-clockwise through the angle $\dfrac{\pi}{2}$ to give $-y + ix$.)

**4.14.** Determine the equations of the tangent and normal lines of the following curves.

a) the cuspidal cubic $y^2 - x^3 = 0$ at the point $(1, 1)$,
$$[-3x + 2y = -1, \ 2x + 3y = 5]$$

b) the tachnodal quartic $x^3 + x^2 - y^4 = 0$ at the point $(-1, 0)$,

c) the cissoid $x^3 + xy^2 - 2y^2 = 0$ at the point $(1, -1)$,
$$[2x + y = 1, \ -x + 2y + 3 = 0]$$

d) Gutschoven's curve $(x^2 + y^2)y^2 - x^2 = 0$ at the point $\left( \frac{1}{\sqrt{2}}, \frac{1}{\sqrt{2}} \right)$.

## 4.6 Arc-length parametrisation

We have shown above that algebraic curves can be parametrised, at least locally. We now consider the arc-length of a parametrised curve. Further given a smooth parametrised curve (which need not be algebraic), we consider the problem of reparametrising it by an arc-length parametrisation. A curve parametrised by arc-length is a unit speed curve.

### 4.6.1 Arc-length

One of the problems which occurred in the history of curves was to calculate arc-length.

We consider polygonal arcs from $\mathbf{r}(t_0)$ to $\mathbf{r}(T)$ all of whose vertices

$$A_1 = \mathbf{r}(t_1), \ A_2 = \mathbf{r}(t_2), \ \ldots, \ A_k = \mathbf{r}(t_k)$$

lie on the curve, where $t_0 < t_1 < \cdots < t_k = T$ (see Figure 4.6). The

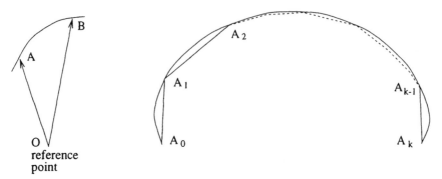

Figure 4.6 *Length of a curve.*

*arc-length* of the smooth curve is the limit, in the case where the limit
exists, of the length of these polygonal arcs as the length of the largest of
the increments $t_{i+1} - t_i$ tends to zero. Given that the curve is a smooth
curve, or more generally that $\mathbf{r}'(t)$ is continuous, it can be shown that
the limit of the lengths of these polygonal arcs exists and is equal to the
quantity given by the following alternative definition.

**Definition 4.25.** The *arc-length* from $t_0$ to $T$ of a curve $t \mapsto \mathbf{r}(t)$, for
which $\mathbf{r}'(t)$ is continuous, is

$$\int_{t_0}^{T} |\mathbf{r}'(t)| dt = \int_{t_0}^{T} \sqrt{x'(t)^2 + y'(t)^2} \ dt.$$

We assume for Definition 4.25 that $t \mapsto \mathbf{r}(t)$ is injective on the domain
$[t_0, T]$ with the possible exception of type 1 and type 2 multiple points as
noted above.

In the remainder of the book we shall use the definition of arc-length
given in Definition 4.25. A proof that this is equivalent to the geometric
definition given above can be found in texts on calculus or analysis; such a
proof can be based on the following consequence of the mean value theorem.

$$\left|\overrightarrow{AB}\right| = |\mathbf{r}(t + \Delta t) - \mathbf{r}(t)| = |\mathbf{r}'(t)|\Delta t + \epsilon.\Delta t \ (\Delta t > 0)$$

where $\epsilon$ tends to 0 as $\Delta t$ tends to 0. Here $A$ and $B$ are the points with
position vectors $\mathbf{r}(t)$ and $\mathbf{r}(t + \Delta t)$ respectively, $\Delta t$ is a small increment,
and $|\mathbf{v}|$ is the length of the vector $\mathbf{v}$.

In the case of a curve parametrised by $x$ for example, that is the graph
of a function $y = g(x)$, we have that the arc-length is given by

$$\int_{x_0}^{X} \sqrt{1 + \left(\frac{dy}{dx}\right)^2} \ dx = \int_{x_0}^{X} \sqrt{1 + \left(\frac{dg}{dx}\right)^2} \ dx.$$

## Exercises

**4.15.** Find the arc-lengths of the following curves.

a) The catenary $y = \cosh x$ from $x = 0$,                    $[\sinh x]$

b) $\mathbf{r}(t) = (3 - 2t^3, 1 + 3t^2)$ from $(3, 1)$ to $(1, 4)$,

c) $\mathbf{r}(t) = (a + 3t^2, -2t^3)$ from $(a, 0)$ to $(a + 3, -2)$.    $\left[2\left(2^{\frac{3}{2}} - 1\right)\right]$

**4.16.** Find the total length of the astroid $\mathbf{r}(\varphi) = (\cos^3 \varphi, \sin^3 \varphi)$.

**4.17.** For the cardioid $t \mapsto \mathbf{r}(t) = (2\cos t + \cos 2t, 2\sin t + \sin 2t)$, show that $|\mathbf{r}'(t)| = \sqrt{8(1 + \cos t)}$. Hence show that the length of the curve from 0 to $t$, for $0 \le t \le \pi$, is $8\sin\dfrac{t}{2}$. Determine the length of the curve from 0 to $t$ for $\pi < t \le 2\pi$; in particular find the length of the whole curve.

$$\left[16 - 8\sin\frac{t}{2}\right]$$

**4.18.** The parabola $y = x^2$ is parametrised by $t \mapsto (t, t^2)$. Prove that $|\mathbf{r}'(t)| = \sqrt{1 + 4t^2}$. Using either the standard technique or the substitution $t = \frac{1}{2}\sinh u$ to evaluate the arc-length integral, show that the arc-length of the parabola from its vertex, is

$$\frac{1}{4}\left(2t\sqrt{1 + 4t^2} + \sinh^{-1} 2t\right) = \frac{1}{4}\left(2t\sqrt{1 + 4t^2} + \log\left(2t + \sqrt{1 + 4t^2}\right)\right).$$

**4.19.** The semi-cubical parabola $y^2 = x^3$ is parametrised by $t \mapsto (t^2, t^3)$. Prove that, for $t \ge 0$, the arc-length from the cusp, is

$$\frac{1}{27}\left((9t^2 + 4)^{3/2} - 8\right).$$

What is the arc-length of the curve from its cusp, as oriented by the direction of its parametrisation, for $t < 0$?

### 4.6.2 Parametrisation by arc-length

Suppose now that we are given an injective smooth parametrisation

$$t \mapsto \mathbf{r}(t).$$

We wish to reparametrise the underlying geometric curve by arc-length. We define the arc-length function

$$s(t) = \int_{t_0}^t |\mathbf{r}'(t)|dt.$$

In the case where the parametrisation $t \mapsto \mathbf{r}(t)$ is smooth and regular, it follows from the Inverse Function Theorem (see §4.7) that $t \mapsto s(t)$ is a smooth bijection $J \to J^1$ between two intervals, which satisfies

$$\frac{ds}{dt} = |\mathbf{r}'(t)| > 0,$$

and which has a smooth inverse $s \mapsto t(s)$. We define the arc-length reparametrisation of the curve by $s \mapsto t(s) \mapsto \mathbf{r}(t(s)) = \mathbf{w}(s)$.

**Theorem 4.26.** *Let $t \mapsto \mathbf{r}(t)$ be a regular smooth curve. Then the curve can be reparametrised by arc-length and the reparametrised curve $s \mapsto \mathbf{w}(s)$ is a regular smooth curve.*

Notice that in the general case the parameter $s$, which is determined by the 'initial' parameter value $t_0$, may take positive and negative values. The theorem extends to the more general case where the parametrisation is continuous and piecewise smooth and regular, the arc length parametrisation is then piecewise smooth.

We note also that

$$\frac{ds}{dt} = |\mathbf{r}'(t)| = \left|\frac{d\mathbf{r}}{dt}\right|.$$

At a point for which $\mathbf{r}'(t) = \mathbf{0}$, if the inverse for $t \mapsto s$ exists, it cannot then be differentiable. If it were differentiable, we should have $\dfrac{dt}{ds}\dfrac{ds}{dt} = 1$ on differentiating $t(s(t)) = t$ by the chain rule, which gives a contradiction since $\left|\dfrac{ds}{dt}\right| = |\mathbf{r}'(t)| = 0$.

In the case of a smooth parametrised curve which is not regular but does satisfy the injectivity condition, the possible smoothness of the arc-length reparametrisation would need more careful analysis.

**Example 4.27.** Parametrise the $x$-axis by $t \mapsto \mathbf{r}(t) = (t^3, 0)$. The curve is smooth but not regular since $\mathbf{r}'(0) = \mathbf{0}$. The associated arc-length parametrisation is $s \mapsto (s, 0)$ which is regular although $s \mapsto t = s^{1/3}$ is not differentiable at 0.

For a regular smooth parametrisation by $t$, the arc-length parametrisation by $s$ is an *equivalent parametrisation*, that is, $s \mapsto t(s)$ is a smooth inverse to the smooth function $t \mapsto s(t)$.

The next theorem is easily proved on replacing $t$ by $s$ in the above definition of arc-length.

**Theorem 4.28.** *Let $s \mapsto \mathbf{r}(s)$, for which $\mathbf{r}'(s)$ is continuous, be parametrised by arc-length, then it is a unit speed curve, that is $|\mathbf{r}'(s)| = 1$.*

## 4.7 Some results in analysis

We collect here some results for reference. They should be referred to when needed. They relate to material in this chapter and elsewhere. Their proofs can be found in texts on calculus or analysis.

### 4.7.1 Functions of one variable

**An important lemma on continuity.**
*Let f be continuous at a and f(a) > 0, then there exists $\epsilon > 0$ such that*
*$f(x) > 0$ for all $x$ satisfying $a - \epsilon < x < a + \epsilon$.*

*Proof\*.* Let $\delta = \frac{1}{2} \mid f(a) \mid > 0$ then there exists $\epsilon > 0$ such that

$$\mid x - a \mid < \epsilon \Rightarrow f(a) - \delta < f(x < f(a) + \delta \Rightarrow f(x) > \frac{1}{2} f(a).$$

The result follows. □

A similar result holds if $f(a) < 0$: in this case $f(x) < 0$ whenever $a - \epsilon < x < a + \epsilon$.

**Taylor's theorem: Young's form.**
*Let $f : \mathbf{R} \supset\!\!\!\rightarrow \mathbf{R}$ and its derivatives up to order $n$ be continuous at the point $a$. Then*

$$f(a + h) = f(a) + hf'(a) + \ldots + \frac{1}{n!} h^n f^{(n)}(a) + R_{n+1}$$

*where $\dfrac{R_{n+1}}{h^n}$ tends to 0 as $h$ tends to 0 (where $n$ is fixed).*

The function $R_{n+1}$ of $f, a$, and $h$ is the *remainder after $(n + 1)$-terms* of the Taylor series; it is the difference between $f(a + h)$ and the polynomial given by the first $(n+1)$-terms of the Taylor series. Other forms of Taylors's theorem give other properties of the remainder $R_{n+1}$. Taylor's Theorem does not tell us that the function is equal to its infinite Taylor series; for that we need to know that $R_{n+1} \rightarrow 0$ as $n \rightarrow \infty$. The convergence of the infinite Taylor series does not guarantee that its limit, for each fixed $h$ near 0, is $f(a + h)$ (see §4.7.2).

**A lemma that in general zeros of functions are isolated.**
*Let $f(a) = 0$, let the $(n + 1)$-st derivative of $f$ be be continuous at the point $a$ ($n \geq 1$), and let*

$$f'(a) = f''(a) = \ldots = f^{(n-1)}(a) = 0, \quad f^{(n)}(a) \neq 0,$$

*then there exists $\epsilon > 0$ such that $f(x) \neq 0$ for $0 < \mid x - a \mid < \epsilon$, that is there is an interval centre $a$ such that the only zero of $f$ within the interval occurs at $a$.*

*Proof\*.* By Taylor's theorem

$$f(x) = \frac{(x - a)^n}{n!} \left( f^{(n)}(a) + \eta(x) \right)$$

where $\eta(x) \rightarrow 0$ as $x \rightarrow a$. By the above lemma there exits $\epsilon > 0$ such that $f^{(n)}(a) + \eta(x)$ has the same sign as $f^{(n)}(a)$ for $\mid x - a \mid < \epsilon$ (and is not zero). The result follows. □

---

\* For the meaning of \* see the preface.

This lemma can be used to show, for example, that non-regular points of a parametrization are in general isolated. It can be used provided not all the derivatives of $\mathbf{r}(t)$ are zero at $t = t_0$.

### 4.7.2 Analyticity

The reader should be aware that smooth functions need not be analytic. A function $f : J \to \mathbf{R}$ defined in a neighbourhood of $a$ in the interval $J$ is *analytic* at $a$ if the infinite Taylor power series converges to $f$ in a neighbourhood of $a$, that is

$$f(a + h) = \sum_{n=0}^{\infty} \frac{1}{n!} h^n f^{(n)}(a)$$

for $|h| < \epsilon$, where $\epsilon > 0$.

In the above study of parametrizations we have only been assuming that the functions are smooth (or the $n$–th derivative is continuous as appropriate).

We note the following theorem.

**Theorem.** *Analytic functions are smooth.*

Smooth functions need not be analytic. For example the function $x(t)$ used to define the curve with a corner in §4.1.4 has all derivatives at 0 equal to zero, and therefore its infinite formal Taylor series at 0 is identically zero. However the function is not equal to zero in a neighbourhood of 0. Therefore it is not analytic at 0. To show that a smooth function is analytic at a point it is not sufficient to show that its infinite formal Taylor series is convergent; it must be shown also that in a neighbourhood of the point the infinite formal Taylor series converges to the function.

So for smooth functions we can use Taylor's theorem (Young's form or the remainder form for example) involving a finite sum, but not generally the infinite Taylor series.

### 4.7.3 Zeros of a function

Let $f : \mathbf{R} \supset\!\!\to \mathbf{R}$ be a function $x \mapsto f(x)$ of one variable defined near $x = a$.

**Definition 4.29.** The function $f$ has a *zero of order* $n \geq 1$ at $a$ if

$$f(a) = f'(a) = f''(a) = \ldots = f^{(n-1)}(a) = 0$$

and $f^{(n)}(a) \neq 0$.

We now show that, in the case where the first $n - 1$ derivatives of a suitable function are zero at a point $a$, we can divide the function by the $n$–th power of $x - a$ and the resulting quotient function will be differentiable at $a$.

**Reduction of zeros theorem.**
Let $x \mapsto f(x)$ be smooth at $x = a$. Then $f$ has a zero of order $n$ at $a$ if, and only if, $f(x) = (x - a)^n \psi(x)$ where $\psi$ is smooth at $a$ and $\psi(a) \neq 0$.

*Sketch of proof\*.* Given that $f$ has a zero of order $n$ at $a$, we have by successive applications of Taylor's theorem

$$f(a + t) = t^n \left( \frac{1}{n!} f^{(n)}(a) + \eta_n(t) \right) = t^n \psi(a + t),$$

$$= t^n \left( \frac{1}{n!} f^{(n)}(a) + \frac{t}{(n+1)!} f^{(n+1)}(a) + t\eta_{(n+1)}(t) \right)$$

$$= t^n \left( \frac{1}{n!} f^{(n)}(a) + \frac{t}{(n+1)!} f^{(n+1)}(a) + \frac{t^2}{(n+2)!} f^{(n+2)}(a) + t^2\eta_{(n+2)}(t) \right)$$

$$= t^n \left( \frac{1}{n!} f^{(n)}(a) + \cdots + \frac{t^k}{(n+k)!} f^{(n+k)}(a) + t^k\eta_{(n+k)}(t) \right),$$

for each $k$, where $\eta_n(t)$, $\eta_{(n+1)}(t)$, $\ldots$, $\eta_{(n+k)}(t)$ all tend to zero as $t$ tends to 0. Thus

$$\psi(a + t) = \frac{1}{n!} f^{(n)}(a) + \eta_n(t)$$

$$= \frac{1}{n!} f^{(n)}(a) + \frac{t}{(n+1)!} f^{(n+1)}(a) + t\eta_{(n+1)}(t)$$

$$= \frac{1}{n!} f^{(n)}(a) + \frac{t}{(n+1)!} f^{(n+1)}(a) + \frac{t^2}{(n+2)!} f^{(n+2)}(a) + t^2\eta_{(n+2)}(t)$$

$$= \frac{1}{n!} f^{(n)}(a) + \cdots + \frac{t^k}{(n+k)!} f^{(n+k)}(a) + t^k\eta_{(n+k)}(t),$$

for each $k$. Therefore

$$\psi(a) = \frac{1}{n!} f^{(n)}(a) \neq 0,$$

$$\psi'(a) = \frac{1}{(n+1)!} f^{(n+1)}(a),$$

and generally

$$\psi^{(k)}(a) = \frac{k!}{(n+k)!} f^{(n+k)}(a).$$

Thus $\psi$ is a smooth function. The result follows readily. □

### 4.7.4 Inverse functions

Let $f : J \to \mathbf{R}$ be a continuous function defined on an interval $J$, then $f(J)$, the set of values taken by $f$, is also an interval $J'$. An *inverse* to the

---

\* For the meaning of \* see the preface.

function $f : J \to J'$ is a function $g : J' \to J$ which satisfies $gf(x) = x$ for each $x$ in $J$ and $fg(y) = y$ for each $y$ in $J'$. Inverse functions, where they exist, are unique. We note the following special form for intervals of the inverse function theorem.

**The inverse function theorem.**
*Let $f : J \to J'$ have continuous n–th derivative $(n \geq 1)$, and let $f'(x) \neq 0$ for each $x$ in $J$. Then there is an inverse function $g : J' \to J$, and $g$ has continuous n–th derivative. In case $f$ is smooth, then $g$ is also smooth.*

### 4.7.5 L'Hôpital's theorem

L'Hôpital's theorem enables us to determine the limit of the quotient of two functions which each tend to the limit zero. For example let $f(x) = x^2$ and $g(x) = x$. Each tends to 0 as $x$ tends to zero, and $\dfrac{0}{0}$ is not meaningful. However $\dfrac{f(x)}{g(x)} = x$ tends to the limit zero as $x$ tends to zero. There are several versions of L'Hôpital's theorem, each having different hypotheses.

**L'Hôpital's theorem.**
*Let $f$ and $g$ be n–times differentiable at $a$, let*

$$f(a) = f'(a) = \cdots = f^{(n-1)}(a) = 0 \quad \text{and}$$
$$g(a) = g'(a) = \cdots = g^{(n-1)}(a) = 0,$$

*and let $g^{(n)}(a) \neq 0$. Then*

$$\lim_{x \to a} \frac{f(x)}{g(x)} = \lim_{x \to a} \frac{f'(x)}{g'(x)} = \cdots = \lim_{x \to a} \frac{f^{(n-1)}(x)}{g^{(n-1)}(x)} = \frac{f^{(n)}(a)}{g^{(n)}(a)}.$$

### 4.7.6 Euler's theorem

We next note a result on homogeneous functions. A function $F : \mathbf{R}^3 \to \mathbf{R}$ is *homogeneous of degree n* if

$$F(\lambda x, \lambda y, \lambda z) = \lambda^n F(x, y, z)$$

for each $x$, $y$, $z$, and $\lambda$.

**Euler's theorem.**
*Let $F : \mathbf{R}^3 \to \mathbf{R}$ be a homogeneous differentiable function of degree $n$, then*

$$nF = x \frac{\partial F}{\partial x} + y \frac{\partial F}{\partial y} + z \frac{\partial F}{\partial z}.$$

*Proof.* The result is immediate on differentiating the equation

$$F(\lambda x, \lambda y, \lambda z) = \lambda^n F(x, y, z),$$

with respect to $\lambda$ and then putting $\lambda = 1$. $\qquad\square$

A similar result holds for homogeneous functions of two variables.

### 4.7.7 Functions of two variables

**Taylor's theorem: Young's form.**
Let $f : \mathbf{R}^2 \supset\rightarrow \mathbf{R}$ *and its partial derivatives up to order $n$ be continuous at $(a, b)$. Then*

$$f(a + h, b + k) = f(a, b) + [\Delta.f]_{(a,b)} + \ldots + \frac{1}{n!}[\Delta^n.f]_{(a,b)} + R_{n+1}$$

*where* $\Delta = h\dfrac{\partial}{\partial x} + k\dfrac{\partial}{\partial y}$, *and* $\dfrac{R_{n+1}}{\|(h,k)\|^n}$ *tends to $0$ as $(h, k)$ tends to $0$ (where $n$ is fixed).*

The function $R_{n+1}$ of $f, a, b, h$, and $k$ is the *remainder after $(n+1)$-terms* of the Taylor series. It is the difference between $f(a + h, b + k)$ and the first $(n+1)$-terms of the Taylor series. Other forms of Taylors's theorem give other properties of $R_{n+1}$. Taylor's theorem does not tell us that the function is equal to its infinite Taylor series; for that, we need to know that $R_{n+1} \to 0$ as $n \to \infty$. As in the case of functions of one-variable, the convergence of the infinite Taylor series does not guarantee that its limit is the original function.

The singular points of an algebraic curve are it precisely those points about which the Taylor expansion of $f(x, y)$ has vanishing linear terms, that is for which $\Delta.f = h\dfrac{\partial f}{\partial x} + k\dfrac{\partial f}{\partial y} = 0$ for each value of $h$ and each value of $k$, that is **grad**$f = \mathbf{0}$ at $(a, b)$. We note, for example, that

$$[\Delta^2.f]_{(a,b)} = h^2 f_{xx} + 2hk f_{xy} + k^2 f_{yy}, \quad \text{and}$$
$$[\Delta^3.f]_{(a,b)} = h^3 f_{xxx} + 3h^2 k f_{xxy} + 3hk^2 f_{xyy} + k^3 f_{yyy},$$

where the partial derivatives are evaluated at the point $(a, b)$.

# 5

# Contact, inflexions, undulations

In this chapter we discuss the 'closeness' of two curves passing through a given point and having a common tangent there. This closeness is measured by contact order. Contact order has several applications. Taking one of the curves to be a line, we show how inflexions and undulations of the second curve can be determined using the method of contact order. In Chapter 6, again using the method of contact order, we show how tangent lines at cusps and other non-regular points of parametric curves can be determined, and how cusps can be classified by their order. Inflexions, undulations, and cusps are 'singularies' of parametrised curves. Contact order can be used to specify the osculating circle or circle of curvature at a point of a parametric curve (see Chapter 9). Also we use contact order in Chapter 15 to give a partial classification of singular points of algebraic curves and to determine the tangent lines at such singular points where appropriate.

## 5.1 Contact

In Chapter 4 we showed that a parametric curve has a tangent line at a regular point and that an algebraic curve can be parametrised near to a non-singular point and has a tangent line at such a point. Two curves are said to be tangent to one another at a point if they have a common tangent line there. We now consider the order of tangency or contact between two curves at a common point. This is a measure of the closeness of the curves near to the point; for example, for all sufficiently small values of $x$, $y = x^6$ is closer to the $x$–axis than is $y = x^2$.

### 5.1.1 Contact between curves

We first consider contact between a parametrised curve or locally paramet-rised curve $\Pi : t \mapsto \mathbf{r}(t)$ and an algebraic curve $\Gamma$ given by $f(x, y) = 0$, where

$f$ is a polynomial, at a point which is both regular for the parametrised curve and non-singular for the algebraic curve. We define

$$\boxed{\gamma(t) = f(x(t), y(t)) = f(\mathbf{r}(t)).}$$

Clearly $\gamma(t_0) = 0$ if, and only if, $\mathbf{r}(t_0)$ lies on the algebraic curve $f(x, y) = 0$.

**Definition 5.1.** Let $\mathbf{r}(t_0)$ be a regular point of $t \mapsto \mathbf{r}(t)$ and a non-singular point of $f(x, y) = 0$. The curves have *n-point contact* or *point-contact of order n* at $\mathbf{r}(t_0)$ if the $n$-th derivative is the first non-vanishing derivative of $\gamma$ at $t_0$, that is $\gamma(t_0) = \gamma'(t_0) = \cdots = \gamma^{(n-1)}(t_0) = 0$ and $\gamma^{(n)}(t_0) \neq 0$. *Transverse intersection* or *simple intersection* is 1–point contact. The curves have $\geq$ *n-point contact* at $\mathbf{r}(t_0)$ if $\gamma(t_0) = \gamma'(t_0) = \cdots = \gamma^{(n-1)}(t_0) = 0$.

The order of point-contact is unchanged if a different parameter is chosen for the parametric curve Π. (See Theorem 5.11.) In the case where both curves are algebraic, we may parametrise either one of them locally by Theorem 4.22, and then use the above analysis locally. The order of point-contact is the same whichever of the curves we parametrise (see Theorem 5.13). Also the order of point contact is unchanged if the space in which the curves lie is changed by rotation, reflexion, or translation of origin (see Theorem 5.10). Notice that the order of point contact at $\mathbf{r}(t_0)$ is the order of the zero of $\gamma(t)$ at $t = t_0$ (see §4.7.3).

For simplicity we use polynomial parametrisations where possible: in the case where $x(t)$ and $y(t)$ are polynomials in $t$, $\gamma(t)$ is also a polynomial in $t$ and the calculation of the derivatives of $\gamma(t)$ is then straightforward. Many curves however cannot be given a polynomial parametrisation. The next proposition is often useful when dealing with polynomial parametrisations.

**Proposition 5.2.** *Let* $t \mapsto \mathbf{r}(t)$ *be a polynomial parametrisation which is regular at* $t = 0$, *and let the algebraic curve* $f(x, y) = 0$ *be non-singular at* $\mathbf{r}(0)$. *Then the order of point-contact of the two curves at* $\mathbf{r}(0)$ *is the degree of the non-zero term of lowest degree in the polynomial* $\gamma(t)$.

*Proof.* Let $\gamma(t) = a_k t^k +$ terms of degree $\geq (k + 1)$, where $a_k \neq 0$. Then $\gamma(0) = \gamma'(0) = \gamma''(0) = \cdots = \gamma^{(k-1)}(0) = 0$ and $\gamma^k(0) = a_k k! \neq 0$.  □

Another measure which is traditionally used is the order of contact. The *order of contact* or *order of tangency* of the two curves at the common point is $(n - 1)$ if the order of point-contact is $n$. Order of contact is used by many authors, and the reader should be aware of this difference of 1 in the values of these two measures. The order of contact of the two curves can be thought of as the number of times the curves are instantaneously tangent at the common point, and the order of point-contact can be thought of as the number of times the two curves instantaneously cross at the common point. In this book we shall use 'order of point-contact' and largely suppress 'order of contact'.

**Mnemonic**

> The curves have point-contact of order $n$ at $t = t_0$ if the first non-vanishing derivative of $\gamma(t)$ is of degree $n$ $(n \geq 1)$. Contact of order $(n-1)$ is point-contact of order $n$. For 2-point contact, $\gamma(t_0) = \gamma'(t_0) = 0$, and $\gamma''(t_0) \neq 0$.

**Worked example 5.3.** Find the order of point-contact, at the origin **0**, of the algebraic curves $y^2 - 2x = 0$ and $(x+1)^2 - y^2 - 1 = 0$.

*Solution.* We parametrise the parabola by $t \mapsto \mathbf{r}(t) = (\tfrac{1}{2}t^2, t)$. Since

$$\mathbf{r}' = (t, 1),$$

the parametrisation is regular. Also for the hyperbola

$$f(x, y) = x^2 + 2x - y^2,$$

we have **grad** $f = 2(x+1, -y) = \mathbf{0}$ if, and only if, $(x, y) = (-1, 0)$, which does not lie on the hyperbola. Therefore the algebraic curve has no singular points. We have

$$\gamma(t) = f(x(t), y(t)) = (\tfrac{1}{2}t^2)^2 + 2(\tfrac{1}{2}t^2) - t^2 = \tfrac{1}{4}t^4,$$

and thus by Proposition 5.2 the curves have 4–point contact at

$$\mathbf{r}(0) = (0, 0).$$

$\square$

**Worked example 5.4.** Write down a parametrisation of the circle

$$x^2 + y^2 = \tfrac{1}{4},$$

and show that this circle has 4–point contact with the parabola

$$f(x, y) = y^2 - x - \tfrac{1}{2} = 0$$

at $\left(-\tfrac{1}{2}, 0\right)$.

*Solution.* The circle can be parametrised by $t \mapsto \mathbf{r}(t) = \left(\tfrac{1}{2}\cos t, \tfrac{1}{2}\sin t\right)$. The point $\left(-\tfrac{1}{2}, 0\right)$ is given by $\mathbf{r}(\pi) = \left(-\tfrac{1}{2}, 0\right)$. The parametrisation is regular since $\mathbf{r}'(t) = \left(-\tfrac{1}{2}\sin t, \tfrac{1}{2}\cos t\right)$ is never zero: it is also smooth. The algebraic curve is non-singular since **grad** $f = (-1, 2y) \neq \mathbf{0}$. Now

$$\gamma(t) = f(x(t), y(t)) = \tfrac{1}{4}\sin^2 t - \tfrac{1}{2}\cos t - \tfrac{1}{2},$$
$$\gamma'(t) = \tfrac{1}{2}\sin t \cos t + \tfrac{1}{2}\sin t = \tfrac{1}{4}\sin 2t + \tfrac{1}{2}\sin t,$$
$$\gamma''(t) = \tfrac{1}{2}\cos 2t + \tfrac{1}{2}\cos t,$$
$$\gamma'''(t) = -\sin 2t - \tfrac{1}{2}\sin t, \quad \text{and}$$
$$\gamma^{(4)}(t) = -2\cos 2t - \tfrac{1}{2}\cos t.$$

Thus $\gamma(\pi) = \gamma'(\pi) = \gamma''(\pi) = \gamma'''(\pi) = 0$ and $\gamma^{(4)}(\pi) = -\frac{3}{2} \neq 0$, and therefore (by Definition 5.1) the two curves have 4–point contact at $\left(-\frac{1}{2}, 0\right)$.

$\square$

### 5.1.2 Contact and tangency

We now describe the conditions for tangency at a common point which is regular for the parametric curve and non-singular for the algebraic curve.

**Proposition 5.5.** *Let* $\mathbf{r}(t_0)$ *be a regular point of* $\Pi : t \mapsto \mathbf{r}(t)$ *and a non-singular point of* $\Gamma : f(x, y) = 0$. *Then* $\gamma'(t_0) = 0$ *if, and only if, the curves have the same tangent line at* $\mathbf{r}(t_0)$.

*Proof.* From the equation $\gamma(t) = f(\mathbf{r}(t))$, we have

$$\gamma' = f_x x' + f_y y' = \mathbf{grad}\, f \cdot \mathbf{r}'\,.$$

Therefore $\gamma'(t_0) = 0$ if, and only if, the non-zero vector $\mathbf{grad}\, f$, which is normal to $f(x, y) = 0$ at $\mathbf{r}(t_0)$, is orthogonal to the non-zero vector $\mathbf{r}'(t_0)$, which is tangent to $t \mapsto \mathbf{r}(t)$ at $\mathbf{r}(t_0)$; that is if, and only if, the two curves have the same tangent line at $\mathbf{r}(t_0)$. $\square$

For this case where $\mathbf{r}(t_0)$ is a regular point for the parametric curve and a non-singular point for the algebraic curve, the hypothesis $\gamma'(t_0) = 0$ is satisfied if, and only if, the curves have $\geq 2$–point contact at $\mathbf{r}(t_0)$.

**Definition 5.6.** A parametric curve and an algebraic curve have *simple tangency* at a common point, which is regular for the parametric curve and non-singular for the algebraic curve, if they have 2–point contact there.

**Worked example 5.7.** Find the points at which the parabola $y^2 = 4x$ and the ellipse $f(x, y) = 4x^2 + y^2 - 8x = 0$ meet and determine those points where the curves are mutually tangent. Is the tangency simple in the latter case?

*Solution.* Parametrise the parabola by $t \mapsto \mathbf{r}(t) = (t^2, 2t)$. Then

$$\gamma(t) = 4t^4 + (2t)^2 - 8t^2 = 4t^4 - 4t^2 = 4t^2(t-1)(t+1) = 0$$

if, and only if, $t = 0$ or $t = \pm 1$. These values of the parameter give the three distinct points where the two curves meet (there are four points of intersection including one iterated point). Now $\mathbf{r}' = 2(t, 1)$, and therefore $\mathbf{r}(t)$ is a regular curve. Also $\mathbf{grad}\, f = (8x - 8, 2y) = 0$ if, and only if, $(x, y) = (1, 0)$, which does not lie on the ellipse. Therefore the ellipse is a non-singular curve. Since $\gamma'(t) = 8(2t^3 - t)$, we have $\gamma'(0) = 0$, and $\gamma'(\pm 1) = \pm 8$. Also $\gamma''(t) = 8(6t^2 - 1)$, and $\gamma''(0) \neq 0$. Thus the two curves have simple tangency at $\mathbf{r} = \mathbf{0}$, and have transverse intersections at $(1, 2)$ and at $(1, -2)$. $\square$

### 5.1.3 Contact between a line and a curve

We now consider the special case where one of the curves is a line. We assume here that the line is given by its algebraic equation, and that the second curve is parametrised, at least locally. We consider more fully in Chapter 15 the alternative case where the line is given by its parametric equation and the curve by its algebraic equation.

The general line $\ell$ has equation

$$f(x,y) = (\mathbf{r} - \mathbf{a}) \cdot \mathbf{n} = 0 \quad (\mathbf{n} \neq \mathbf{0})$$

and the parametric curve has equation $t \mapsto \mathbf{r}(t)$. All points on $\ell$ are non-singular points, since $\mathbf{grad}\, f = \mathbf{n} \neq \mathbf{0}$. The line and the curve intersect at precisely those points with parameter values $t$ for which $\gamma(t) = f(\mathbf{r}(t)) = 0$. Assuming that the line and the curve intersect, we can choose $\mathbf{a} = \mathbf{r}(t_0)$ to be one of the points of intersection. We have

$$\gamma(t) = (\mathbf{r}(t) - \mathbf{r}(t_0)) \cdot \mathbf{n},$$
$$\gamma'(t) = \mathbf{r}'(t) \cdot \mathbf{n},$$

and generally

$$\gamma^{(k)}(t) = \mathbf{r}^{(k)}(t) \cdot \mathbf{n} \quad (k \geq 1).$$

Thus $\gamma'(t_0) = 0$ at a regular point $\mathbf{r}(t_0)$ if, and only if, $\mathbf{r}'(t_0) \cdot \mathbf{n} = 0$, that is if, and only if, $\ell$ is tangent to the curve. The next proposition is immediate.

**Proposition 5.8.** *Let $\mathbf{r}(t_0)$ be a regular point of $t \mapsto \mathbf{r}(t)$. Then all lines through $\mathbf{r}(t_0)$, other than the tangent line, have one-point contact at $\mathbf{r}(t_0)$ with the parametrised curve, and the tangent line has $m$–point contact where $m \geq 2$.*

**Worked example 5.9.** Find the order of point-contact which the lines through $\mathbf{0}$ have at the origin with the curve $t \mapsto \mathbf{r}(t) = (t, t + t^4)$.

*Solution.* We have $\mathbf{r}' = (1, 1 + 4t^3) \neq \mathbf{0}$ and therefore the curve is regular. For the line $py - qx = 0$ we have

$$\gamma(t) = p(t + t^4) - qt = t(p - q) + t^4 p.$$

Thus the line $y = x$ (given by $p = q = 1$) has 4–point contact with the curve at the origin and all other lines have 1–point contact. □

### Exercises

**5.1.** Find the orders of point-contact of the following pairs of algebraic curves:

a) $y = x^3$ and $y = 0$ at $(0,0)$, [order is 3]

b) $x^2 + y^2 = x^3$ and $x = 1$ at $(1,0)$,

c) $y^2 + y^3 = x^4$ and $y = -1$ at $(0,-1)$, [order is 4]

d)   $(x^2 + y^2 - 2x)^2 - 16(x^2 + y^2) = 0$ and $x = -2$ at $(-2, 0)$,

e)   the acnodal cubic $x^2 + y^2 = x^3$ and $(x - \frac{3}{2})^2 + y^2 = \frac{1}{4}$ at $(1, 0)$.
[order is 3]

**5.2.** Show that the parabola $y^2 = 4ax$ has 2–point contact with the circle $x^2 - x + y^2 = 0$ at the origin except in the case $a = \frac{1}{4}$, in which case it has 4–point contact.

**5.3.** Show that the parabola $y^2 = 4ax$ has 2–point contact with the hyperbola $(x + 1)^2 - y^2 - 1 = 0$ at the origin except in the case $a = \frac{1}{2}$, in which case it has 4–point contact.

**5.4.** Find $\gamma(t) = f(\mathbf{r}(t))$ where $\mathbf{r}(t) = (2 + 4\cos t, 4\sin t)$ is the circle and $f(x, y) = (x^2 + y^2 - 2x)^2 - 16(x^2 + y^2) = 0$ is the limaçon. Deduce that the curves meet in two points, having 2–point contact at each.

**5.5.** Find the parameter values of the points of intersection of the parabola $\mathbf{r}(t) = (-1 + t^2, t)$ and the nodal cubic $y^2 - 4x^2 - 4x^3 = 0$. Show that the curves have 4–point contact at $(-1, 0)$ and two simple intersections.
$$[0, \pm\sqrt{2}]$$

## 5.2 Invariance of point-contact order*

In defining the order of point-contact we made a number of implicit or explicit choices. We now show that the order of point-contact does not depend on these choices. The reader for whom this is a first course in geometry may wish to omit this section.

We first show that the order of point-contact does not depend on the choice of origin or axes.

**Theorem 5.10.** *The order of point-contact between an algebraic curve and a smooth parametrised curve at a common point, which is non singular for the algebraic curve and regular for the parametrised curve, is unchanged if the space in which they lie is changed by rotations, reflexions, or translations of origin.*

*Proof.* Let the (smooth) transformation $F : \mathbf{R}^2 \rightarrow \mathbf{R}^2$ be $(u, v) = F(x, y)$, with (smooth) inverse transformation $(x, y) = F^{-1}(u, v)$. Then the algebraic curve becomes $g(u, v) = 0$ where $g(u, v) = f\left(F^{-1}(u, v)\right)$, and the parametric curve becomes $\mathbf{w}(t) = F(\mathbf{r}(t))$. Thus

$$g(\mathbf{w}(t)) = f\left(F^{-1}(F(\mathbf{r}(t)))\right) = f(\mathbf{r}(t)) = \gamma(t).$$

This shows that $\gamma$ is unchanged by the transformation, and therefore the order of point-contact is unchanged by the transformation.  □

In §5.1.1 we defined the order of point-contact between two curves where one of them is given as a parametric curve and the other is given as an

---

* For the meaning of * see the preface.

algebraic curve. The question arises as to whether the order of point-contact is the same if the position is reversed, that is if the first curve is given algebraically and the second parametrically. We now answer this question in the affirmative. Although we consider contact between two algebraic curves, the proof we give is valid more generally for contact between two smooth Cartesian curves at a common non-singular point where each can be given a regular smooth local parametrisation.

We first show that the order of point-contact as previously defined does not change if the parametrised curve is given a different parametrisation. Let $f(x, y) = 0$ be an algebraic curve which is non-singular at $(a, b)$, and let a second curve have a smooth local parametrisation $t \mapsto \mathbf{r}(t)$ near $(a, b)$, with $\mathbf{r}(t_0) = (a, b)$, which is regular at $t_0$. In §5.1.1 we defined the order of point-contact of the two curves to be the order of the first non-vanishing derivative of $\gamma(t) = f(\mathbf{r}(t))$ at $t_0$.

**Theorem 5.11.** *The order of point-contact between an algebraic curve and a smooth parametrised curve at a common point, which is non singular for the algebraic curve and regular for the parametrised curve, is independent of the parametrisation.*

*Proof.* It is sufficient to prove this in the case where the second parametrisation is that of arc-length. Since reparametrising the curve in opposite direction by $t \mapsto \mathbf{r}(-t)$ clearly does not change the order of point-contact, we can assume that $\dfrac{ds}{dt} > 0$, and use the arc-length reparametrisation formula given in §4.6.2. Let $s \mapsto \mathbf{v}(s) = \mathbf{r}(t(s))$ be this arc-length parametrisation with $\mathbf{v}(s_0) = (a, b)$, and let

$$\omega(s) = f(\mathbf{v}(s)) = f(\mathbf{r}(t(s))) = \gamma(t(s)),$$

where $s \mapsto t(s)$ is the local inverse of the function $t \mapsto s(t)$. We now prove that the orders of the first non-vanishing derivatives of $\gamma(t)$ and $\omega(s)$ at $t_0$ and $s_0 = t(s_0)$ respectively are equal. We have

$$\omega(s) = \gamma(t(s))$$

$$\omega' = \gamma' \frac{dt}{ds}$$

$$\omega'' = \gamma'' \frac{dt}{ds} + \gamma' \left( \frac{dt}{ds} \right)^2, \quad \text{and generally}$$

$$\omega^{(k)} = \gamma^{(k)} \frac{dt}{ds} + \gamma^{(k-1)} B_{k-1}^k + \gamma^{(k-2)} B_{k-2}^k + \cdots + \gamma' B_1^k,$$

where the $B_j^i$ are functions of $t$. Since $\dfrac{dt}{ds} \neq 0$, we have easily by induction that $\omega = \omega' = \cdots = \omega^{(m)} = 0$ and $\omega^{(m+1)} \neq 0$ at $s_0$ if, and only if, $\gamma = \gamma' = \cdots = \gamma^{(m)} = 0$ and $\gamma^{(m+1)} \neq 0$ at $t_0$. Alternatively this can easily be proved using matrix theory since, for $k \geq 1$,

$$
\begin{pmatrix} \omega' \\ \omega'' \\ \vdots \\ \omega^{(k)} \end{pmatrix} = A \begin{pmatrix} \gamma' \\ \gamma'' \\ \vdots \\ \gamma^{(k)} \end{pmatrix}
$$

where $A$ is a triangular matrix whose determinant is $\left(\dfrac{dt}{ds}\right)^k \neq 0$.    □

**Remark 5.12.** The condition in this theorem that the parametrisations are regular at the common point cannot be removed. For example we showed in Example 5.3 that the parabola $t \mapsto (\frac{1}{2}t^2, t^3)$ and the hyperbola $x^2 + 2x - y^2 = 0$ have 4–point contact at $\mathbf{0}$. The parabola can also be parametrised by $t \mapsto (\frac{1}{2}t^6, t^9)$ apparently giving 12–point contact at $\mathbf{0}$; but, for this parametrisation, $\mathbf{0}$ is not a regular point. By restricting, for the moment, consideration to regular points we avoid such difficulties. It can however similarly be shown that point-contact is invariant for all parametrisations equivalent to a given one, that is where the parameters $t$ and $\tau$ are connected by a smooth or suitably differentiable function $\tau = \tau(t)$ where $\dfrac{d\tau}{dt} \neq 0$, at least locally at the point in question.

Next we show that the order of point-contact of two algebraic curves is well defined, in that it is independent of which of the two curves is parametrised. Let the algebraic curves $f(x, y) = 0$ and $g(x, y) = 0$ be non-singular at a common point $(a, b)$ and have regular local parametrisations, respectively $t \mapsto \mathbf{r}(t)$ and $\tau \mapsto \mathbf{v}(\tau)$, near $(a, b)$, with $\mathbf{r}(t_0) = (a, b)$ and $\mathbf{v}(\tau_0) = (a, b)$. In §5.1.1 we defined the order of point-contact of the two curves to be the order of the first non-vanishing derivative of $\gamma(t) = g(\mathbf{r}(t))$ at $t_0$. Let $\sigma(\tau) = f(\mathbf{v}(\tau))$. We must show that the order of the first non-vanishing derivative of $\sigma(\tau)$ at $\tau_0$ is equal to the order of the first non-vanishing derivative of $\gamma(t)$ at $t_0$.

**Theorem 5.13.** *The order of point-contact of two algebraic curves $f(x, y)$ and $g(x, y)$ at a common point at which both curves are non-singular is independent of which of the curves is parametrised.*

*Proof.* One-point contact occurs at a common point if, and only if, the non-zero vectors $\mathbf{grad}\, f$ and $\mathbf{grad}\, g$ are not parallel there. In this case the result is immediate. We consider next the case where the curves have a common tangent line at the common point. After making a rotation and translation if necessary, we can assume further that the common point is the origin $\mathbf{0}$ and that the common tangent line is $y = 0$. Let the algebraic curves in this new position be $f(x, y) = 0$ and $g(x, y) = 0$. The curves are both non-singular at $\mathbf{0}$, and hence the partial derivatives $f_y(\mathbf{0})$ and $g_y(\mathbf{0})$ are both non-zero. Therefore, by Theorem 4.22, the curves can be parametrised near the origin by the $x$–coordinate, with regular parametrisations $x \mapsto \mathbf{r}(x) = (x, R(x))$ and $x \mapsto \mathbf{v}(x) = (x, V(x))$ respectively. Define $\gamma(x) = g(\mathbf{r}(x))$ and $\omega(x) = f(\mathbf{v}(x))$. We need to show that the orders of the first non-vanishing derivatives of $\gamma(x)$ and $\omega(x)$ are equal at $x = 0$. Let $W(x) = R(x) - V(x)$. We shall prove that the order of the first non-vanishing derivative of $\gamma(x)$ is equal to the order of the first non-vanishing derivative of $W(x)$. The theorem will then follow by symmetry. We have

$$\gamma(x) = g(\mathbf{r}(x)) = g(\mathbf{v}(x) + (0, R(x) - V(x)))$$
$$\gamma' = \mathbf{grad}\, g \cdot \mathbf{v}' + g_y \cdot (R' - V')$$
$$= g_y \cdot W', \text{ since } g(\mathbf{v}(x)) \text{ is identically zero,}$$
$$\gamma'' = g_y \cdot W'' + A_1^2 \cdot W', \text{ and generally}$$
$$\gamma^{(k)} = g_y \cdot W^{(k)} + A_{k-1}^k \cdot W^{(k-1)} + \cdots + A_1^k \cdot W',$$

where the $A_j^i$ are functions of $x$. Since $g_y \neq 0$, we now have easily by induction that $W = W' = \cdots = W^{(m)} = 0$ and $W^{(m+1)} \neq 0$ at $x = 0$ if, and only if, $\gamma = \gamma' = \cdots = \gamma^{(m)} = 0$ and $\gamma^{(m+1)} \neq 0$ at $x = 0$. Alternatively this can also be proved using the triangular matrix connecting $W, W', \ldots, W^{(m)}$ and $\gamma, \gamma', \ldots, \gamma^{(m)}$ as in the proof of Theorem 5.11. $\qquad\square$

**Remark 5.14.** The local parametrisation by the $x$–coordinate which we have considered in the proof of this theorem corresponds for a curve in general position to projecting a point $(x, y)$ of the curve near $(a, b)$ orthogonally onto the tangent line at $(a, b)$ and giving $(x, y)$ the parameter value corresponding to the directed distance in a specified direction along the tangent line from $(a, b)$ to the projected point. We could use this local parametrisation to prove Theorem 5.13 directly, without first moving the point of contact to the origin. It also follows from the proof that the order of point-contact of the two curves is equal to the order of point-contact at $0$ between the curve $x \mapsto (x, R(x) - V(x))$ and its tangent line $y = 0$. Given two parametric curves which have the same tangent at a common point, we can move the origin to the common point, and rotate the curves until the $x$–axis becomes the common tangent. Using the inverse function theorem, we can reparametrise the curves by the $x$–coordinate as $x \mapsto (x, R(x))$ and $x \mapsto (x, V(x))$. We can then define the order of point contact to be the order of the first non-vanishing derivative of $R(x) - V(x)$.

## Exercise

**5.6.** Show that two curves which each have point-contact of order $n$ with a third curve at a given point have point-contact of order $\geq n$ with each other at the point.

(*Hint:* Assume that the common point is at the origin, the common tangent is the $x$–axis, and the curves are parametrised by $x$.)

## 5.3 Inflexions and undulations

We have shown in §5.1.3 that a line through a regular point of a parametrised curve is tangent to the curve if, and only if, the line has $\geq$ 2–point contact with the curve at the point. For the tangent line to have higher point-contact than 2 imposes conditions on the shape of the curve near the point. We now investigate some of these conditions. Simple inflexions correspond to 3–point contact, simple undulations correspond to 4–point contact, and for $n$–point contact determines whether the curve crosses the tangent line or stays on one side of it depending on whether $n$ is odd or even.

Let $\ell$ to be the tangent line at a regular point $\mathbf{a} = \mathbf{r}(t_0)$ of the parametric curve $t \mapsto \mathbf{r}(t)$: we define $\mathbf{n} = (-y'(t_0), x'(t_0)) \neq \mathbf{0}$ to be the normal vector obtained by rotating the tangent vector $\mathbf{r}'(t_0)$ counter-clockwise through an angle $\dfrac{\pi}{2}$. The equation of the line $\ell$ is $(\mathbf{r} - \mathbf{a}) \cdot \mathbf{n} = 0$, and thus we have

$$\gamma(t) = f(\mathbf{r}(t)) = (\mathbf{r}(t) - \mathbf{r}(t_0)) \cdot \mathbf{n} = (\mathbf{r}(t) - \mathbf{r}(t_0)) \cdot (-y'(t_0), x'(t_0)).$$

We shall use the order of contact between the curve and its tangent line to describe specific properties of the curve itself.

### 5.3.1 Simple inflexions

A simple inflexion occurs where the curve 'instantaneously' crosses its tangent line three times at the point of contact (see Figure 5.1).

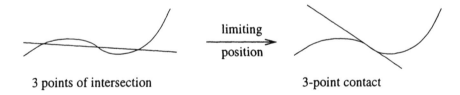

limiting
position $\longrightarrow$

3 points of intersection                    3-point contact

Figure 5.1 *Simple inflexion.*

**Definition 5.15.** A *point of simple inflexion* of a parametric curve is a regular point $\mathbf{r}(t_0)$ on the curve at which the tangent line has 3–point contact, that is second-order contact. A *simple inflexion* of a regular curve is that part of the curve close to a point of simple inflexion.

Notice how the point at which the inflexion occurs is distinguished from the inflexion itself. 'Inflexion' describes the 'shape' of the curve near the point. Compare this definition of inflexion with the 'point at which a local maximum occurs' and the 'local maximum' itself; the latter relates to the shape and inclination of the curve near the specific point. The 'shape' of a local maximum at $x = a$ is 'concave downwards'and 'lying below the line $y = f(a)$ near $a$'.

**Worked example 5.16.** Show that the curve $\mathbf{r}(t) = (t, t+t^3)$ has a simple inflexion at the origin.

*Solution.* We have $\mathbf{r}'(t) = (1, 1 + 3t^2)$ and $\mathbf{r}'(0) = (1, 1)$. Therefore the tangent line at $(0, 0)$ is $y - x = 0$, and

$$\gamma(t) = y(t) - x(t) = (t + t^3) - t = t^3.$$

The first non-vanishing derivative of $\gamma$ at the origin has order 3. Equivalently $\gamma$ has a zero of order 3 at $t = 0$. Thus the curve has a simple inflexion at the origin.                                                                    □

We now give a criterion for points of inflexions which is potentially more useful than the original definition because it enables us more readily to locate points of inflexion on a general curve. First we need to define linear points. These occur where the rate of change of the tangent vector $\mathbf{r}'(t)$ is parallel to the tangent vector itself.

**Definition 5.17.** A *linear point* of a parametric curve $t \mapsto \mathbf{r}(t)$ is a regular point $\mathbf{r}(t_0)$ at which $\mathbf{r}'(t_0)$ and $\mathbf{r}''(t_0)$ are parallel, that is at which

$$x'y'' = y'x''.$$

**Criterion 5.18.** *The regular point $\mathbf{r}(t_0)$ is a point of simple inflexion if, and only if, $\mathbf{r}'(t_0)$ and $\mathbf{r}''(t_0)$ are parallel and $\mathbf{r}'''(t_0)$ is not parallel to $\mathbf{r}'(t_0)$ (this allows the possibility that $\mathbf{r}''(t_0) = 0$).*

*Proof.* We have

$$\gamma(t) = (\mathbf{r}(t) - \mathbf{r}(t_0)) \cdot \mathbf{n},$$

where $\mathbf{n} = (-y'(t_0), x'(t_0))$, and therefore

$$\gamma'(t_0) = \mathbf{r}'(t_0) \cdot \mathbf{n} = 0,$$

since $\mathbf{r}'(t_0)$ is a tangent vector. Also

$$\gamma''(t_0) = \mathbf{r}''(t_0) \cdot \mathbf{n} = 0$$

if, and only if, $\mathbf{r}''(t_0)$ is orthogonal to $\mathbf{n}$, and therefore if, and only if, $\mathbf{r}'(t_0)$ and $\mathbf{r}''(t_0)$ are parallel. Again

$$\gamma'''(t_0) = \mathbf{r}'''(t_0) \cdot \mathbf{n} \neq 0$$

if, and only if, $\mathbf{r}'''(t_0)$ is not orthogonal to $\mathbf{n}$, and therefore if, and only if, $\mathbf{r}'(t_0)$ and $\mathbf{r}'''(t_0)$ are not parallel. The result follows easily.            □

The reader should be familiar with the criteria for two vectors to be parallel (see §0.6), namely that $(a, b)$ is parallel to $(c, d)$ if, and only if,

$ad = bc$, or, equivalently, that one vector is a multiple of the other. The linear points on the parametric curve are given by solving the equation $x'y'' = y'x''$. To determine the points of simple inflexion, we then consider whether or not $\mathbf{r}'''$ is parallel to $\mathbf{r}'$ for each linear point, or we find the order of point-contact at each linear point.

**Worked example 5.19.** Show that $t \mapsto \mathbf{r}(t) = (t - 3t^2, 2t + 3t^3)$ has two linear points and that both are points of simple inflexion.

*Solution.* The curve is regular since $\mathbf{r}'(t) = (1 - 6t, 2 + 9t^2) \neq \mathbf{0}$. Now $\mathbf{r}'(t)$ and $\mathbf{r}''(t) = (-6, 18t)$ are parallel if, and only if,

$$3t(1 - 6t) + (2 + 9t^2) = -9t^2 + 3t + 2 = -(3t - 2)(3t + 1) = 0,$$

that is if, and only if, $t = \frac{2}{3}$ or $t = -\frac{1}{3}$. These values of $t$ give the two linear points. Also $\mathbf{r}'(\frac{2}{3}) = (-3, 6)$ and $\mathbf{r}'(-\frac{1}{3}) = (3, 3)$ are not parallel to $\mathbf{r}''' = (0, 18)$. Therefore the two linear points are both points of simple inflexion.  □

### 5.3.2 Simple undulations

A simple undulation occurs where the curve 'instantaneously ' crosses its tangent line four times at the point of contact (see Figure 5.2).

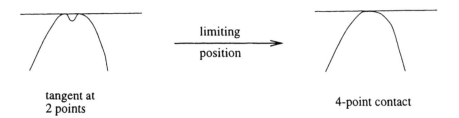

tangent at
2 points

limiting
position

4-point contact

Figure 5.2 *Simple undulation.*

**Definition 5.20.** A *point of simple undulation* of a parametric curve is a regular point $\mathbf{r}(t_0)$ on the curve at which the tangent line has 4–point contact, that is third-order contact. A *simple undulation* is that part of a curve close to a point of simple undulation.

**Worked example 5.21.** Show that the curve $\mathbf{r}(t) = (t + t^5, t + t^4)$ has a simple undulation at the origin.

*Solution.* We have $\mathbf{r}'(t) = (1 + 5t^4, 1 + 4t^3)$ and $\mathbf{r}'(0) = (1, 1)$. Therefore the tangent line at $(0, 0)$ is $y - x = 0$, and

$$\gamma(t) = y(t) - x(t) = (t + t^5) - (t + t^4) = -t^4 + t^5.$$

The first non-vanishing derivative of $\gamma$ at the origin has order 4. Equivalently $\gamma$ has a zero of order 4 at $t = 0$. Thus the curve has a simple undulation at the origin. $\qquad\square$

We now give a criterion for points of simple undulation which can be used to locate such points on a general curve. Again these points are initially located as linear points.

**Criterion 5.22.** *The regular point* $\mathbf{r}(t_0)$ *is a point of simple undulation if, and only if,* $\mathbf{r}'(t_0)$, $\mathbf{r}''(t_0)$ *and* $\mathbf{r}'''(t_0)$ *are parallel, and* $\mathbf{r}^{(4)}(t_0)$ *is not parallel to* $\mathbf{r}'(t_0)$. *(This allows the possibility that one or both of* $\mathbf{r}''(t_0)$ *and* $\mathbf{r}'''(t_0)$ *is zero.)*

*Proof.* As in the proof of Criterion 5.18, we have

$$\gamma(t_0) = \gamma'(t_0) = \gamma''(t_0) = \gamma'''(t_0) = 0$$

if, and only if, $\mathbf{r}'(t_0)$, $\mathbf{r}''(t_0)$ and $\mathbf{r}'''(t_0)$ are parallel. Now

$$\gamma^{(4)}(t_0) = \mathbf{r}^{(4)}(t_0) \cdot \mathbf{n} \neq 0$$

if, and only if, $\mathbf{r}^{(4)}(t_0)$ is not orthogonal to $\mathbf{n}$, and therefore if, and only if, $\mathbf{r}'(t_0)$ and $\mathbf{r}^{(4)}(t_0)$ are not parallel. The result follows easily. $\qquad\square$

Since $\mathbf{r}'(t_0) \neq 0$ at a point of undulation, in practice we show that $\mathbf{r}'$ and $\mathbf{r}''$ are parallel, that $\mathbf{r}'$ and $\mathbf{r}'''$ are parallel, and that $\mathbf{r}'$ and $\mathbf{r}^{(4)}$ are not parallel. The points of simple undulation, if they exist, may be located by finding the linear points and then determining which of these is a point of simple undulation, either by using Criterion 5.22 or by determining the order of point contact between the curve and its tangent line.

### 5.3.3 Higher order inflexions and undulations

Points of inflexion and points of undulation are not all simple in the above sense. We now consider higher inflexions and higher undulations.

**Definition 5.23.** A *point of higher inflexion* at a regular point of a parametric curve is a point $\mathbf{r}(t_0)$ on the curve at which the tangent line has odd–point contact ($\geq 5$). A *point of higher undulation* at a regular point of a parametric curve is a point $\mathbf{r}(t_0)$ on the curve at which the tangent line has even–point contact ($\geq 6$).

The shapes of the curve near such points are similar to the corresponding shapes in the simple cases in that the curve crosses its tangent line at an inflexion and does not cross its tangent line at an undulation. However, near to the point of tangency, the curves will be closer to the tangent line than in the case of simple inflexion or simple undulation. Compare 'simple' as used here with the 'simple' local minimum at the origin of $x \mapsto x^2$ and the 'higher' local minimum at the origin of $x \mapsto x^4$.

Again we give a criterion which can be used for locating such points: the proof is similar to the special cases given in Criterion 5.18 and Criterion 5.22.

**Criterion 5.24.** *At the regular point* $\mathbf{r}(t_0)$ *the curve* $t \mapsto \mathbf{r}(t)$ *has n–point contact with its tangent line if, and only if, the first derivative, at* $t = t_0$, *of* $\mathbf{r}(t)$ *which is not parallel to* $\mathbf{r}'(t_0)$ *is the n–th derivative* $\mathbf{r}^{(n)}(t_0)$ $(n \geq 3)$.

As a general procedure to find higher inflexions and undulations, we first find the linear points, then determine the type of singularity either by finding the order of point-contact with the tangent line at each linear point, or using Criterion 5.24 at each linear point.

**Worked example 5.25.** Show that the curve $y = x + x^6$ has a point of undulation at the origin, and no other linear point.

*Solution.* The curve is parametrised as a graph by $\mathbf{r}(x) = (x, x + x^6)$. We have $\mathbf{r}' = (1, 1 + 6x^5) \neq \mathbf{0}$ and therefore the curve is regular. Also $\mathbf{r}'' = (0, 30x^4) = 0$ if, and only if, $x = 0$. Therefore the only linear point is at $x = 0$. Now $\mathbf{r}'(0) = (1, 1)$, the next four derivatives are zero at $x = 0$, and $\mathbf{r}^{(6)} = (0, 720)$ which is not parallel to $(1, 1)$. Thus the first derivative of $\mathbf{r}(x)$ which is not parallel to $\mathbf{r}'(0)$ is that of even order 6 and therefore we have a higher undulation.  □

**Remark 5.26.** More general definitions of inflexion and undulation are given as follows. The curve has an *inflexion at* $t = t_0$ if $x'y'' - y'x''$ is zero at $t_0$ and has different signs immediately before and after $t_0$. The curve has an *undulation at* $t = t_0$ if $x'y'' - y'x''$ is zero at $t_0$ and has the same sign immediately before and after $t_0$. The simple and higher inflexions and undulations we have previously considered are special cases of these more general ones.

### 5.3.4 Special case: curve parametrised by x

We now consider the case where $x$ is chosen as a parameter, or at least as a local parameter. In this case the conditions for an inflexion take on a special form. Let the curve be $x \mapsto \mathbf{r}(x) = (x, f(x))$; thus, the curve is the graph of the function $x \mapsto f(x)$. Since $\mathbf{r}' = (1, f')$, the curve is regular at all points for which $f'$ exists.

We recall that a *critical point* (or *stationary point*) of a function $x \mapsto f(x)$ is a point $x_0$ at which $\dfrac{dy}{dx} = f'(x_0) = 0$. See Figure 5.3.

**Criterion 5.27.** *The point* $\mathbf{r}(x_0)$ *is a linear point if, and only if,*

$$f''(x_0) = 0.$$

*The point* $\mathbf{r}(x_0)$ *is a point of inflexion if, and only if,* $f''(x_0) = 0$ *and the first non-vanishing derivative of* $f(x)$ *of order* $\geq 3$ *is of odd order* $n \geq 3$. *The inflexion is simple for* $n = 3$, *and is a higher inflexion for* $n \geq 5$. *The point* $\mathbf{r}(x_0)$ *is a point of undulation if, and only if,* $f''(x_0) = 0$ *and the first non-vanishing derivative of* $f(x)$ *of order* $\geq 3$ *is of even order* $n \geq 4$. *The undulation is simple for* $n = 4$, *and is a higher undulation for* $n \geq 6$.

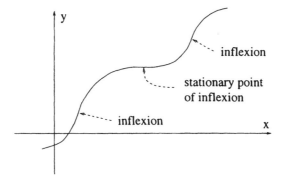

Figure 5.3 *Inflexions of a graph.*

*Proof.* We have

$$\mathbf{r}'(x) = (1, f'(x)),$$
$$\mathbf{r}''(x) = (0, f''(x)),$$

and more generally

$$\mathbf{r}^{(k)}(x) = (0, f^{(k)}(x))$$

for $k \geq 2$. The results are now immediate. □

Similar results to those above apply if a curve is parametrised by the coordinate $y$ rather than by $x$.

The reader should compare carefully the statement of this criterion with that of the general case in Criterion 5.24. This criterion refers to the first *non-vanishing* derivative of $f(x)$, whereas Criterion 5.24 refers to the first derivative of $\mathbf{r}(t)$ *not parallel* to $\mathbf{r}'$.

As a general procedure to find inflexions and undulations, we first find the linear points, then determine the type of singularity either by finding the order of point-contact with the tangent line at each linear point, or using Criterion 5.27 at each linear point.

**Worked example 5.28.** Show that $y = x^3$ has a stationary point of simple inflexion at the origin.

*Solution.* This is immediate since $\dfrac{dy}{dx} = 3x^2$ and $\dfrac{d^2y}{dx^2} = 6x$ are both 0 at $x = 0$ and $\dfrac{d^3y}{dx^3} = 6 \neq 0$ at $x = 0$. □

**Worked example 5.29.** Show that $y = x + x^4$ has a simple undulation at the origin, which is not a stationary point.

*Solution.* This is immediate since $\dfrac{dy}{dx} = 1 + 4x^3$, $\dfrac{d^2y}{dx^2} = 12x^2$, and

$$\frac{d^3y}{dx^3} = 24x.$$

At the origin $\dfrac{d^2y}{dx^2} = \dfrac{d^3y}{dx^3} = 0$, and $\dfrac{d^4y}{dx^4} = 24 \neq 0$, but $\dfrac{dy}{dx} = 1 \neq 0.$　　□

**Worked example 5.30.** Show that $y = \dfrac{1-x}{x^2+1}$ has three points of simple inflexion.

*Solution.* Differentiating gives

$$y' = \frac{x^2 - 2x - 1}{(x^2+1)^2}, \quad \text{and} \quad y'' = \frac{-(x+1)(x-2+\sqrt{3})(x-2-\sqrt{3})}{(x^2+1)^3}.$$

Thus $y'' = 0$ if, and only if, $x = -1, 2 \pm \sqrt{3}$; these values give three linear points. By Lemma 8.4, we see easily that $y''' \neq 0$ for each of these three values of $x$, and therefore they give three points of simple inflexion.　　□

**Remark 5.31.** A more general definition of inflexion and undulation in this case of a curve parametrised by $x$ is given as follows. Let $f(x)$ be a function with continuous second derivative for which $f''(x_0) = 0$ and $x_0$ is an isolated zero of $f''$. We define $x_0$ to be a point of inflexion if $f''$ changes sign at $x_0$, and to be a point of undulation if $f''$ has the same sign immediately before and after $x_0$. It can be shown, using Taylor's theorem, that this definition is equivalent to those given above in the cases which they cover.

### 5.3.5 $\dfrac{d^2y}{dx^2}$ the second derivative*

To determine the inflexions of a curve parametrised by $t$, we find the linear points and use the result of Criterion 5.18. However, for completeness, we note here how to determine $\dfrac{d^2y}{dx^2}$ for such a curve. Implicit in what follows is the existence of a local parametrisation giving $y$ as a function of $x$: the condition for this is that $\mathbf{r}'(t_0)$ is not parallel to the $y$–axis, that is that $\dfrac{dx}{dt} \neq 0$ at $t = t_0$. For, by the inverse function theorem, the smooth function $t \mapsto x(t)$ has a smooth local inverse $x \mapsto t(x)$ near $t = t_0$ $(n \geq 2)$, and $x \mapsto (x, y(t(x)))$ is the required smooth local parametrisation. Differentiating $y(x(t))$ and $\dfrac{dy}{dx}$ with respect to $t$ we have, by the chain rule,

$$y' = \frac{dy}{dx}\,x' \quad \text{and} \quad \frac{y''}{x'} - \frac{x''y'}{(x')^2} = \frac{d}{dt}\left(\frac{y'}{x'}\right) = \frac{d}{dt}\left(\frac{dy}{dx}\right) = \frac{d^2y}{dx^2}\,x'.$$

* For the meaning of * see the preface.

These two equations can be used to determine $\dfrac{d^2y}{dx^2}$ and $\dfrac{dy}{dx}$ in terms of the derivatives of $x$ and $y$ with respect to $t$. Thus

$$\frac{dy}{dx} = \frac{y'}{x'} \quad \text{and} \quad \frac{d^2y}{dx^2} = \frac{x'y'' - x''y'}{(x')^3}.$$

The higher derivatives $\dfrac{d^ny}{dx^n}$ can be calculated in a similar way.

**Exercises**

**5.7.** Find $\dfrac{d^3y}{dx^3}$ in terms of the derivatives of $x(t)$ and $y(t)$.

In the following exercises, only consider linear points, inflexions, and undulations at regular points.

**5.8.** Find the linear points of $t \mapsto \mathbf{r}(t) = (t - t^2, 4t + 3t^3)$.

**5.9.** Show that $t \mapsto \mathbf{r}(t) = (t^2 - t^3, 2t^2 - t^4)$ has no linear points.

**5.10.** Show that $t \mapsto \mathbf{r}(t) = (\frac{1}{2}t^2, \frac{1}{4}t^4 + \frac{1}{5}t^5)$ has one linear point and that it is a point of simple inflexion.

**5.11.** Show that $t \mapsto \mathbf{r}(t) = (-2t^2 + \frac{2}{3}t^3, \frac{1}{2}t^3 - \frac{1}{4}t^2)$ has two linear points, each of which is a point of simple inflexion.                     [1, 3]

**5.12.** Show that $t \mapsto \mathbf{r}(t) = (\frac{3}{2}t^2 - t^3, \frac{10}{9}t^2 - t^4)$ has two linear points, each of which is a point of simple inflexion.

**5.13.** Find the points of inflexion of the curves
  a)  $\mathbf{r}(t) = (1 + t + t^2, 1 + t + t^2 + t^3)$,                 [0, −1; simple, 3]
  b)  $\mathbf{r}(t) = (2t + t^4, 1 + t^3)$.                              [0, 1; simple, 3]
In each case determine whether or not the inflexion is simple and state the order of point-contact the curve has with its tangent line at the point of contact.

**5.14.** Show that the curve $\mathbf{r}(t) = (\frac{4}{3}t + t^2, 3 + t^4)$ is regular. Find its inflexions and undulations.

**5.15.** Show that the curve $\mathbf{r}(t) = (-t + t^3, t^4)$ is regular. Find its inflexions and undulations.                     [±1, simple inflexions; 0, simple undulation]

**5.16.** Show that the curve $\mathbf{r}(t) = (t^6, -\frac{8}{5}t + t^2)$ has a higher undulation at the origin. Find its inflexions.

**5.17.** Show that each of the following curves has a point of simple inflexion at the origin
  a)  $y = x \cos x$,
  b)  $y = \tan x$,
  c)  $y = x^2 \log(1 - x)$.

**5.18.** Show that $y = xe^x$ has a point of simple inflexion at

$$\left(-2, -\frac{2}{e^2}\right).$$

**5.19.** Show that each of the following curves has two points of simple inflexion.

> a)  $12y = x^4 - 16x^3 + 42x^2 + 12x + 1$,                                [1, 7]
> b)  $y = x - 15x^2 + x^6$.                                                       [±1]

**5.20.** Show that each of the following curves has points of simple inflexion as indicated

> a)  $y = \dfrac{x}{(x+9)^2}$.             $x = 18$,
>
> b)  $y = \dfrac{1}{x^2 + 3}$             $x = \pm 1$,
>
> c)  $y = \dfrac{x^3}{x^2 + 3}$             $x = 0, \pm 3$.

**5.21.** Show that the curve $y = x + x^8$ has a point of higher undulation at the origin.

---

Solve the following three exercises by finding the order of point-contact between the algebraic curve $f(x, y) = 0$ and its tangent line $\mathbf{r}(t) = t(p, q)$, where $(-q, p) = \mathbf{grad}\, f$.

**5.22.** Show that the cubic curve $y = axy + by^2 + cx^3$   $(c \neq 0)$ has a point of simple inflexion at the origin.

**5.23.** Show that the quartic curve $y = ax^4 + bx^2 y^2 + cy^4$   $(a \neq 0)$ has a point of simple undulation at the origin.

**5.24.** Show that the quintic curve $y = xy^2 + 3x^3 y - x^5 + x^4 y$ has a point of inflexion at the origin, but that it is not simple inflexion.

---

**5.25.** Show that 'linear point', 'point of simple inflexion' and 'point of simple undulation' are independent of the parametrisation.
(*Hint:* Compare a given parametrisation with the arc-length parametrisation.)

## 5.4 Geometrical interpretation of $n$–point contact

We note the following criteria (see Figure 5.4) for two algebraic curves at a common point which is non-singular for both, or for an algebraic curve and a parametric curve at a common point which is non-singular for the algebraic curve and regular for the parametric curve.

> a)  **1–point contact:** the curves do not have the same tangent line.
> b)  **$\geq$ 2–point contact:** the curves touch, that is they have the same tangent line.
> c)  **3–point contact with one curve a line:** the second curve has a point of simple inflexion at the point of contact.

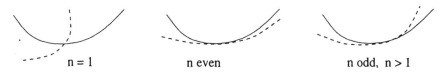

n = 1    n even    n odd, n > 1

Figure 5.4 *n–point contact of two curves.*

d)  **3–point contact with one curve a circle:** the circle is the circle of curvature (osculating circle) of the second curve at the point of contact – it is uniquely determined (see Chapter 9).

e)  **4–point contact with one curve a line:** the second curve has an undulation at the point of contact.

f)  *n*–**point contact** – *n* even: the curves do not 'cross' at the point, that is each stays on one side of the other locally. Where one curve is a line and $n \geq 6$, the second has a higher undulation.

g)  *n*–**point contact** – *n* odd: the curves do 'cross' at the point, that is the first curve crosses from one side of the second curve to the other side of the second curve. Where one curve is a line and $n \geq 5$, the second has a higher inflexion.

The limiting positions of the diagrams in Figure 5.5 illustrate the order of point-contact between a (regular) 'geometric' curve and a line. At 'singularities' of parametric or algebraic curves, the relationship between the order and tangency, for example, can be different. We shall discuss such singularities later; they include cusps and nodes.

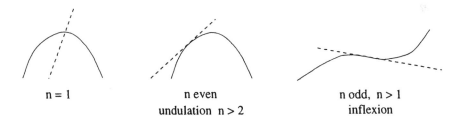

n = 1    n even    n odd, n > 1
        undulation n > 2    inflexion

Figure 5.5 *n–point contact of a line and a curve.*

## 5.5  An analytical interpretation of contact

By definition, a smooth curve $t \mapsto \mathbf{r}(t)$, regular at $t_0$, and an algebraic curve $f(x, y) = 0$, non-singular at $\mathbf{r}(t_0)$, have *n*–point contact at $\mathbf{r}(t_0)$ if

$$\gamma(t_0) = \gamma'(t_0) = \gamma''(t_0) = \ldots = \gamma^{(n-1)}(t_0) = 0 \text{ and } \gamma^{(n)}(t_0) \neq 0,$$

where $\gamma(t) = f(\mathbf{r}(t))$. The next proposition (see the reduction of zeros theorem in §4.7.3) gives an analytical interpretation of the order of point-contact.

**Proposition 5.32.** *Let $t \mapsto \mathbf{r}(t)$ be a smooth curve which is regular at $t = t_0$, and let the algebraic curve $f(x, y) = 0$ be non-singular at $\mathbf{r}(t_0)$, then the order of point contact of the two curves at $\mathbf{r}(t_0)$ is $n \geq 1$ if, and only if, $\gamma(t) = (t - t_0)^n \psi(t)$ where $\psi$ is smooth at $t = t_0$ and $\psi(t_0) \neq 0$.*

In the case where $\gamma(t)$ is a polynomial, we can also write

$$\gamma(t) = b_n(t - t_0)^n + \text{ (finitely many) terms of higher order in } (t - t_0).$$

A similar but infinite expansion applies in the case where $\gamma$ is analytic, but is not in general valid for smooth functions.

Proposition 5.32 tells us that, close to a (regular) point of $n$–point contact with its tangent line, the curve $t \mapsto \mathbf{r}(t)$ approximates an $n$-th power curve; indeed it approximates the curve $y = cx^n$ after the latter has been rotated and translated to an appropriate position, where $c$ is a non-zero constant. The form of the curve near the origin after such a rotation and translation is

$$y = x^n \zeta(x)$$

where $\zeta(0) = c \neq 0$.

# 6

# Cusps, non-regular points

In this chapter we give further applications of the method of contact order; in particular, we show how tangent lines at cusps and other non-regular points of parametric curves can be determined, and how quadratic cusps can be classified by their order.

In Chapter 4 we showed that parametric curves have tangents at regular points and that algebraic curves have tangents and can be parametrised locally at non-singular points. We consider here, for parametrised curves, the existence of tangents at non-regular points and in particular at cusps. In Chapter 5, we showed that the tangent line to a parametric curve at a regular point is the unique line having $\geq 2$–point contact with the curve at the point, all other lines having 1–point contact. We consider here the order of contact between a line and a parametrised curve at certain non-regular points: all the lines through such points have point-contact of order $\geq 2$. In the case of a quadratic cusp, the cuspidal tangent line is the unique line having $\geq 3$–point contact with the curve at the point, all other lines having 2–point contact.

## 6.1 Cusps

Most non-regular points of parametric curves which we consider in this book are cusps, which we now define. Recall that a non-regular value of $t$ is a value for which $\mathbf{r}'(t) = \mathbf{0}$.

**Definition 6.1.** A *quadratic cusp* is a point with non-regular value $t_0$ for which $\mathbf{r}''(t_0) \neq \mathbf{0}$. An *ordinary cusp* is a point with non-regular value $t_0$ for which $\mathbf{r}''(t_0) \neq \mathbf{0}$, and $\mathbf{r}'''(t_0)$ is not parallel to $\mathbf{r}''(t_0)$. A cusp is *not ordinary* if $\mathbf{r}''(t_0)$ and $\mathbf{r}'''(t_0)$ are parallel.

**Definition 6.2.** A *branch* near a cusp $\mathbf{r}(t_0)$ is either that part of the curve near $t_0$ for which $t \geq t_0$ or that part of the curve near $t_0$ for which $t \leq t_0$.

A standard example of an ordinary cusp is given by the usual parametrisation of the semi-cubical parabola $y^2 = x^3$ (see Figure 3.1). For many named curves, the cusps which occur are all ordinary.

**Worked example 6.3.** Prove the semi-cubical parabola $\mathbf{r}(t) = (t^2, t^3)$ has an ordinary cusp at $t = 0$.

*Solution.* We have $\mathbf{r}' = (2t, 3t^2)$, $\mathbf{r}'' = (2, 6t)$ and $\mathbf{r}''' = (0, 6)$. Thus $\mathbf{r}' = 0$ if, and only if, $t = 0$, and therefore the curve has precisely one non-regular point, and that is at $t = 0$. Since $\mathbf{r}''(0) = (2, 0) \neq 0$, the non-regular point is a cusp. Since $\mathbf{r}''(0) = (2, 0)$ and $\mathbf{r}'''(0) = (0, 6)$ are not parallel, the cusp is ordinary.                                                            □

**Worked example 6.4.** Show that the curve $\mathbf{r} = (t^2, t^5)$ has a cusp at $t = 0$, but that it is not an ordinary cusp.

*Solution.* We have $\mathbf{r}' = (2t, 5t^4)$, $\mathbf{r}'' = (2, 20t^3)$ and $\mathbf{r}''' = (0, 60t^2)$. Thus $\mathbf{r}' = 0$ if, and only if, $t = 0$, and therefore the curve has precisely one non-regular point: since $\mathbf{r}''(0) = (2, 0) \neq 0$, this is a cusp. Since $\mathbf{r}''(0) = (2, 0)$ and $\mathbf{r}'''(0) = (0, 0)$ are parallel, the cusp is not ordinary.             □

A parametric cusp can have coincident branches.

**Example 6.5.** The parametrisation $t \mapsto \mathbf{r}(t) = (t^2, t^4)$ (for $t$ in $R$) covers twice the curve $\{y = x^2, x \geq 0\}$, which is part of a parabola. This parametric curve has two coincident branches and there is a cusp at the origin corresponding to $t = 0$. Similarly $t \mapsto \mathbf{r}(t) = (t^2, 0)$ (for $t$ in $R$) consists of two coincident copies of the half-line $\{x \geq 0, y = 0\}$.

In general the cusps we consider do not have coincident branches.

## 6.2 Tangents at cusps

At a regular point $\mathbf{r}(t_0)$ the tangent is parallel to $\mathbf{r}'(t_0) \neq 0$: the slope of the tangent is $\dfrac{y'(t_0)}{x'(t_0)}$ and the slope of the normal is $-\dfrac{x'(t_0)}{y'(t_0)}$. If $x'(t_0)$, for example, is zero, we adopt the convention that the slopes of the tangent and normal are $\infty$ and $0$ respectively. We now consider tangents at cusps. At a cusp $\mathbf{r}(t_0)$ we have $\mathbf{r}'(t_0) = \mathbf{0}$; in this case $\dfrac{y'(t_0)}{x'(t_0)} = \dfrac{0}{0}$ is not meaningful. Therefore we must define the tangent line at a cusp in a different way from the way in which it is defined at a regular point. For this we use the following result.

**Proposition 6.6.** *Let* $t \mapsto \mathbf{r}(t)$ *have a cusp at* $t = t_0$, *then the tangent line* $L_t$ *to the curve at a regular point* $t$ *moves to a limiting position* $L_{t_0}$ *as* $\mathbf{r}(t)$ *moves along the curve to the limiting position* $\mathbf{r}(t_0)$. *Moreover the line* $L_{t_0}$ *is parallel to* $\mathbf{r}''(t_0) \neq \mathbf{0}$.

*Proof\**. Since $\mathbf{r}'(t_0) = \mathbf{0}$ and $\mathbf{r}''(t_0) \neq \mathbf{0}$, we have, by L'Hôpital's theorem (see §4.7.5),

$$\lim_{t \to t_0} \frac{y'(t)}{x'(t)} = \frac{y''(t_0)}{x''(t_0)}.$$

The slope of the line $L_t$ is $\dfrac{dy}{dx} = \dfrac{y'(t)}{x'(t)}$, and therefore, as we move along the curve and reach the cusp, $L_t$ becomes the line $L_{t_0}$ with slope $\dfrac{y''(t_0)}{x''(t_0)}$. This line is parallel to $\mathbf{r}''(t_0)$. □

**Remark\* 6.7.** The vector $\mathbf{r}''(t_0)$ is parallel to the limiting direction of the vector $\mathbf{r}'(t)$ as $t$ tends to $t_0$ from above and is parallel to the limiting direction of $\mathbf{r}'(t)$ as $t$ tends to $t_0$ from below even though $\mathbf{r}'$ tends to 0 as $t$ tends to $t_0$. Indeed one can show, using L'Hôpital's theorem that the unit vector $\dfrac{\mathbf{r}'}{|\mathbf{r}'|}$ tends to a limit $\mathbf{v}$ as $t$ tends to $t_0$ from above, and tends to $-\mathbf{v}$ as $t$ tends to $t_0$ from below, where $\mathbf{v}$ is parallel to the cuspidal tangent vector. Notice however that there is no continuous non-zero tangent vector defined in the direction of the parametrisation near a cusp since the curve 'reverses direction' as it passes through the cusp.

In view of Proposition 6.6, we can now define the tangent line at a cusp of a parametrised curve.

**Definition 6.8.** The *cuspidal tangent line* at a cusp $\mathbf{r}(t_0)$ is the line passing through $\mathbf{r}(t_0)$ and parallel to the vector $\mathbf{r}''(t_0)$.

The equation of the cuspidal tangent line is thus

$$s \mapsto \mathbf{v}(s) = \mathbf{r}(t_0) + s\mathbf{r}''(t_0), \quad \text{or}$$

$$(\mathbf{v} - \mathbf{r}(t_0)) \cdot (-y''(t_0), x''(t_0)) = 0.$$

**Worked example 6.9.** Find the cuspidal tangent line, at the cusp given by $t = 0$, of the semicubical parabola $\mathbf{r}(t) = (t^2, t^3)$.

*Solution.* We have $\mathbf{r}'(t) = (2t, 3t^2) = t(2, 3t)$, and $\mathbf{r}''(t) = (2, 6t)$. Thus the cuspidal tangent line is parallel to $\mathbf{r}''(0) = (2, 0)$ and therefore parallel to $(1, 0)$. Therefore its equation is $s \mapsto \mathbf{v}(s) = s(1, 0)$, or $y = 0$. The reader should note also the following short cut in this case. Since $\mathbf{r}'(t) = t(2, 3t)$ is parallel to $(2, 3t)$ for non-zero values of $t$, the cuspidal tangent is parallel to the non-zero limit of $(2, 3t)$ as $t$ tends to 0, namely to the vector $(2, 0)$. □

---

\* For the meaning of \* see the preface.

## 6.3 Contact between a line and a curve at a cusp

We considered in §5.1.3 contact between a line and a parametric curve at a regular point, and showed that all lines through the point have 1–point contact, except the tangent line which has $k$–point contact where $k \geq 2$. We now consider the corresponding result for contact between a line and a parametrised curve at a quadratic cusp (cf. Proposition 5.8).

**Proposition 6.10.** *Let $\mathbf{r}(t_0)$ be a quadratic cusp of $t \mapsto \mathbf{r}(t)$. Then all lines through $\mathbf{r}(t_0)$, other than the cuspidal tangent line, have 2–point contact at $\mathbf{r}(t_0)$ with the parametrised curve, and the cuspidal tangent line has $k$–point contact where $k \geq 3$.*

*Proof.* We have $\mathbf{r}'(t_0) = \mathbf{0}$, and $\mathbf{r}''(t_0) \neq \mathbf{0}$. The general line $\ell$ has equation

$$\varphi(x, y) = (\mathbf{r} - \mathbf{a}) \cdot \mathbf{n} = 0 \ (\mathbf{n} \neq \mathbf{0}).$$

We consider

$$\gamma(t) = (\mathbf{r}(t) - \mathbf{r}(t_0)) \cdot \mathbf{n}.$$

Now

$$\gamma'(t_0) = \mathbf{r}'(t_0) \cdot \mathbf{n} = 0$$

and

$$\gamma''(t_0) = \mathbf{r}''(t_0) \cdot \mathbf{n}.$$

Thus $\gamma''(t_0) = 0$ if, and only if, $\mathbf{r}''(t_0)$ is perpendicular to $\mathbf{n}$, that is if, and only if, the line $\ell$ is the cuspidal tangent line. $\quad\square$

Figure 6.1 *Non-tangent lines through a cusp.*

All lines through the quadratic cusp other than the cuspidal tangent line have 2–point contact with the curve at the cusp. For any line having 2–point contact,

$$(\mathbf{r}(t) - \mathbf{r}(t_0)) \cdot \mathbf{n} = \gamma(t) = (t - t_0)^2 \psi(t)$$

by §4.7.3, where $\psi$ is continuous at $t_0$ and $\psi(t_0) \neq 0$, and we have, by §4.7.1, that $\psi(t)$ has constant sign for $|\, t - t_0 \,| < \epsilon$ (where $\epsilon$ is small). Therefore $\gamma(t)$

has constant sign for $0 <| t - t_0 | < \epsilon$. Now one side of the line $(\mathbf{r} - \mathbf{a}) \cdot \mathbf{n} = 0$ is given by $(\mathbf{r} - \mathbf{a}) \cdot \mathbf{n} > 0$ and the other side of the line is given by $(\mathbf{r} - \mathbf{a}) \cdot \mathbf{n} < 0$. Thus the curve close to $t_0$ stays on one side of any non-tangent line; this gives an indication of possible shapes of the curve near to a cusp (see Figure 6.1).

The cuspidal tangent line is $(\mathbf{r} - \mathbf{r}(t_0)) \cdot (-y''(t_0), x''(t_0)) = 0$. We define

$$\gamma(t) = (\mathbf{r}(t) - \mathbf{r}(t_0)) \cdot (-y''(t_0), x''(t_0)),$$

in order to determine the order of contact between the curve and the cuspidal tangent line at the cusp. Since $t = t_0$ gives a cusp, we have

$$\gamma(t_0) = \gamma'(t_0) = \gamma''(t_0) = 0,$$

and

$$\gamma^{(n)}(t_0) = \mathbf{r}^{(n)}(t_0) \cdot (-y''(t_0), x''(t_0)) \quad (n \geq 3).$$

**Definition 6.11.** A quadratic cusp has *order* $k$ if the first non-vanishing derivative of $\gamma$ at $t = t_0$ is that of order $k + 2$. A *higher (quadratic) cusp* is a cusp of order $\geq 2$.

We show below that a cusp has order one if, and only if, it is an ordinary cusp.

In practice, in case the cusp is at $t = 0$, it is often simpler to use the fact that the first non-vanishing derivative of $\gamma$ at $t = 0$ is that of order $k + 2$ if, and only if, $\gamma$ has a zero of order $k + 2$ at $t = 0$, that is if, and only if,

$$\gamma(t) = t^{k+2}\psi(t),$$

where $\psi(0) \neq 0$ (see §4.7.3). In case $\gamma(t)$ is a polynomial, the order of the cusp at $t = 0$ is $m - 2$ where $m$ is the degree of the term of $\gamma(t)$ of lowest degree $\geq 1$.

**Worked example 6.12.** Show that $\mathbf{r}(t) = (t^2, t^5)$ has a cusp of order 3 at $(0, 0)$.

*Solution.* We have $\mathbf{r}'(t) = (2t, 5t^4) = 0$ if, and only if, $t = 0$. Also

$$\mathbf{r}''(t) = (2, 20t^3) \neq 0,$$

and therefore $\mathbf{r}(0) = \mathbf{0}$ is a cusp. The cuspidal tangent line is parallel to $(1, 0)$ and has equation $y = 0$. Thus $\gamma(t) = t^5$ has a zero of order 5 and therefore the cusp has order $5 - 2 = 3$. $\qquad\square$

The following criterion is analogous to Criterion 5.24.

**Criterion 6.13.** *The quadratic cusp* $\mathbf{r}(t_0)$ *has order* $k$ *if, and only if, the first derivative at* $t_0$ *of* $\mathbf{r}(t)$ *not parallel to* $\mathbf{r}''(t_0)$ *is the* $(k + 2)$*-nd derivative* $\mathbf{r}^{(k+2)}(t_0)$ *(k $\geq 1$).*

*Proof.* Since $\gamma^{(n)}(t) = \mathbf{r}^{(n)}(t) \cdot (-y''(t_0), x''(t_0))$ for $n \geq 3$, the first non-vanishing derivative of $\gamma(t)$ is that of order $k + 2$ if, and only if, the first derivative of $\mathbf{r}(t)$ at $t_0$ not parallel to $\mathbf{r}''(t_0)$ is that of order $k + 2$. $\qquad\square$

**Corollary 6.14.** *A cusp has order one if, and only if, it is an ordinary cusp.*

**Worked example 6.15.** $\mathbf{r}(t) = (t^2 + t^3, t^4)$ has a cusp of order 2 at $(0,0)$.

*Solution.* We have $\mathbf{r}' = (2t + 3t^2, 4t^3) = 0$ if, and only if, $t = 0$. Also

$$\mathbf{r}'' = (2 + 6t, 12t^2) \neq \mathbf{0},$$

therefore there is a cusp at the origin. Now $\mathbf{r}''(0) = (2,0)$, $\mathbf{r}'''(0) = (6,0)$, and the first derivative not parallel to $\mathbf{r}''(0)$ is $\mathbf{r}^{(4)}(0) = (0,24)$. Therefore the cusp has order $4 - 2 = 2$.                                            □

**Proposition 6.16.** *At a cusp of even order the curve stays on one side of the cuspidal tangent line close to the cusp, and thus the two branches at the cusp lie on the same side of the cuspidal tangent line near the cusp. At a cusp of odd order the curve crosses the cuspidal tangent line at the cusp, and thus the two branches at the cusp lie on opposite sides of the cuspidal tangent line near the cusp.*

*Proof.* Let the cusp be of order $k$. We have

$$\gamma(t) = (\mathbf{r}(t) - \mathbf{r}(t_0)) \cdot (-y''(t_0), x''(t_0)) = (t - t_0)^{(k+2)}\psi(t)$$

where $\psi$ is continuous at $t_0$ and $\psi(t_0) \neq 0$. By the §4.7.1, $\psi$ has constant sign for $0 < |t - t_0| < \epsilon$ (where $\epsilon$ is small). Therefore, in case $k$ is odd, $\gamma(t)$ has opposite signs for $t_0 - \epsilon < t < t_0$ and for $t_0 < t < t_0 + \epsilon$. In case $k$ is even, $\gamma(t)$ has the same sign for $t_0 - \epsilon < t < t_0$ and for $t_0 < t < t_0 + \epsilon$.   □

This proposition is illustrated in Figure 6.2.

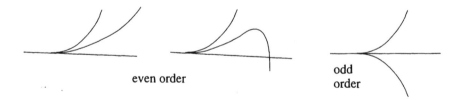

even order                                        odd order

Figure 6.2 *Cusp and cuspidal tangent line.*

## 6.4 Higher singularities*

A *higher singularity* of a parametric curve is a point $\mathbf{r}(t_0)$ where

$$\mathbf{r}'(t_0) = \mathbf{r}''(t_0) = \mathbf{0}.$$

* For the meaning of * see the preface.

We can define a limiting tangent line in the case where

$$\frac{d^s \mathbf{r}}{dt^s}(t_0) = 0 \quad \text{for } s = 1, \ldots, n-1, \quad \text{and} \quad \frac{d^n \mathbf{r}}{dt^n}(t_0) \neq 0 \quad (n \geq 3).$$

Again using L'Hôpital's theorem we have in this case

$$\lim_{t \to t_0} \frac{y'(t)}{x'(t)} = \frac{y^{(n)}(t_0)}{x^{(n)}(t_0)}.$$

The equation of the tangent line at such a point $\mathbf{r}(t_0)$ is

$$s \mapsto \mathbf{v}(s) = \mathbf{r}(t_0) + s\mathbf{r}^{(n)}(t_0) \quad \text{(for } s \text{ in } \mathbf{R}\text{), or}$$

$$(\mathbf{v} - \mathbf{r}(t_0)) \cdot (-y^{(n)}(t_0), x^{(n)}(t_0)) = 0.$$

**Worked example 6.17.** Show that the curve $t \mapsto \mathbf{r}(t) = (t^3, t^4)$ (for $t$ in $\mathbf{R}$) has a non-regular point at the origin. Find the tangent line there.

*Solution.* We have

$$\mathbf{r}'(t) = (3t^2, 4t^3),$$
$$\mathbf{r}''(t) = (6t, 12t^2),$$

and

$$\mathbf{r}'''(t) = (6, 24t).$$

Thus $\mathbf{r}'(0) = \mathbf{0}$, $\mathbf{r}''(0) = \mathbf{0}$, and $\mathbf{r}'''(0) = (6,0) \neq \mathbf{0}$. The point is not a cusp, and the tangent line is $s \mapsto \mathbf{v}(s) = s(1,0)$ or $y = 0$. $\quad\square$

The real algebraic curve of Example 6.17 with equation $y^3 = x^4$ can also be written as the graph of $y = x^{\frac{4}{3}}$. The $x$-coordinate gives a parametrisation of the curve, which is smooth except at $x = 0$: the derivative $\frac{dy}{dx} = \frac{4}{3}x^{\frac{1}{3}}$ is 0 at $x = 0$, but $\frac{d^2y}{dx^2}$ is not defined at $x = 0$. Clearly the critical point given by $x = 0$ is a local minimum, and the shape of the curve near $x = 0$ is similar to that of the curve $y = x^2$. The non-regular point of $\mathbf{r}(t) = (t^3, t^4)$ is not a cusp in the sense of two branches of the curve coming together at a sharp point. The reader should compare this example with the semi-cubical parabola $y^2 = x^3$, parametrised by $\mathbf{r}(t) = (t^2, t^3)$, which consists of two graphs (two branches) $y = \pm x^{\frac{3}{2}}$ ($x \geq 0$), joined together at a cusp. Also the real part of the curve $y^4 = x^5$, parametrised by $\mathbf{r}(t) = (t^4, t^5)$, consists of two real branches $y = \pm x^{\frac{5}{4}}$ ($x \geq 0$), joined together at a sharp point. Near the non-regular point the parametric curve has the shape of a cusp. However, regarded as a complex algebraic curve, $y^4 = x^5$ is also satisfied by $y = \pm ix^{\frac{5}{4}}$, so the singularity of this algebraic curve is theoretically somewhat different to that of the algebraic curve $y^2 = x^3$. In the present chapter we have considered cusps of parametric curves. We shall mention cusps of algebraic curves again in Chapter 15.

A point of a curve which is regular under one parametrisation can be non-regular under another. For example the parametrisation $t \mapsto \mathbf{r}(t) = (t, t^2)$ (for $t$ in $\mathbf{R}$) of the parabola $y = x^2$ is regular, whereas the parametrisation $t \mapsto \mathbf{r}(t) = (t^3, t^6)$ (for $t$ in $\mathbf{R}$) of the same algebraic curve is not regular at $t = 0$. So a non-regular value of a parametrisation may exist because the parametrisation is not sufficiently 'lean' rather than because it signifies a 'singularity' of the curve itself. We avoided this difficulty when considering contact between two curves in Chapter 5 by considering only the case where the parametrised curve is regular, at least near the point of contact. At cusps of parametrised curves we cannot do this, but we avoided the same difficulty in the case of cusps by considering only the case where the parametrisation satisfies $\mathbf{r}''(t_0) \neq \mathbf{0}$ at the cusp. This allows us to consider many examples of quadratic cusps, including ordinary cusps, and to avoid the need to fit into the theory parametrisations such as $\mathbf{r}(t) = (t^6, t^9)$ for the semi-cubical parabola, where the 'degree of singularity' is increased by the parametrisation itself. However it also means that we do not discuss, as cusps, higher non-regular points having cusp-like shapes such as that at the origin of the parametrisation $\mathbf{r}(t) = (t^4, t^5)$ of the real algebraic curve $y^4 = x^5$.

We next note an example of a smooth parametrised curve for which, at one point, no tangent can be defined using these analytical methods.

**Example 6.18.** The smooth curve with a corner, considered in Example 4.7, has no tangent at $\mathbf{0}$. Notice in this example that the derivatives of all orders are zero; that is,

$$\left.\frac{d^n \mathbf{r}}{dt^n}\right|_{t=0} = \mathbf{0}, \quad \text{for all } n \geq 0.$$

The two half-lines with the usual arc-length parametrisations do, of course, each have limiting tangent directions as the cusp is approached, though the arc-length parametrisation is not differentiable at the origin. This curve does not have two branches coming together at a sharp point as does a cusp, but has two 'branches' coming together at a 'blunt' point at an angle of $\dfrac{\pi}{4}$.

## Exercises

**6.1.** Show that each of the following curves has two cusps and find them
   a)  $\mathbf{r}(t) = (-2t^2 + t^4, -3t + t^3)$,
       $[t = \pm 1, 3x - 4y = 14, 3x + 4y = 14, \text{both ordinary}]$
   b)  $\mathbf{r}(t) = (5t^2 - 2t^5, 3t^2 - 2t^3)$,
   c)  $\mathbf{r}(t) = (5t^2 - 2t^5, 2t^2 - t^4)$,
       $[t = 0, 1; 2x = 5y, 4x - 15y = -3, \text{order 2, simple}]$
For each cusp determine the cuspidal tangent line, and determine whether or not the cusp is ordinary. Find the order of each cusp.

**6.2.** Let $\mathbf{r}(t) = (t^2 - t^3, t^2 + t^4)$. Show that this parametric curve has exactly one non-regular point and that this is a cusp. Determine the cuspidal tangent line, and find the order of the cusp. Determine the point $P$ on the curve, not at the cusp, at which the tangent is parallel to the $y$-axis. Show that there are precisely two points at which the curve has an inflexion, and that precisely one of these, $Q$, has parameter value in the range $-1 \le t \le 1$.

**6.3.** Show that the curve $\mathbf{r}(t) = (-2t^2 + \frac{2}{3}t^3, \frac{1}{2}t^3 - \frac{1}{4}t^4)$ has an ordinary cusp at the origin. Write down the equation of the cuspidal tangent line. Show that the curve has two simple inflexions and determine the tangent lines at these. $\qquad [y = 0; t = 1, 3; 3x + 12y = 7, 9x - 4y = 27]$

**6.4.** Show that the curve $\mathbf{r}(t) = (t + t^2, t^2 + \frac{4}{3}t^3)$ has 3-point contact with the parabola $y = x^2$ at $t = 0$. Show also that $t \mapsto \mathbf{r}(t)$ has precisely one non-regular point, and that this is an ordinary cusp. Write down the equation of the cuspidal tangent line. Show further that the curve has no inflexions. Find all the points of intersection of the curve with the coordinate axes and with the parabola $y = x^2$.

Draw the curve $t \mapsto \mathbf{r}(t)$ and the parabola on the same diagram at a scale of 1 unit = 20 cm for $-1.1 \le t \le 0.2$.

**6.5.** Let $\mathbf{r}(t) = (t^2 - t^3, 2t^2 - t^4)$ for $t$ in $\mathbf{R}$. Show that the parametric curve has exactly one non-regular point and that this is a cusp. Determine the cuspidal tangent line, and find the order of the cusp. Determine the point $P$ on the curve, not at the cusp, at which the tangent is parallel to the $y$-axis; and determine the points $Q$ and $R$ on the curve, not at the cusp, at which the tangent is parallel to the $x$-axis.

Show that the curve has no point of inflexion.

Plot the curve for $-\sqrt{2} \le t \le \sqrt{2}$ using 4 cm as 1 unit. Indicate the points $P$, $Q$ and $R$ and plot about 11 other points. Draw the cuspidal tangent line.

(Use the portrait mode with the $y$-axis 5 cm from the left side.)
$$[t = 0, y = 2x, 1; t = \tfrac{2}{3}, \pm 1; 3t^2 - 4t + 3 = 0]$$

**6.6.** Let $\mathbf{r}(t) = (\frac{3}{2}t^2 - t^3, \frac{10}{9}t^2 - t^4)$ for $t$ in $\mathbf{R}$. Show that the parametric curve has exactly one non-regular point and that this is a cusp. Determine the cuspidal tangent line, and find the order of the cusp. Determine the point on the curve, not at the cusp, at which the tangent is parallel to the $y$-axis; and determine the points on the curve, not at the cusp, at which the tangent is parallel to the $x$-axis.

Show that the curve has two points of inflexion and find them.

Plot and draw the curve for $-0.6 \le t \le 1.2$ using 20 cm as 1 unit. Plot, in the given range, any inflexions, points where the tangents are parallel to either axis, the point at which the curve crosses the $x$-axis, and plot about 10 other points. Draw the cuspidal tangent line.

(Use the landscape mode with the $y$-axis at the left side and the $x$-axis 10 cm from the bottom.)

**6.7.** Let $\mathbf{r}(t) = \left(3t^2 - 2t^3,\, 2t^2 - t^4\right)$ for $t$ in $\mathbf{R}$. Prove that

$$\mathbf{r}'(t) = t(1 - t)\left(6,\, 4(1 + t)\right).$$

Deduce that the parametric curve has exactly two non-regular points and show that these are both cusps. Determine the cuspidal tangent lines, and show that both cusps are ordinary. Determine the point on the curve, not at the cusp, at which the tangent is parallel to the $x$-axis. Show that there are no points on the curve at which the tangent is parallel to the $y$-axis. Show further that, for each vector $(a, b)$ with $a \neq 0$, there is precisely one point on the curve at which the tangent or cuspidal tangent is parallel to $(a, b)$, and write down the parameter value of that point.

Plot and draw the curve for $-1 \le t \le 1.5$ using 5 cm as 1 unit. First draw (in colour) the cuspidal tangent lines. Plot the points given by

$$t = -1, -0.5, 0, 0.5, 1, \frac{4}{3}, 1.5$$

and about 7 other points.

(Use the landscape mode with the $y$-axis 1 cm from the left side and the $x$-axis 9 cm from the bottom.)

$[t = 0, 1; 2x = 3y, 4x - 3y = 1; \mathbf{r}''(1) = (-6, -8), \mathbf{r}'''(1) =$
$(-12, -24); t = \dfrac{3b}{2a} - 1]$

**6.8.** Let $\mathbf{r}(t) = \left(10t^2,\, 5t^2 + 20t^4 + 16t^5\right)$ for $t$ in $\mathbf{R}$. Show that the parametric curve has exactly one non-regular point and that this is a cusp. Determine the cuspidal tangent line, and find the order of the cusp. Determine the point $P$ on the curve, not at the cusp, at which the tangent is parallel to the cuspidal tangent line. Show that there is exactly one point $Q$ at which the curve has an inflexion, and determine the point $Q$.

Plot and draw the curve for $-\frac{5}{4} \le t \le \frac{3}{4}$ using 1 cm as 1 unit. First draw (in colour) the cuspidal tangent line. Plot the points given by

$$t = -\tfrac{5}{4}, -1, -0.9, -\tfrac{3}{4}, -\tfrac{2}{3}, 0, \tfrac{2}{3}, \tfrac{3}{4}$$

and about 6 other points. Indicate the points $P$ and $Q$.

(Use the landscape mode with the $y$-axis 1 cm from the left side and the $x$-axis 1 cm from the bottom.)

# 7

# Curvature

---

The (signed) curvature of a curve parametrised by its arc-length is the rate of change of direction of the tangent vector. The absolute value of the curvature is a measure of how sharply the curve bends. Curves which bend sharply will have a large absolute curvature. Curves which bend slowly, which are almost straight lines, will have a small absolute curvature. Straight lines themselves have zero curvature. Curves which swing to the left have positive curvature and curves which swing to the right have negative curvature. The curvature of the direction of a road will affect the maximum speed at which vehicles can travel without skidding, and the curvature in the trajectory of an aeroplane will affect whether the pilot will suffer "blackout" as a result of the g–forces involved. In this chapter we give the basic definitions and properties of curvature. In Chapter 8 we consider some applications of the curvature of parametrised curves, and in Chapter 9 we consider the circle of curvature.

## 7.1 Cartesian coordinates

### 7.1.1 General parameters

Let $t \mapsto \mathbf{r}(t)$ be a regular smooth parametrised curve[†]. The parameter $t$ will not in general be arc-length. Since the curve is regular and smooth, the parametrisation will determine a direction along the curve as $t$ increases. An *oriented curve* is a curve with such a direction along it. We wish to measure the rate of change of direction of the tangent vector in a way which is intrinsic to curve, that is, in a way which is not dependent on a particular parametrisation of the curve, but only on the orientation of the curve (direction along the curve) determined by the parametrisation. We

---

[†] Often we use $t$ as parameter. The reader should however become familiar with the use of other parameters such as $s$, usually the arc-length, and $\varphi$, often used in case the parameter is an angle.

measure this rate of change of direction with respect to the arc-length $s$. Let $\psi$ be the angle measured counter-clockwise from the direction of the initial half-line (the positive $x$-axis) to the tangent vector (see Figure 7.1). Notice that[†]

$$\frac{d\mathbf{r}}{ds} = (\cos\psi, \sin\psi).$$

We assume in general that the direction of the positive $y$–axis is given by

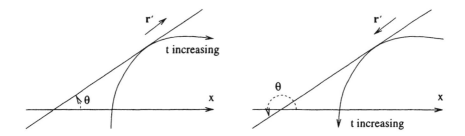

Figure 7.1 *Curvature.*

a counter-clockwise rotation through an angle $\dfrac{\pi}{2}$ from the direction of the positive $x$–axis.

**Definition 7.1.** The *curvature* or *signed curvature* $\kappa$ of a smooth curve $t \mapsto \mathbf{r}(t)$ at a regular point is the rate of change of direction at that point of the tangent line with respect to arc-length, that is

$$\boxed{\kappa = \frac{d\psi}{ds},}$$

where arc-length $t \mapsto s(t)$ is measured along the curve so that $s$ increases as $t$ increases. The *absolute curvature* of the curve at a point is the absolute value $|\kappa|$ of the curvature at that point.

In general the arc-length parametrisation of the curves we consider is not given by a simple formula and, in order to calculate the curvature of a specific parametrised curve, we need to determine a formula for curvature in terms of the given parameter. As a first step to this we give in the following lemma a formula for curvature in terms of the rate of change of direction of the tangent vector with respect to the parameter $t$ by which the curve is described.

[†] We use $\psi$ for the polar angle of $\mathbf{r}'$, and generally use $\theta$ or $\varphi$ when considering the polar angle of $\mathbf{r}$.

**Lemma 7.2.** *The curvature of a curve* $t \mapsto \mathbf{r}(t)$ *at a regular point is*

$$\kappa = \kappa(t) = \frac{d\psi}{dt} \Bigg/ \left|\frac{d\mathbf{r}}{dt}\right| = \frac{\dfrac{d\psi}{dt}}{\sqrt{\left(\dfrac{dx}{dt}\right)^2 + \left(\dfrac{dy}{dt}\right)^2}} = \frac{d\psi}{dt} \Bigg/ \frac{ds}{dt} = \frac{d\psi}{ds} \,,$$

*where arc-length* $t \mapsto s(t)$ *is measured in the direction given by t increasing.*

*Solution\*.* We need to prove the last of the equalities. Since the arc-length is measured in terms of $t$ increasing, we have at regular points $\dfrac{ds}{dt} > 0$ and therefore $\dfrac{ds}{dt} = |\mathbf{r}'|$. For a regular (smooth) curve $t \mapsto \mathbf{r}(t)$, we also have that $t \mapsto s(t)$ is smooth (see §4.6.2). By the inverse function theorem (see §4.7), in this case of a regular smooth curve, the function $t \mapsto s(t)$ has an inverse $s \mapsto t(s)$ which is also smooth. Therefore $\psi$ can be regarded as a function of $s$ by $s \mapsto \psi(t(s))$. The last of the equalities in the definition now follows from the chain rule $\dfrac{d\psi}{dt} = \dfrac{d\psi}{ds}\dfrac{ds}{dt}$, on regarding $\psi$ as a function of $s$ on the one hand and as a function of $t$ on the other, where $s = s(t)$. $\qquad\square$

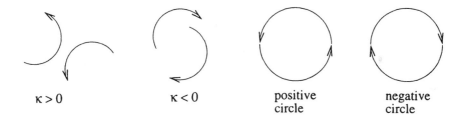

| $\kappa > 0$ | $\kappa < 0$ | positive circle | negative circle |

Figure 7.2 *Sign of the curvature.*

The curvature can be positive, negative or zero. The angle $\psi$ is measured counterclockwise. Therefore the curvature of a parametrised curve is *positive* $\left(\kappa = \dfrac{d\psi}{ds} > 0\right)$ if the curve swings to the left as $t$ increases, that is $\psi$ increases as $s$ increases. Similarly the curvature is *negative* $\left(\kappa = \dfrac{d\psi}{ds} < 0\right)$ if the curve swings to the right as $t$ increases, that is $\psi$ decreases as $s$ increases (see Figure 7.2). Thus the *sign* of the curvature at a point depends on the direction along which the curve is parametrised at the point. If the curve is reparametrised in the opposite direction then the sign of the curvature becomes the opposite of what it had been.

---

\* For the meaning of \* see the preface.

**Mnemonic**

A curve with positive curvature swings to the left.

We next give a formula for curvature which is the one normally used in calculations.

**Theorem 7.3.** *At a regular point of a smooth curve* $t \mapsto \mathbf{r}(t)$, *the curvature satisfies*

$$\kappa = \frac{x'\, y'' - y'\, x''}{((x')^2 + (y')^2)^{\frac{3}{2}}} = \frac{x'\, y'' - y'\, x''}{|\mathbf{r}'|^3} = -\frac{\mathbf{r}' \cdot (-y'', x'')}{|\mathbf{r}'|^3}.$$

*Proof.* Differentiating $\tan \psi = \dfrac{y'}{x'}$ with respect to $t$, we have, for $x' \neq 0$,

$$\frac{x'\, y'' - y'\, x''}{(x')^2} = \sec^2 \psi \cdot \psi' = (1 + \tan^2 \psi) \cdot \psi' = \left(1 + \left(\frac{y'}{x'}\right)^2\right) \cdot \psi',$$

and the result follows easily on multiplying each side by $(x')^2$. At any (regular) point at which $x' = 0$, we cannot use the above argument, but, since then $y' \neq 0$, we differentiate instead $\cot \psi = \dfrac{x'}{y'}$ to get the same formula for $\kappa$. $\square$

**Worked example 7.4.** Show that the curvature of straight lines is zero.

*Solution.* We parametrise the line by $\mathbf{r}(t) = \mathbf{a} + t\mathbf{b}$ where $\mathbf{b} \neq \mathbf{0}$. We have $\mathbf{r}' = \mathbf{b}$, $\mathbf{r}'' = \mathbf{0}$, and $\kappa = \dfrac{-\mathbf{b} \cdot \mathbf{0}}{|\mathbf{b}|^3} = 0$. $\square$

**Worked example 7.5.** Show that the curvature of a positively described (counter-clockwise) circle of radius $R$ is $\dfrac{1}{R}$, and that the curvature of a negatively described (clockwise) circle of radius $R$ is $-\dfrac{1}{R}$.

*Solution.* We parametrise the circle having centre $\mathbf{a}$ and radius $R$ in the counter-clockwise direction by $\mathbf{r}(t) = \mathbf{a} + R(\cos t, \sin t)$. Therefore we have $\mathbf{r}' = R(-\sin t, \cos t)$, and $\mathbf{r}'' = -R(\cos t, \sin t)$. Therefore

$$\kappa = \frac{x'\, y'' - y'\, x''}{((x')^2 + (y')^2)^{\frac{3}{2}}} = \frac{R^2(\sin^2 t + \cos^2 t)}{(R^2 \sin^2 t + R^2 \cos^2 t)^{\frac{3}{2}}} = \frac{1}{R}.$$

If the circle is oriented in the clockwise direction, by

$$\mathbf{r}(t) = \mathbf{a} + R(\cos t, -\sin t)$$

for example, we obtain in a similar way that the curvature is $-\dfrac{1}{R}$. $\square$

By Example 7.5, we have the following memory-aid for the sign of the curvature of a curve.

**Mnemonic**

A positively (counter-clockwise) described circle has positive curvature.

**Worked example 7.6.** Find the curvature of the curve $x = t^3 - t$, $y = t^2$.

*Solution.* We have $\mathbf{r}' = (3t^2 - 1, 2t)$, and $\mathbf{r}'' = (6t, 2)$. Therefore

$$\kappa = \frac{x'\,y'' - y'\,x''}{((x')^2 + (y')^2)^{\frac{3}{2}}} = \frac{(3t^2 - 1).2 - 2t.6t}{((3t^2 - 1)^2 + (2t)^2)^{\frac{3}{2}}} = -\frac{6t^2 + 2}{(9t^4 - 2t^2 + 1)^{\frac{3}{2}}}.$$

$\square$

Notice that Theorem 7.3 tells us that

$$\frac{x'\,y'' - y'\,x''}{((x')^2 + (y')^2)^{\frac{3}{2}}}$$

is invariant under change of rectangular axes (rotation and/or translation), since the curvature $\dfrac{d\psi}{ds}$ is clearly invariant under such a change. If the axes are changed by a reflexion, the curvature simply changes sign. The curvature is also independent of the choice of parametrisation in a fixed direction (see Exercise 7.20). We subsequently consider the derivatives of the curvature. The differentiability is given by the following corollary to Theorem 7.3.

**Corollary 7.7.** *At a regular point of a smooth parametrisation* $t \mapsto \mathbf{r}(t)$ *($n \geq 2$), the curvature* $t \mapsto \kappa(t)$ *is a smooth function.*

*Proof.* The curvature is given by

$$\kappa = \frac{x'\,y'' - y'\,x''}{((x')^2 + (y')^2)^{\frac{3}{2}}}.$$

The numerator is a polynomial in the smooth functions $x'$, $y'$, $x''$, and $y''$, and therefore is a smooth function; similarly $(x')^2 + (y')^2$ is smooth. Also $X \to X^{\frac{3}{2}}$ is a smooth function of $X$ at points for which $X > 0$. The result follows since the composite of two smooth functions is smooth, and the quotient of a smooth function by a non-zero smooth function is also smooth. $\square$

### 7.1.2 Coordinates as parameter

An important special case of Theorem 7.3 is the formula for the curvature of a graph. The graph of a function $y = g(x)$ is usually assumed to be parametrised using $x$ as parameter, that is by $\mathbf{r}(x) = (x, g(x))$. With this parametrisation the graph of $y = g(x)$ has the orientation given by $x$ increasing. Similarly the curve given by $x = h(y)$ is usually assumed to be parametrised using $y$ as parameter, that is by $\mathbf{r}(y) = (h(y), y)$, and has then the orientation given by $y$ increasing.

**Corollary 7.8.** *The graph, parametrised by the coordinate $x$, of a smooth function $y = g(x)$ has curvature*

$$\kappa(x) = \frac{\dfrac{d^2 y}{dx^2}}{\left(1 + \left(\dfrac{dy}{dx}\right)^2\right)^{\frac{3}{2}}} = \frac{g''(x)}{(1 + (g'(x))^2)^{\frac{3}{2}}}.$$

*Proof.* We have $\mathbf{r}' = (1, g'(x))$ and $\mathbf{r}'' = (0, g''(x))$. Thus

$$\kappa = \frac{x' y'' - y' x''}{|\mathbf{r}'|^3} = \frac{1 \cdot g''(x) - g'(x) \cdot 0}{(1 + (g'(x))^2)^{\frac{3}{2}}}.$$

$\square$

For a curve parametrised by the $x$–coordinate, the curvature is positive where the concavity of the curve is "upward", and is negative where the concavity is "downward"; moreover, in this case of a curve parametrised by the $x$–coordinate, the curvature has the same sign as $\dfrac{d^2 y}{dx^2}$ by Corollary 7.8.

Compare the formula of Corollary 7.8 carefully with the formula for the curvature for general parametrisations. Of course, by Theorem 4.22, any algebraic curve may be parametrised locally, near a non-singular point, by either the $x$ or the $y$ coordinate, or by both. Each such local parametrisation will give a local orientation along the curve, determined by $x$ increasing or $y$ increasing respectively, and hence will determine a sign for the curvature. For some curves the curvature will have the opposite sign and the curve will have the opposite local orientation for increasing $x$ to those given by increasing $y$.

**Worked example 7.9.** Find the curvature of the graph of $y = x^3$ parametrised by $x$.

*Solution.* We have $\dfrac{dy}{dx} = 3x^2$, and $\dfrac{d^2 y}{dx^2} = 6x$. Therefore

$$\kappa = \frac{6x}{(1 + 9x^4)^{\frac{3}{2}}}.$$

$\square$

**Worked example 7.10.** Find the point of greatest absolute curvature of the graph of $y = \log x$.

*Solution.* The domain of parametrisation is $x > 0$. We have

$$\frac{dy}{dx} = \frac{1}{x}, \quad \text{and}$$

$$\frac{d^2 y}{dx^2} = -\frac{1}{x^2}.$$

Therefore

$$\kappa = -\frac{x}{(1 + x^2)^{\frac{3}{2}}}, \quad \text{and}$$

$$\frac{d\kappa}{dx} = \frac{2x^2 - 1}{(1 + x^2)^{\frac{5}{2}}} = \frac{2\left(x - \frac{1}{\sqrt{2}}\right)\left(x + \frac{1}{\sqrt{2}}\right)}{(1 + x^2)^{\frac{5}{2}}}.$$

The critical point is at $x = \dfrac{1}{\sqrt{2}}$, and the required point, which gives a minimum value of the signed curvature corresponding to the parametrisation by $x$ and a maximum value of the absolute curvature, is $\left(\dfrac{1}{\sqrt{2}}, -\dfrac{1}{2}\log 2\right)$.

The curve is not defined at $x = -\dfrac{1}{\sqrt{2}}$. □

## 7.1.3 A short cut

We note the following technique which is sometimes useful in shortening the calculation of curvature and in determining whether or not a cusp is ordinary. Recall that $(-v', u')$ is obtained by rotating $(u', v')$ counterclockwise through the angle $\dfrac{\pi}{2}$, in the case where $(u', v') \neq 0$. Let $\mathbf{Z}(t) = (X(t), Y(t))$ satisfy $\mathbf{Z}' = f(t) \cdot (u(t), v(t))$. Suppose we wish to calculate

$$\mathbf{Z}' \cdot (-Y'', X'').$$

We have

$$\mathbf{Z}'' = f'(t) \cdot (u(t), v(t)) + f(t) \cdot (u'(t), v'(t)).$$

Therefore

$$\mathbf{Z}' \cdot (-Y'', X'') = \mathbf{Z}' \cdot (f(t) \cdot (-v'(t), u'(t))),$$

since $(u'(t), v'(t)) \cdot (-v'(t), u'(t)) = 0$.

Notice that we have differentiated

$$f(t) \cdot (u(t), v(t))$$

as a product of a scalar function and a vector function, rather than writing it in the form $(f(t).u(t), f(t).v(t))$.

**Exercises**

**7.1.** Find the curvature at regular points of the following curves:

a)   The rectangular hyperbola $t \mapsto \mathbf{r}(t) = \left(at, \dfrac{a}{t}\right)$.

$$\left[ \frac{2a^2 t^4}{a^3(1+t^4)^{\frac{3}{2}}}; \text{ none, none} \right]$$

b)   The cuspidal cubic $t \mapsto \mathbf{r}(t) = (t^2, t^3)$.

c)   The crunodal cubic $t \mapsto \mathbf{r}(t) = (t^2 - 1, t(t^2 - 1))$.

$$\left[ \frac{6t^2 + 2}{(9t^4 - 2t^2 + 1)^{\frac{3}{2}}}; \text{ none, none} \right]$$

d)   The acnodal cubic $t \mapsto \mathbf{r}(t) = (1 + t^2, t(1 + t^2))$.

In each case find the non-regular points and the points at which the curvature is zero.

**7.2.** Prove that the ellipse $x = a\cos\varphi, y = b\sin\varphi$ $(0 < b < a)$ has curvature

$$\kappa = \frac{ab}{\left(a^2 \sin^2\varphi + b^2 \cos^2\varphi\right)^{\frac{3}{2}}}.$$

**7.3.** Show that the right-hand branch $x = a\sec\varphi, y = b\tan\varphi$ of the hyperbola $\dfrac{x^2}{a^2} - \dfrac{y^2}{b^2} = 1$ has curvature

$$\kappa = - \frac{ab}{\left(a^2 \tan^2\varphi + b^2 \sec^2\varphi\right)^{\frac{3}{2}}}.$$

Find the curvature on the left-hand branch $x = -a\sec\varphi, y = b\tan\varphi$ .

**7.4.** Show that the parabola $x = at^2, y = 2at$ has curvature

$$\kappa = - \frac{1}{2a(1 + t^2)^{\frac{3}{2}}}.$$

**7.5.** Show that the curvature of the cycloid $x = a(t - \sin t), y = a(1 - \cos t)$ is

$$-\frac{1}{4a\left|\sin\dfrac{t}{2}\right|}.$$

**7.6.** Show that the curvature of the curve

$$t \mapsto \mathbf{r}(t) = (a\sin 2t\,(1 + \cos 2t), a\cos 2t\,(1 - \cos 2t))$$

is $4a\cos 3t$.

**7.7.** The curvature of the ellipse given by $x = a\cos\varphi, y = b\sin\varphi$ $(0 < b < a)$ is

$$\kappa = \frac{ab}{\left(a^2 \sin^2\varphi + b^2 \cos^2\varphi\right)^{\frac{3}{2}}}.$$

Show that

$$a^2 \sin^2 \varphi + b^2 \cos^2 \varphi = (a^2 - b^2) \sin^2 \varphi + b^2$$

is strictly monotone increasing in the interval $0 \le \varphi \le \dfrac{\pi}{2}$ and hence deduce that there are precisely four points on the ellipse at which the curvature of the ellipse is $\dfrac{1}{a}$. (This is the curvature of the circle of radius $a$ which has, as a diameter, the major axis of the ellipse.)

**7.8.** Show that the curve $r(t) = a(\sin 2bt, \sin bt)$ is regular and that its tangent is parallel to the $y$–axis if, and only if, $bt = \pm\dfrac{\pi}{4}$ or $\pm\dfrac{3\pi}{4}$. Find the point where the tangent is parallel to the $x$–axis. Determine the curvature at those points where the tangent is parallel to either axis.

**7.9.** Find the curvature at the origin of the curve $y = \sin ax^2$ parametrised by the $x$ coordinate. $\qquad\qquad\qquad\qquad\qquad\qquad\qquad [\kappa = a]$

**7.10.** Find the curvature at the origin of the curve $y = x^2 - x^3$ parametrised by the $x$ coordinate.

**7.11.** Show that the curvature of the catenary $y = a\cosh\dfrac{x}{a}$ parametrised by the $x$ coordinate is $\dfrac{a}{y^2}$ .

**7.12.** Find the curvature of a curve given by $x = h(y)$ and parametrised by the $y$–coordinate.

**7.13.** A curve which can be parametrised near the origin by a smooth function $y = f(x)$ has the $x$–axis as tangent at the origin. Prove that the curvature at the origin is given by Newton's formula:

$$\kappa(0) = \lim_{x \to 0} \frac{2y}{x^2} \, .$$

Obtain a similar formula in case the curve is parametrised near the origin by a smooth function $x = g(y)$ and has the $y$–axis as tangent at the origin. (*Hint:* Use L'Hôpital's theorem – see §4.7.5.)

$$\left[ \kappa(0) = -\lim_{y \to 0} \frac{2x}{y^2} \right]$$

**7.14.** Find the curvature at the origin of the quartic curve

$$3x^4 + 2y^4 + 4xy^2 + xy - x^2 + 2y = 0.$$

(*Hint:* Use the local parametrisation theorem (Theorem 4.22) to show that there is a local parametrisation $y = f(x)$ near the origin, divide the algebraic equation by $x^2$, and use Exercise 7.13.) $\qquad\qquad\qquad [\kappa = 1]$

**7.15.** Show that the curvature of the ellipse $\dfrac{x^2}{a^2} + \dfrac{y^2}{b^2} = 1$ parametrised locally by $x$ is $\kappa = \pm\dfrac{ab}{(a^2 - e^2x^2)^{\frac{3}{2}}}$, where $e$ is the eccentricity. Indicate on a diagram which parts of the ellipse have positive curvature in this case and which have negative curvature.

**7.16.** Find the maximum distance from the origin of points on the curve

$$t \mapsto \mathbf{r}(t) = \left( a \sin t - b \sin \frac{at}{b} , a \cos t - b \cos \frac{at}{b} \right) .$$

Show that the curvature at such a point of maximum distance is $\dfrac{a+b}{4ab}$ .

**7.17.** Show that the curvature of the hyperbola $\dfrac{x^2}{a^2} - \dfrac{y^2}{b^2} = 1$ parametrised by $x$ is $\kappa = \pm \dfrac{ab}{(e^2 x^2 - a^2)^{\frac{3}{2}}}$ , where $e$ is the eccentricity. Indicate on a diagram which parts of the hyperbola have positive curvature in this case and which have negative curvature.

**7.18.** Show that the curvature of a general curve satisfies

$$\kappa = \frac{\mathbf{r}'' \cdot \mathbf{n}}{|\mathbf{r}'|^3} ,$$

where $\mathbf{n} = (-y', x')$ is the (usual) normal.

**7.19.** Let $s \mapsto \mathbf{r}(s)$ be a curve parametrised by arc-length. Prove that $\mathbf{r}'$ and $(-y'', x'')$ are parallel. Deduce that $|\kappa(s)| = |\mathbf{r}''(s)|$. Prove further that $\mathbf{r}'' \cdot \mathbf{r}'' + \mathbf{r}' \cdot \mathbf{r}''' = 0$ and deduce that $\kappa(s)^3 = x'' y''' - y'' x'''$.

**7.20.** Show that the curvature of a curve $t \mapsto \mathbf{r}(t)$ is unchanged by rotation or translation of the coordinate axes. Prove that it is also left unchanged by a change of parameter given by $t = g(\tau)$ where $g$ is a smooth function and $g'(\tau) > 0$ for all relevant values of $\tau$.

**7.21.** Let $t \rightarrow \mathbf{r}(t)$ be a regular parametrised curve and let $t \rightarrow \mathbf{n}(t)$ be a smooth function such that $\mathbf{n}(t)$ is one of the two unit normals to the curve at $\mathbf{r}(t)$. Let the curvature, for this choice of normal, be defined by

$$\mathbf{n}'(t) + \kappa(t)\mathbf{r}'(t) = \mathbf{0}.$$

Deduce that $\mathbf{n} \cdot \mathbf{r}'' = \kappa \mathbf{r}' \cdot \mathbf{r}'$ for each value of $t$, and that $\mathbf{n} \cdot \mathbf{r}''' = 3\kappa \mathbf{r}' \cdot \mathbf{r}''$ at $t$ if, and only if, $\kappa'(t) = 0$. Prove also that if $\mathbf{n}' \cdot \mathbf{n}' = 1$ for all $t$, then $\kappa'(t) = 0$ if, and only if, $\mathbf{r}' \cdot \mathbf{r}'' = 0$.

## 7.2 Curves given by polar equation

We now develop a technique which we can use to find the curvature of curves given by a polar equation. Let

$$\mathbf{w} = (\cos \varphi, \sin \varphi),$$

be the parametrisation of the unit circle $x^2 + y^2 = 1$ where the parameter $\varphi$ in this case is the polar angle $\theta$, then we have

$$\mathbf{w}' = (-\sin\varphi, \cos\varphi), \text{ and}$$
$$\mathbf{w}'' = -(\cos\varphi, \sin\varphi).$$

The vectors $(\cos\varphi, \sin\varphi)$ and $(-\sin\varphi, \cos\varphi)$ satisfy

$$\begin{aligned}
(\cos\varphi, \sin\varphi) \cdot (\cos\varphi, \sin\varphi) &= 1, \\
(-\sin\varphi, \cos\varphi) \cdot (-\sin\varphi, \cos\varphi) &= 1, \text{ and} \\
(\cos\varphi, \sin\varphi) \cdot (-\sin\varphi, \cos\varphi) &= 0,
\end{aligned} \tag{7.1}$$

that is they are orthogonal unit vectors or orthonormal vectors.

A curve given by a polar equation $r = r(\theta)$, has the parametric equation

$$\mathbf{r} = r(\varphi)(\cos\varphi, \sin\varphi),$$

according to our convention of replacing $\theta$, the polar angle, by $\varphi$. Then we have by the chain rule

$$\mathbf{r}' = r'(\varphi)(\cos\varphi, \sin\varphi) + r(\varphi)(-\sin\varphi, \cos\varphi),$$
$$\mathbf{r}'' = r''(\varphi)(\cos\varphi, \sin\varphi) + 2r'(\varphi)(-\sin\varphi, \cos\varphi) - r(\varphi)(\cos\varphi, \sin\varphi),$$

and

$$(-y'', x'') = r''(\varphi)(-\sin\varphi, \cos\varphi) - $$
$$2r'(\varphi)(\cos\varphi, \sin\varphi) - r(\varphi)(-\sin\varphi, \cos\varphi).$$

Notice how $\mathbf{r}$, $\mathbf{r}'$, $\mathbf{r}''$, and $(-y'', x'')$ are expressed as linear combinations of the two orthogonal unit vectors $(\cos\varphi, \sin\varphi)$ and $(-\sin\varphi, \cos\varphi)$. These two vectors vary by rotating as $\varphi$ increases. The above technique of writing the successive derivatives of $\mathbf{r}$ as linear combinations of these two rotating orthogonal vectors considerably shortens and simplifies the calculation needed to determine the curvature of curves given by their polar equation. We now have easily, on using Equations 7.1,

$$|\mathbf{r}'|^2 = \{r'(\varphi)\}^2 + \{r(\varphi)\}^2, \text{ and}$$
$$-\mathbf{r}' \cdot (-y'', x'') = r^2 + 2(r')^2 - r\,r''.$$

The following theorem is obtained by substituting these into the formula for curvature given in Theorem 7.3.

**Theorem 7.11.** *The curvature of a curve given by a polar equation $r = r(\theta)$ is*

$$\kappa(\theta) = \frac{r^2 + 2\left(\dfrac{dr}{d\theta}\right)^2 - r\dfrac{d^2r}{d\theta^2}}{\left(r^2 + \left(\dfrac{dr}{d\theta}\right)^2\right)^{\frac{3}{2}}}.$$

The parameter used for this result is $\theta$ and therefore the curve has the orientation given by increasing $\theta$. As usual the sign of the curvature is positive if the curve turns to the left as $\theta$ increases. Notice that the formula for $\kappa(\theta)$ is unchanged when $r$ is replaced by $-r$ and $\theta$ is replaced by $\theta + \pi$. Therefore the formula is valid where $r = r(\theta)$ is given by the polar equation, and is equally valid where $r$ is the polar coordinate $(r \geq 0)$ of the point given by the polar equation; recall that the polar coordinates are either $(r(\theta), \theta)$ or $(-r(\theta), \theta + \pi)$, where $r$ is given by the polar equation.

We now use the above technique as a model for calculating the curvature in a specific example. It is suggested that the reader calculate the curvature of curves given by their polar equation directly in this way rather than by learning and using the formula of Theorem 7.11. We stress that direct calculations are far more complicated if the above technique involving orthogonal vectors is not used.

**Worked example 7.12.** Find the curvature of the equi-angular spiral

$$\mathbf{r}(\varphi) = e^{a\varphi}(\cos\varphi, \sin\varphi),$$

where $a$ is a real number.

*Solution.* We have

$$\mathbf{r}' = ae^{a\varphi}(\cos\varphi, \sin\varphi) + e^{a\varphi}(-\sin\varphi, \cos\varphi),$$
$$\mathbf{r}'' = a^2 e^{a\varphi}(\cos\varphi, \sin\varphi) + 2ae^{a\varphi}(-\sin\varphi, \cos\varphi) - e^{a\varphi}(\cos\varphi, \sin\varphi),$$

and

$$(-y'', x'') = a^2 e^{a\varphi}(-\sin\varphi, \cos\varphi) - 2ae^{a\varphi}(\cos\varphi, \sin\varphi) - e^{a\varphi}(-\sin\varphi, \cos\varphi).$$

Therefore $\mathbf{r}' \cdot (-y'', x'') = -2a^2 e^{2a\varphi} + a^2 e^{2a\varphi} - e^{2a\varphi}$ (deduce this directly using the fact that $(\cos\varphi, \sin\varphi)$ and $(-\sin\varphi, \cos\varphi)$ are orthogonal unit vectors). Hence the curvature is

$$\kappa = \frac{(1 + a^2)e^{2a\varphi}}{(1 + a^2)^{\frac{3}{2}} e^{3a\varphi}} = \frac{e^{-a\varphi}}{\sqrt{1 + a^2}}.$$

□

## Exercises

**7.22.** For the following curves use the technique described in this section to show that the curvature is as given:

a)   The circle $r = a\cos\theta$, $\kappa = \dfrac{2}{a}$.

b)   The Archimedean spiral $r = a\theta$, $\kappa = \dfrac{2a^2 + r^2}{(a^2 + r^2)^{\frac{3}{2}}}$.

c)   The cardioid $r = a(1 + \cos\theta)$, $\kappa = \dfrac{3}{4a}\left|\sec\dfrac{\theta}{2}\right|$.

**7.23.** Calculate the curvature of the curves of Exercise 7.22 by using the formula of Theorem 7.11.

**7.24.** Show that the astroid $\varphi \mapsto \mathbf{r}(\varphi) = (a\cos^3\varphi, a\sin^3\varphi)$ has curvature at regular points given by

$$\kappa = -\frac{2}{3a|\sin 2\varphi|}.$$

**7.25.** Show that the curvature of $r = a\cos 2\theta$ is

$$\frac{3\sin^2 2\theta + 5}{a(3\sin^2 2\theta + 1)^{\frac{3}{2}}}.$$

**7.26.** Show that the curvature of the curve

$$x(t) = \int_0^t \frac{\cos u}{\sqrt{u}}\,du, \qquad y(t) = \int_0^t \frac{\sin u}{\sqrt{u}}\,du$$

is $\kappa = -\sqrt{t}$. Sketch the curve.

**7.27.** Show that the curvature of $r\cos 2\theta = a$ at $r = a$ is

$$\kappa = -\frac{3}{a}.$$

**7.28.** Show that the curvature of $r^2 = a^2(1 + \cos\theta)$ at $\theta = 0$ is

$$\kappa = \frac{5}{4\sqrt{2}\,a}.$$

**7.29.** For the sinusoidal spirals $r^m = a^m \cos m\theta$, prove that

$$\frac{dr}{d\theta} = -r\tan m\theta, \text{ and that } \frac{d^2 r}{d\theta^2} = r(\tan^2 m\theta - m\sec^2 m\theta).$$

Deduce that

$$\kappa = \frac{(m+1)|r|^{m-1}}{a^m}.$$

(Special cases of this general curve for various values of rational $m$ include the line ($m = -1$), the circle ($m = 1$), the parabola $\left(m = -\frac{1}{2}\right)$, the cardioid $\left(m = \frac{1}{2}\right)$, the rectangular hyperbola ($m = -2$), and Bernoulli's lemniscate ($m = 2$).)

## 7.3 Curves in the Argand diagram

For certain curves, especially ones given by a polar equation, the representation of points $\mathbf{r} = (x, y)$ of the plane by the complex number $z = x + iy$ can very considerably shorten and simplify the calculation of the curvature of the curve.

Let the parametrised curve $J \mapsto \mathbf{C}$ in the Argand diagram be given by $t \mapsto z(t) = x(t) + iy(t)$. Recall that multiplying by $i$ in the Argand diagram corresponds to rotating counter-clockwise through an angle $\dfrac{\pi}{2}$. The next theorem, which is is equivalent to Theorem 7.3, follows on using the bijection between the plane $\mathbf{R}^2$ and the Argand diagram $\mathbf{C}$ which associates the vector $\mathbf{r} = (x, y)$ in the plane to the vector $z = x + iy$ in the Argand diagram: we simply replace the vector $\mathbf{r}$ in the formula for curvature in Theorem 7.3 by the vector $z$, and recall that the scalar product in $\mathbf{C}$ of $z$ and $w = u + iv$ is given by $(z, w) \mapsto \mathrm{re}\,(z\overline{w})$, and thus the scalar product of $z$ and $iw$ is $-\mathrm{im}\,(z\overline{w})$.

**Theorem 7.13.** *At a regular point of a smooth parametrisation the curvature satisfies*

$$\boxed{\kappa = -\frac{\mathrm{im}\,(z'\,\overline{z''})}{|z'|^3}.}$$

Note especially the negative signs which occur in the last formulae of Theorem 7.3 and in Theorem 7.13. Also when calculating curvature we use the formula $|z'|^2 = z' \cdot \overline{z'}$, which again can shorten the calculations for curves related to circles.

The technique we use in calculations is similar to that given in §7.2. Let

$$w(\varphi) = e^{i\varphi},$$

be the parametrisation of the unit circle $|z| = 1$ where the parameter $\varphi$ is the polar angle $\theta$, then we have

$$w' = ie^{i\varphi}, \text{ and}$$

$$w'' = -e^{i\varphi}.$$

The vectors $e^{i\varphi}$ and $ie^{i\varphi}$ are orthogonal unit vectors, or orthonormal vectors, in the Argand diagram, that is they satisfy $|e^{i\varphi}| = 1$, $|ie^{i\varphi}| = 1$, and

$$\mathrm{re}\,\left(e^{i\varphi} \cdot \overline{ie^{i\varphi}}\right) = \mathrm{re}\,(-i) = 0.$$

The condition for complex numbers $z$ and $w$ to be orthogonal is that $\mathrm{re}\,(z.\overline{w}) = 0$, or, equivalently, that multiplying one by $i$ gives a scalar (i.e., real) multiple of the second. A curve given by a polar equation $r = r(\theta)$, has the parametric equation

$$z(\varphi) = r(\varphi)e^{i\varphi}$$

according to our convention of replacing $\theta$ by $\varphi$. Then we have

$$z' = r'e^{i\varphi} + rie^{i\varphi},$$
$$z'' = (r'' - r)e^{i\varphi} + 2r'ie^{i\varphi}.$$

Notice how $z$, $z'$, and $z''$ are expressed as linear combinations of the two orthogonal unit vectors $e^{i\varphi}$ and $ie^{i\varphi}$, which rotate as $\varphi$ increases. We now have easily

$$|z'|^2 = \{r'(\varphi)\}^2 + \{r(\varphi)\}^2, \text{ and}$$
$$-\mathrm{im}\,(z'\,\overline{z''}\,) = r^2 + 2(r')^2 - r\,r''.$$

Theorem 7.11 again follows from Theorem 7.13 and these last two equations. However, we use the above technique, which is similar to that of §7.2, for calculating the curvature in specific examples.

**Worked example 7.14.** Show that the equi-angular spiral $z = e^{(a+i)\varphi}$, where $a$ is real, has curvature $\kappa = \dfrac{e^{-a\varphi}}{\sqrt{1+a^2}}$.

*Solution.* We have

$$z' = (a+i)e^{(a+i)\varphi},$$
$$z'' = (a+i)^2 e^{(a+i)\varphi},$$
$$|z'| = |(a+i)e^{i\varphi}e^{a\varphi}| = \sqrt{1+a^2}e^{a\varphi},$$
$$z'\,\overline{z''} = (a+i)e^{(a+i)\varphi}\,\overline{(a+i)^2 e^{(a+i)\varphi}}$$
$$= (a+i)e^{(a+i)\varphi}(a-i)^2 e^{(a-i)\varphi} = (a-i)(1+a^2)e^{2a\varphi}, \text{ and}$$
$$-\mathrm{im}\,(z'\,\overline{z''}) = (1+a^2)e^{2a\varphi}.$$

Hence the curvature $\kappa = \dfrac{(1+a^2)e^{2a\varphi}}{(1+a^2)^{\frac{3}{2}}e^{3a\varphi}} = \dfrac{e^{-a\varphi}}{\sqrt{1+a^2}}.$ $\qquad\square$

### Exercises

**7.30.** Use the technique of this section to show that the curvature is as given for the following curves (in each case $a$ is a real number):

a) The circle $z = (a\cos\varphi)\,e^{i\varphi}$, $\kappa = \dfrac{2}{a}$.

b) The Archimedean spiral $z = a\varphi e^{i\varphi}$, $\kappa = \dfrac{2+\varphi^2}{a(1+\varphi^2)^{\frac{3}{2}}}$.

c) The cardioid $z = a(1+\cos\varphi)e^{i\varphi}$, $\kappa = \dfrac{3}{4a}\left|\sec\dfrac{\varphi}{2}\right|$.

## 7.4 An alternative formula

As well as being an important theoretical result which we shall use later, the following theorem gives a useful alternative way of calculating the curvature, especially where points of the plane are represented by complex numbers in the Argand diagram.

**Theorem 7.15.** *At a regular point of smooth parametrised curve in the Argand diagram we have*

$$\frac{d}{dt}\left(\frac{z'(t)}{|z'(t)|}\right) = i\kappa z'(t).$$

*In particular, in the case where the parametrisation is by the arc-length $s$, we have*

$$z''(s) = i\kappa z'(s).$$

*Proof.* Since $z' = |z'|e^{i\psi}$, where $\psi = \psi(t)$ as in §7.1.1 is the angle measured counter-clockwise from the initial half-line to the tangent vector $z' = z'(t)$ or equivalently is the polar angle of $z'$, we have, on using Lemma 7.2,

$$\frac{d}{dt}\left(\frac{z'}{|z'|}\right) = \frac{d}{dt}\left(e^{i\psi}\right) = ie^{i\psi}\psi' = ie^{i\psi}|z'|\kappa = i\kappa z',$$

as required.                                                                    □

Notice that $\dfrac{z'(t)}{|z'(t)|}$ is the unit (length) tangent vector, and that $iz'$ is a normal vector. The derivative of the unit tangent vector is therefore orthogonal to the tangent vector provided that the curvature is not zero. In the case of arc-length parametrisation, $z'(s)$ is already a unit tangent vector for each $s$ and $iz'(s)$ is the corresponding unit normal vector.

**Worked example 7.16.** Use Theorem 7.15 to show that the curvature at all points of a circle of radius $R$ is either $\dfrac{1}{R}$ or $-\dfrac{1}{R}$, depending on the orientation.

*Solution.* Consider the circle oriented counter-clockwise by the parametrisation $z = z_0 + Re^{it}$ $(0 \leq t \leq 2\pi)$. This has centre $z_0 = x_0 + iy_0$ and radius $R$. We have

$$z' = iRe^{it}, \quad \frac{z'}{|z'|} = ie^{it}, \quad \text{and} \quad \frac{d}{dt}\left(\frac{z'}{|z'|}\right) = -e^{it} = \frac{1}{R}iz'.$$

Thus, by Theorem 7.15, the curvature at all points of this circle of radius $R$ (which is oriented counter-clockwise) is $\dfrac{1}{R}$. Similarly for circles

$$z = z_0 + Re^{-it} \quad (0 \leq t \leq 2\pi)$$

oriented clockwise the curvature is $-\dfrac{1}{R}$ at all points.                                         □

The equivalent of Theorem 7.15 for curves in $\mathbf{R}^2$ is the following corollary.

**Corollary 7.17.** *At a regular point of a smooth parametrised curve in $\mathbf{R}^2$ we have*

$$\boxed{\frac{d}{dt}\left(\frac{1}{|\mathbf{r}'(t)|}\mathbf{r}'(t)\right) = \kappa \cdot (-y'(t), x'(t)).}$$

*In particular, in the case where the parametrisation is by the arc-length $s$, we have*

$$\boxed{\mathbf{r}''(s) = \kappa\,(-y'(s), x'(s)) \quad and \quad (-y''(s), x''(s)) = -\kappa \mathbf{r}'(s).}$$

Note that in the latter case, $\mathbf{r}'(s)$ is a unit tangent vector for each $s$ and $(-y'(s), x'(s))$ is the corresponding unit normal vector.

It is not necessary to calculate the curvature itself to find points at which the curvature is zero. The following corollary can save unnecessary calculation.

**Corollary 7.18.** *At a regular point of a smooth parametrisation the curvature is zero if, and only if,*

$$x'y'' - y'x'' = \mathbf{r}' \cdot (-y'', x'') = -\operatorname{im}\left(z'\,\overline{z''}\right) = 0.$$

*Equivalently the curvature is zero if, and only if,*

$$\frac{d}{dt}\left(\frac{1}{|\mathbf{r}'|}\mathbf{r}'\right) = \frac{d}{dt}\left(\frac{z'}{|z'|}\right) = 0.$$

In order to use this result to find points at which the curvature is zero, we must check that the points we find are regular points. The curvature is not normally defined at non-regular points.

**Exercises**

**7.31.** Use Theorem 7.13 to show that the Archimedean spiral $z = a\varphi e^{i\varphi}$ has curvature $\kappa = \dfrac{\varphi^2 + 2}{a(\varphi^2 + 1)^{\frac{3}{2}}}$.

**7.32.** Use Theorem 7.15 to show that the equi-angular spiral $z = e^{(a+i)t}$ has curvature $\dfrac{e^{-at}}{\sqrt{1 + a^2}}$.

**7.33.** Let $\mathbf{n} = \dfrac{1}{|\mathbf{r}'|}(-y', x')$ be the standard unit normal vector. Show that

$$\mathbf{n}' = -\kappa \mathbf{r}'.$$

**7.34.** Show that the line $z = z_0 + tw$ $(w \neq 0)$ has curvature $\kappa = 0$.

# 8

# Curvature: applications

In this chapter we give applications of the curvature of parametrised curves. First we show how the curvature and its derivatives can be used to determine inflexions, vertices, and undulations. Then we show how the curvature of algebraic curves can be determined. Finally we show how cusps can be classified by considering the limiting curvature at the cusp, together with the sign of the curvature at points on each side of the cusp.

## 8.1 Inflexions of parametric curves at regular points

We defined a point of simple inflexion at a regular point of a parametrised curve to be a point at which the tangent has three-point contact, or, equivalently, a point at which $\mathbf{r}''$ is parallel to $\mathbf{r}'$ and $\mathbf{r}'''$ is not parallel to $\mathbf{r}'$ (see §5.3.1). Recall that $\mathbf{r}''$ is 'parallel to $\mathbf{r}'$ includes the possibility that $\mathbf{r}'' = \mathbf{0}$. We now give a criterion for determining inflexions using the curvature and its derivatives.

**Criterion 8.1.** *A regular point of a twice differentiable parametrisation is a point of simple inflexion if, and only if, $\kappa = 0$ and $\kappa' \neq 0$.*

*Proof.* This result follows easily from Theorem 7.3. First $(-y'', x'')$ is obtained from $\mathbf{r}''$ by a counter-clockwise rotation through the angle $\dfrac{\pi}{2}$. We have $\kappa = 0$ if, and only if, $\mathbf{r}' \cdot (-y'', x'') = 0$, that is if, and only if, $\mathbf{r}''$ is a parallel to $\mathbf{r}'$. Also

$$\kappa' = \frac{d}{dt}\left(-\frac{\mathbf{r}' \cdot (-y'', x'')}{|\mathbf{r}'|^3}\right)$$

$$= -\frac{\mathbf{r}'' \cdot (-y'', x'')}{|\mathbf{r}'|^3} - \frac{\mathbf{r}' \cdot (-y''', x''')}{|\mathbf{r}'|^3} - \mathbf{r}' \cdot (-y'', x'')\frac{d}{dt}\left(\frac{1}{|\mathbf{r}'|^3}\right).$$

The first term on the right is always 0, the third term is 0 whenever $\kappa$ is 0, and the second term is 0, if, and only if, $\mathbf{r}'''$ is a parallel to $\mathbf{r}'$. $\qquad\square$

More generally a point of inflexion is a point $(a, b)$ where $\kappa = 0$ and where $\kappa$ has opposite signs on the curve close to $(a, b)$ on either side of $(a, b)$, that is it is positive along the curve on one side of $(a, b)$ and negative along the curve on the other side of $(a, b)$.

**Example 8.2.** On the line $\mathbf{r} = (a, b) + t(p, q)$ we have $\mathbf{r}' = (p, q)$ and $\mathbf{r}'' = \mathbf{0}$. Thus $\kappa = \kappa' = 0$ at all points on the line. Therefore at every point on a line we have $\kappa = 0$; but no point on a line is a point of simple inflexion, since $\kappa'$ is zero at all such points.

**Worked example 8.3.** Prove that the acnodal cubic $y^2 = x^3 - x^2$ has precisely two points of simple inflexion and find them.

*Solution.* We parametrise the curve excluding its isolated point, by

$$\mathbf{r} = (x, y) = (t^2 + 1, \ t(t^2 + 1))$$

for real values of $t$. We have

$$\mathbf{r}' = (2t, \ 3t^2 + 1),$$
$$\mathbf{r}'' = (2, \ 6t),$$
$$(-y'', x'') = (-6t, \ 2),$$
$$(\mathbf{r}')^2 = 9t^4 + 10t^2 + 1, \text{ and}$$
$$-\mathbf{r}' \cdot (-y'', x'') = 6t^2 - 2.$$

Thus the curvature is given by

$$\kappa = \frac{6t^2 - 2}{(9t^4 + 10t^2 + 1)^{\frac{3}{2}}} \ ,$$

which is zero if, and only if, $t = \pm\frac{1}{\sqrt{3}}$ . Writing

$$t^2 - \tfrac{1}{3} = \left(t - \tfrac{1}{\sqrt{3}}\right)\left(t + \tfrac{1}{\sqrt{3}}\right),$$

we have, by Lemma 8.4 below,

$$\kappa'\left(\tfrac{1}{\sqrt{3}}\right) = \left. \frac{6\left(t + \tfrac{1}{\sqrt{3}}\right)}{(9t^4 + 10t^2 + 1)^{\frac{3}{2}}} \right|_{t=\frac{1}{\sqrt{3}}} \neq 0,$$

Similarly $\kappa'\left(-\tfrac{1}{\sqrt{3}}\right) \neq 0$. Therefore the acnodal cubic has precisely two points of simple inflexion, at $t = \pm\frac{1}{\sqrt{3}}$ .    □

### 8.1.1  A useful lemma

The following lemma can save a great deal of work in calculating the derivative of a function at point at which the function is zero, and in determining the sign of such a derivative. One use is to show easily (in suitable cases) which inflexions are simple. Another is to determine whether a critical point is a maximum or a minimum.

**Lemma 8.4.** *Let $f(x) = g(x)h(x)$ where $g$ and $h$ are differentiable at $a$ and $g(a) = 0$. Then $f'(a) = g'(a)h(a)$.*

By this lemma, in order to determine $f'(a)$, or simply the sign of $f'(a)$, it is not necessary to differentiate the whole of the formula for $f$, but simply that 'factor' which is zero at $x = a$.

*Proof.* We have

$$f'(a) = g'(a)h(a) + g(a)h'(a)$$

by the rule for differentiating a product, and the result is immediate. □

**Example 8.5.** Let $f(x) = (x - 2)\tan^{-1}\left(\dfrac{x^2 + x + 1}{x^4 + 1}\right)$, then (immediately) we have

$$f'(2) = \tan^{-1}\left(\frac{4 + 2 + 1}{16 + 1}\right) = \tan^{-1}\left(\frac{7}{17}\right).$$

**Exercises**

**8.1.** Find the curvature of the following curves and determine their inflexions

a) $\left(\frac{1}{2}t^2, t + \frac{1}{3}t^3\right)$

$$\left[\frac{t^2 - 1}{(t^4 + 3t^2 + 1)^{\frac{3}{2}}}; \pm 1, \text{ both simple}\right]$$

b) $\left(t + \frac{1}{2}t^2, \frac{1}{3}t^3\right)$

**8.2.** Let $f(x) = g(x)h(x)$ where $g$ and $h$ are twice differentiable at $a$ and $g(a) = g'(a) = 0$. Show that $f''(a) = g''(a)h(a)$.

**8.3.** Let $f(x) = g(x)h(x)$ where $g$ and $h$ are $n$-times differentiable at $a$ and $g(a) = g'(a) = \cdots = g^{(n-1)}(a) = 0$. Show that $f^n(a) = g^n(a)h(a)$.

## 8.2 Vertices and undulations at regular points

We defined a point of simple undulation at a regular point of a parametrised curve to be a point at which the tangent has four-point contact, or, equivalently a point at which $\mathbf{r}'$, $\mathbf{r}''$, and $\mathbf{r}'''$, are parallel and $\mathbf{r}^{(4)}$ is not parallel to $\mathbf{r}'$ (see §5.3.1). We now give an alternative criterion for undulation in terms of the curvature.

**Criterion 8.6.** *A regular point of a three times differentiable parametrisation is a point of simple undulation if, and only if, $\kappa = \kappa' = 0$ and $\kappa'' \neq 0$.*

*Proof.* By Criterion 8.1 we know that $\kappa = 0$ if, and only if, $\mathbf{r}'$ and $\mathbf{r}''$ are parallel, that $\kappa = \kappa' = 0$ if, and only if, $\mathbf{r}'$, $\mathbf{r}''$, and $\mathbf{r}'''$ are parallel, and that in general

$$\kappa' = -\frac{\mathbf{r}' \cdot (-y''', x''')}{|\mathbf{r}'|^3} - (\mathbf{r}' \cdot (-y'', x'')) \frac{d}{dt}\left(\frac{1}{|\mathbf{r}'|^3}\right).$$

At a point where $\mathbf{r}'$, $\mathbf{r}''$ and $\mathbf{r}'''$ are parallel, we have, on using Lemma 8.4,

$$\kappa'' = -\frac{\mathbf{r}' \cdot (-y^{(4)}, x^{(4)})}{|\mathbf{r}'|^3}.$$

Thus given that $\kappa = \kappa' = 0$, we have that $\kappa'' \neq 0$ if, and only if, $\mathbf{r}^{(4)}$ is not parallel to $\mathbf{r}'$.                                                                    □

More generally a point of undulation is a point $(a, b)$ where $\kappa = 0$ and where $\kappa$ has the same sign on the curve on either side of $(a, b)$, that is either it is positive along the curve on both sides of $(a, b)$ or negative along the curve on both sides of $(a, b)$.

**Definition 8.7.** A *vertex* is a (regular) point for which $\kappa' = 0$.

Thus points, at which the curvature $\kappa(t)$ has a local maximum, a local minimum or a stationary point of inflexion, are vertices.

Note carefully the difference between simple inflexion, vertex, and simple undulation using the curvature criteria. Note also the difference between vertex and corner; a vertex is a regular point and a corner is not.

**Worked example 8.8.** Show that, on circles and lines, all points are vertices.

*Solution.* For a line we have $\kappa = 0$ and for a circle we have that $\kappa$ is constant. In each case $\kappa' = 0$.                                                           □

**Worked example 8.9.** Find the curvature of $y = x^4$. Deduce that the curve has a simple undulation at the origin, and, additionally, two vertices close to the origin. Show that there is a strict local maximum of the curvature at each of these two vertices.

*Solution.* We parametrise the curve by $x \mapsto (x, x^4) = \mathbf{r}(x)$. We have

$$\kappa = \frac{x' y'' - y' x''}{|\mathbf{r}'|^3} = \frac{12x^2}{(16x^6 + 1)^{3/2}}, \text{ and}$$

$$\kappa' = \frac{24x(1 - 56x^6)}{(16x^6 + 1)^{5/2}}.$$

Also at the point where $x = 0$, we have by Lemma 8.4,

$$\kappa''(0) = \left. \frac{24(1 - 56x^6)}{(16x^6 + 1)^{5/2}} \right|_{x=0} = 24.$$

Therefore $\kappa(0) = \kappa'(0) = 0$ and $\kappa''(0) \neq 0$. Thus there is a simple undulation at $x = 0$. Also there are two vertices and the values taken by $x$ at these are the two real sixth roots of $\frac{1}{56}$, which are $\approx \pm.51$; thus the vertices are $\approx (\pm.51, .07)$. Using Lemma 8.4, we readily see that a strict local maximum of the curvature occurs at each of the vertex points, since, at the vertex points,

$$\kappa'' = \frac{24x(-56.6x^5)}{(16x^6 + 1)^{5/2}}\Big|_{x=\pm\frac{1}{\sqrt[6]{56}}} < 0.$$

□

**Worked example 8.10.** Let $\mathbf{r}(t) = (3t^2 - 1, \, t\,(3t^2 - 1)\,)$. Show that the curve $t \mapsto \mathbf{r}(t)$ is regular, and find its curvature. Show that the curve has no inflexions and only one vertex. Find the points at which the tangents are horizontal and the points at which the tangents are vertical. Find the slopes of the tangent lines to the curve at the point $\mathbf{r} = \mathbf{0}$. Find the local maxima and minima of $y$ regarded as a function of $x$. Write down the algebraic equation of the curve.

*Solution.* We have $\mathbf{r}' = (6t, 9t^2 - 1)$, $\mathbf{r}'' = (6, 18t)$, and $|\mathbf{r}'|^2 = (9t^2 + 1)^2$. The parametrisation is regular, since $\mathbf{r}' \neq \mathbf{0}$. The curvature is

$$\kappa = \frac{x'\,y'' - y'\,x''}{|\mathbf{r}'|^3} = \frac{6(9t^2 + 1)}{(9t^2 + 1)^3} = \frac{6}{(9t^2 + 1)^2} \,, \text{ and}$$

$$\kappa' = \frac{6.18t.(-2)}{(9t^2 + 1)^3} \,.$$

Thus there are no inflexions and a vertex only at $t = 0$: at $t = 0$ we have (using Lemma 8.4)

$$\kappa'(0) = 6.18.(-2) < 0,$$

and thus the curvature has a local maximum where $t = 0$, that is at the point $(-1, 0)$. The tangent is horizontal (i.e., $y' = 0$ and $x' \neq 0$) if, and only if, $t = \pm\frac{1}{3}$, and is vertical (i.e., $x' = 0$ and $y' \neq 0$) if, and only if, $t = 0$. Also $\mathbf{r} = \mathbf{0}$ if, and only if, $t = \pm\frac{1}{\sqrt{3}}$. At $t = \pm\frac{1}{\sqrt{3}}$ we have

$$\frac{dy}{dx} = \frac{y'}{x'} = \pm\frac{1}{\sqrt{3}}.$$

Now

$$\frac{dy}{dx} = \frac{y'}{x'} = \frac{9t^2 - 1}{6t} = 0$$

if, and only if, $t = \pm\frac{1}{3}$. Also

$$\frac{d^2y}{dx^2}\,x' = \frac{d}{dt}\left(\frac{dy}{dx}\right),$$

and again using Lemma 8.4, we see that $\dfrac{d^2y}{dx^2} \neq 0$ at $t = \pm\frac{1}{3}$, and therefore the local maxima and minima of $y$, regarded as a function of $x$, are at $(-\frac{2}{3}, \pm\frac{2}{9})$. For large values of $t$, the curve is 'close' to the semi-cubical parabola $x^3 = 3y^2$ parametrised by $t \mapsto (3t^2, 3t^3)$. The algebraic equation of the curve is given by

$$y^2 = x^2 t^2 = \frac{x^2(x+1)}{3},$$

that is by $3y^2 = x^2(x+1)$: so it is a cubic curve. $\qquad\square$

**Worked example 8.11.** Show that the ellipse $\mathbf{r} = (a\cos t, b\sin t)$ has no inflexions and has precisely four vertices.

*Solution.* We have

$$\kappa = \frac{ab}{\left(a^2\sin^2 t + b^2\cos^2 t\right)^{\frac{3}{2}}}, \quad \text{and}$$

$$\kappa' = -\frac{3ab(a^2-b^2)\sin t \cos t}{(a^2\sin^2 t + b^2\cos^2 t)^{5/2}}.$$

Thus the ellipse has no inflexions and has precisely four vertices. The parameter values at which these vertices occur are $t = 0, \dfrac{\pi}{2}, \pi, \dfrac{3\pi}{2}$. Using Lemma 8.4, we readily see that $\kappa'' \neq 0$ at each of these vertices. Therefore the curvature has a local maximum or a local minimum at each vertex. $\qquad\square$

This example illustrates a general property of simple closed curves, that is that such curves have at least four vertices.

**Theorem 8.12.** *(Four-vertex theorem) A simple closed curve has at least four vertices; at least two correspond to local maximum values of $\kappa$ and at least two correspond to local minimum values of $\kappa$.*

A *simple* curve here means a regular curve with no multiple points except the end points, that is $\mathbf{r}(t_0) = \mathbf{r}(T)$. The proof of this theorem is hard and beyond the scope of this book. The proof can be found in texts on algebraic curves, or algebraic geometry.

**Exercises**

**8.4.** Show that the curve $t \mapsto \mathbf{r}(t) = (t^2 - 1, t(t^2 - 1))$ is regular, and find its curvature. Show that $\kappa' = 0$ if, and only if, $9t^5 + 4t^3 - 1 = 0$, and deduce that the curve has no inflexions and precisely three vertices. Find the points at which the tangents are horizontal and the points at which the tangents are vertical. Find the slopes of the tangent lines to the curve at the point $\mathbf{r} = \mathbf{0}$. Find the local maxima and minima of $y$ regarded as a function of $x$. Write down the algebraic equation of the curve.

$$\left[\frac{6t^2 + 2}{(9t^4 - 2t^2 + 1)^{\frac{3}{2}}}; 0, t^2 = \tfrac{1}{18}\left(\sqrt{52} - 4\right); 0, \pm\tfrac{1}{\sqrt{3}}; \pm 1; \pm\tfrac{1}{\sqrt{3}}; y^2 = x^2(1+x)\right]$$

**8.5.** Find the curvature of $(t + \tfrac{1}{2}t^2, \tfrac{1}{2}t^2)$. Show that the curve has no inflexions and has precisely one vertex. Write down the algebraic equation of the curve.

**8.6.** Prove that the curvature of $r = 2 + \cos\theta$ is

$$\kappa = \frac{6(1 + \cos\theta)}{(5 + 4\cos\theta)^{\frac{3}{2}}}.$$

Show that the curve has a simple undulation at $\theta = -\pi$, and find the vertices.

(*Hint:* Use Exercise 8.2.)

$$\left[0, \frac{2\pi}{3}, \pi, \frac{4\pi}{3}\right]$$

## 8.3 Curvature of algebraic curves*

In general algebraic curves cannot be parametrised in a simple way, that is by using a simple formula which gives a parametrisation of the whole curve. Therefore the formulae for the curvature which we have considered previously cannot be used. However algebraic curves can be parametrised locally near a non-singular point, though not in general using a simple formula. We now use this result to give a parameter-free formula for the curvature of algebraic curves.

Given a polynomial $f(x, y)$ let $J = \begin{pmatrix} f_{xx} & f_{xy} \\ f_{xy} & f_{yy} \end{pmatrix}$ be the Hessian matrix.

**Theorem 8.13.** *The curvature of the algebraic curve* $f(x, y) = 0$ *at a non-singular point* $(x, y)$ *on the curve is*

$$\kappa = \pm \frac{(f_y, -f_x) J \begin{pmatrix} f_y \\ -f_x \end{pmatrix}}{(f_x^2 + f_y^2)^{\frac{3}{2}}} = \pm \frac{f_y^2 f_{xx} - 2f_x f_y f_{xy} + f_x^2 f_{yy}}{(f_x^2 + f_y^2)^{\frac{3}{2}}}$$

$$= \pm \frac{f_y^2 f_{xx} - 2f_x f_y f_{xy} + f_x^2 f_{yy}}{|\mathbf{grad}\, f|^3},$$

*where the sign* $-$ *is taken in case the motion along the curve is in the direction of the vector* $(f_y, -f_x)$ *and where the sign* $+$ *is taken in case the motion along the curve is in the direction of the vector* $-(f_y, -f_x)$.

*Proof.* By Theorem 4.4, the algebraic curve can be given a regular local parametrisation near a non-singular point by $t \mapsto \mathbf{r}(t)$. Differentiating

$$f(x(t), y(t)) = 0,$$

using the chain rule, we obtain

$$\mathbf{r}' \cdot (f_x, f_y) = 0.$$

Therefore $\mathbf{r}'$ is orthogonal to $(f_x, f_y)$, and we have $\mathbf{r}' = \lambda(f_y, -f_x)$ where $\lambda \neq 0$ is a function of $t$, and $(-y', x') = \lambda(f_x, f_y)$. On differentiating once

---

* For the meaning of * see the preface.

more we obtain

$$\mathbf{r}'' \cdot (f_x, f_y) + (x', y') \, J \begin{pmatrix} x' \\ y' \end{pmatrix} = 0.$$

Therefore we have

$$x'\, y'' - y'\, x'' = \mathbf{r}'' \cdot (-y', x') = \lambda \mathbf{r}'' \cdot (f_x,\ f_y) = -\lambda^3 (f_y, -f_x)\, J \begin{pmatrix} f_y \\ -f_x \end{pmatrix}.$$

Since $|\mathbf{r}'| = |\lambda| \, |(f_y, -f_x)|$, the result follows.                    □

**Corollary 8.14.** *The algebraic curve has a point of inflexion at a non-singular point* $(a, b)$ *if, and only if,*

$$f_y^2 f_{xx} - 2 f_x f_y f_{xy} + f_x^2 f_{yy}$$

*is zero at* $(a, b)$ *and changes sign as* $(x, y)$ *moves through* $(a, b)$ *along the curve.*

Thus in order to find all points of inflexion of an algebraic curve we first determine the points where the curvature is zero by finding the simultaneous solutions of the equations

$$f(x, y) = 0, \text{ and}$$
$$f_y^2 f_{xx} - 2 f_x f_y f_{xy} + f_x^2 f_{yy} = 0,$$

and then find those solutions which are non-singular points. Next we determine whether the curvature changes sign as we move along the curve past the point of zero curvature. If the curvature changes sign, the point is an inflexion. If the sign of the curvature is the same immediately before and after the point, the point is an undulation. In practice we can check on the change of sign if, for example, the curve can be parametrised near the point by $x$, and

$$f_y^2 f_{xx} - 2 f_x f_y f_{xy} + f_x^2 f_{yy}$$

becomes a function of $x$ after substituting from $f(x, y) = 0$. Such a local parametrisation by $x$ exists by Theorem 4.4 provided that the tangent at the point is not parallel to the $y$-axis.

**Worked example 8.15.** Calculate the curvature of the hyperbola

$$x^2 - 3y^2 = 1$$

at $(2, 1)$.

*Solution.* Let $f = x^2 - 3y^2 - 1$. We have $f_x = 2x$, $f_y = -6y$, $f_{xx} = 2$, $f_{xy} = 0$, and $f_{yy} = -6$. Therefore

$$\kappa = \pm \frac{(-6y)^2 . 2 + (4x)^2 . (-6)}{(4x^2 + 36y^2)^{\frac{3}{2}}} = \pm \frac{9y^2 - 12x^2}{(x^2 + 9y^2)^{\frac{3}{2}}}.$$

In particular

$$\kappa(2, 1) = \pm \frac{-3}{\sqrt{13}}.$$

□

**Worked example 8.16.** Find the points of inflexion of the acnodal cubic

$$f(x,y) = y^2 + x^2 - x^3 = 0.$$

*Solution.* We have $f_x = 2x - 3x^2$, $f_y = 2y$, $f_{xx} = 2 - 6x$, $f_{xy} = 0$, and $f_{yy} = 2$. Eliminating $y$ from

$$f(x,y) = y^2 + x^2 - x^3 = 0, \text{ and}$$

$$f_y^2 f_{xx} - 2f_x f_y f_{xy} + f_x^2 f_{yy} = 4y^2(2 - 6x) + 2(2x - 3x^2)^2$$
$$= 8y^2 - 24xy^2 + 8x^2 - 24x^3 + 18x^4,$$

we have, on the curve, that

$$f_y^2 f_{xx} - 2f_x f_y f_{xy} + f_x^2 f_{yy} = 8x^3 - 6x^4.$$

The origin is a singular point. Therefore there are precisely two points of zero curvature. They are $\left(\frac{4}{3}, \pm\frac{4}{3\sqrt{3}}\right)$. At these points

$$\mathbf{grad}\, f = (2x - 3x^2, 2y)$$

is not parallel to the $y$–axis, and therefore the curve can be parametrised locally by $x$. Also

$$8x^3 - 6x^4 = 6x^3 \left(\frac{4}{3} - x\right)$$

changes sign as we move along the curve past each of the points and therefore the curvature also changes sign. Thus the two points are points of inflexion. (Near points of undulation the curvature does not change sign.)

□

**Exercises**

**8.7.** Calculate the curvature of the ellipse $\dfrac{x^2}{4} + y^2 = 1$ at $(2,0)$ using Theorem 8.13. Compare your result with that given by using the parametrisation $t \mapsto (2\cos t, \sin t)$. $\left[\mp\frac{3}{8}\right]$

**8.8.** Show that the curvature of $x^4 + y^2 = 2a(x+y)$ at the origin is

$$\pm\frac{1}{2\sqrt{2}a}.$$

**8.9.** Show that the curvature of $y^2(a-x) = x^2(a+x)$ at $(-a,0)$ is $\pm\dfrac{4}{a}$.

**8.10.** Show that the curvature of $x^2 y = x^2 + y^2$ at non-singular points is

$$\pm\frac{2x^3 y^4 (y^2 - 3x^2)}{(4y^6 + x^2(y^2 - x^2)^2)^{\frac{3}{2}}}.$$

**8.11.** Show that the curvature of $x^3 - xy^2 - y = 0$ at points where $y$ has a local maximum or minimum value is $\pm 2^{-\frac{1}{2}} 3^{\frac{5}{4}}$.

## 8.4 Limiting curvature of algebraic curves at cusps

As a further application of curvature, we show how curvature can be used to classify the cusps of algebraic curves. We divide cusps into three types, ceratoid cusps, cusps with zero limiting curvature, and rhamphoid cusps (see Figure 8.1). Although the curvature is not defined at a cusp, in that the formula for curvature is not meaningful at a cusp, it is nevertheless useful to consider the limit of the curvature at a point as the point approaches the cusp. The following analysis of types of cusps does not extend to non-algebraic curves in general.

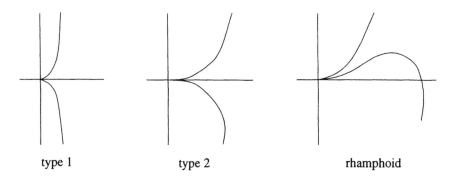

type 1                          type 2                          rhamphoid

Figure 8.1 *Limiting curvature at cusps.*

### 8.4.1 Type 1: Ceratoid cusps (limiting radius of curvature zero).

A point $z(t_0)$ is a *ceratoid cusp* if it is a cusp for which $|\kappa(t)|$ tends to $\infty$ along both branches as $t$ tends to $t_0$, or equivalently for which the radius of curvature tends to zero as $t$ tends to $t_0$. For ceratoid cusps the curve appears to move away from the cuspidal tangent line 'fairly rapidly' as it leaves the cusp, since near the cusp the curvature is very large and therefore the curve bends sharply.

**Example 8.17.** For the semi-cubical parabola $x^3 = y^2$ parametrised by $t \mapsto \mathbf{r}(t) = (t^2, t^3)$ for all real $t$, we have

$$\mathbf{r}' = (2t, 3t^2),$$
$$\mathbf{r}'' = (2, 6t), \text{ and}$$
$$\mathbf{r}''' = (0, 6).$$

The curvature of the curve is given by

$$\kappa = \frac{6t^2}{(4t^2 + 9t^4)^{\frac{3}{2}}} = \frac{6}{|t|(4 + 9t^2)^{\frac{3}{2}}}.$$

In this calculation we have used the formula $\sqrt{t^2} = |t|$. We have that $\kappa$ tends to $\infty$ as $t$ tends to 0, and therefore the limiting radius of curvature is zero. The semi-cubical parabola has an ordinary cusp at $\mathbf{r}'(0) = \mathbf{0}$, since $\mathbf{r}''(0) = (2,0)$ and $\mathbf{r}'''(0) = (0,6)$ are linearly independent.

### 8.4.2 Type 2: limiting radius of curvature infinite

The second type of cusp is a cusp at a point $z(t_0)$ for which $\kappa(t)$ tends to 0 along both branches as $t$ tends to $t_0$, or equivalently for which the limiting radius of curvature is infinite. In this case the curve appears to linger near the cuspidal tangent line as it leaves the cusp, since the curvature is close to zero near the cusp.

**Example 8.18.** For the curve $(2\,x)^5 = (5\,y)^2$, parametrised by

$$t \mapsto \mathbf{r}(t) = \left(\tfrac{1}{2}t^2, \tfrac{1}{5}t^5\right)$$

for all real $t$, we have

$$\mathbf{r}' = (t, t^4),$$
$$\mathbf{r}'' = (1, 4t^3), \text{ and}$$
$$\mathbf{r}''' = (0, 12t^2).$$

The curvature of the curve is given by

$$\kappa = \frac{3t^4}{(t^2 + t^8)^{\frac{3}{2}}} = \frac{3|t|}{(1 + t^6)^{\frac{3}{2}}}.$$

We have that $\kappa$ tends to 0 as $t$ tends to 0, and the limiting radius of curvature is infinite. This cusp is not ordinary since $\mathbf{r}'' = (1,0)$ and $\mathbf{r}''' = (0,0)$ are not linearly independent at it. In this calculation we have used the formula $\sqrt{t^2} = |t|$. Recall that the curvature is positive as the curve swings to the left; this is indicated in the figure.

Cusps ($\mathbf{r}'' \neq \mathbf{0}$), for which, in moving through the cusp, the curve crosses the cuspidal tangent line from one side to the other, are necessarily of Type 1 or of Type 2 (ie. $|\kappa| = \infty$ or $\kappa = 0$ in the limit). For Type 1 and Type 2 cusps the curvature has the same sign, close to the cusp, on each side of the cusp.

### 8.4.3 Type 3: Rhamphoid cusps

Rhamphoid cusps are cusps for which the curve, close to the cusp, lies on one side of the cuspidal tangent line. For such cusps the curvature has different signs, close to the cusp, on each side of the cusp. These cusps can also have limiting radii of curvature other than 0 or $\infty$.

**Example 8.19.** The curve $(y - x^2)^2 = x^5$, or $y = x^2 \pm x^{\frac{5}{2}}$, parametrised by $t \mapsto \mathbf{r}(t) = (t^2, t^4 + t^5)$ for all real $t$, satisfies

$$\mathbf{r}' = (2t, 4t^3 + 5t^4),$$
$$\mathbf{r}'' = (2, 12t^2 + 20t^3), \text{ and}$$
$$\mathbf{r}''' = (0, 24t + 60t^2).$$

The curvature of the curve is given by

$$\kappa = \frac{(16t^3 + 30t^4)}{(4t^2 + (4t^3 + 5t^4)^2)^{\frac{3}{2}}} = \frac{t}{|t|} \frac{(16 + 30t)}{(4 + (4t^2 + 5t^3)^2)^{\frac{3}{2}}}.$$

In this calculation we have used the formula $\sqrt{t^2} = |t|$. We have $|\kappa|$ tends to 2 as $t$ tends to 0, $\kappa$ tends to $+2 > 0$ as $t > 0$ tends to 0, and $\kappa$ tends to $-2 < 0$ as $t < 0$ tends to 0. Also we have the slope

$$\frac{dy}{dx} = \frac{y'}{x'} = \frac{t^3(4 + 5t)}{2}.$$

This curve has a point of simple inflexion where $\kappa = 0$, that is, where $t = -\frac{8}{15}$. Also, regarded as the graph of a function of $x$, the curve has a horizontal tangent at $t = -\frac{4}{5}$. Again the cusp is not ordinary since $\mathbf{r}'' = (2, 0)$ and $\mathbf{r}''' = (0, 0)$ are not linearly independent at it.

**Drawing cusps.** Note the different shapes of these cusps depending on the different limiting radii of curvature, and on whether or not the cusp is rhamphoid.

### Exercises

**8.12.** For each of the following curves, show that there is a cusp at the origin, and classify the cusp by finding the limiting radius of curvature and determining whether or not it is rhamphoid.

   i) $(\frac{1}{2}t^2, \frac{1}{3}t^3)$,                                           [$\infty$; type 1, ordinary]

   ii) $(\frac{1}{2}t^2, \frac{1}{4}t^4 + \frac{1}{5}t^5)$,

   iii) $(t^2, t^5)$,                                                         [0; type 2]

   vi) $(\frac{1}{2}t^2, \frac{1}{6}t^6 + \frac{1}{6}t^6)$.

**8.13.** Show that $\mathbf{r}(t) = (t^2, t^n)$ $(n \geq 7)$ has, at the origin, a type 2 cusp if $n$ is odd.

**8.14.** Show that $\mathbf{r}(t) = (t^2, t^n + t^{n+1})$ $(n \geq 6)$ has, at the origin, a rhamphoid cusp if $n$ is even.

**8.15.** Use L'Hôpital's theorem (see §4.7.5) to show that, at a cusp $\mathbf{r}(t_0)$ of a parametrised curve $t \mapsto \mathbf{r}(t)$,

$$\lim_{t \to t_0} \frac{x'y'' - y'x''}{(x')^2 + (y')^2} = \frac{x''y''' - y''x'''}{2(x'')^2 + 2(y'')^2}.$$

Deduce that $\lim_{t \to t_0} |\kappa(t)| = \infty$ at a simple cusp.

We give here some more general exercises which can include curve sketching or plotting if desired.

**8.16.** Let $\mathbf{r}(t) = (\sqrt{3}(t^2 - 1),\ t\,(t^2 - 1)\,)$. Show that the curve $t \mapsto \mathbf{r}(t)$ is regular. Prove that the curvature of $t \mapsto \mathbf{r}(t)$ at $\mathbf{r}(t)$, is given by

$$\kappa = \frac{2\sqrt{3}}{(3t^2 + 1)^2},$$

and show that the curve has no inflexions and only one vertex. Show further that the curvature $\kappa$ does not have an inflexion at the vertex of the curve. Find the points at which the tangents are horizontal and the points at which the tangents are vertical. Hence sketch the curve, noting the points with horizontal tangents and the points with vertical tangents, and the local maxima and minima of $y$. Write down the algebraic equation of the curve.

$$\left[ t = 0, \kappa''(0) \neq 0 : \pm \tfrac{1}{\sqrt{3}}; 0 : 9y^2 - 3x^2 - \sqrt{3}x^3 = 0 \right]$$

**8.17.** Show that the curve $z(t) = (\cos t, \cos t \sin t)$ is regular and has curvature

$$\kappa = \frac{\cos t\,(1 + 2\sin^2 t)}{\left(1 - 3\sin^2 t + 4\sin^4 t\right)^{\frac{3}{2}}}.$$

Hence find the inflexions of the curve. Show that every point of $z(t)$ belongs to the algebraic curve

$$y^2 = x^2 - x^4.$$

By writing $x^2 - x^4 = x^2(1 - x^2)$, show conversely that every point of the algebraic curve has the form $z(t)$ for some $t$. Show that the algebraic curve has a singular point at $(0,0)$ and no other singular point. Sketch the curve, indicating the positions of the inflexions. $\left[ n\pi + \tfrac{\pi}{2}; |x| \leq 1,\ \text{thus } x = \cos t \right]$

**8.18.** Show that the limaçon with polar equation

$$r = \sqrt{2} + \cos\theta$$

has inflexions where $\cos\theta = -\tfrac{2\sqrt{2}}{3}$ and vertices where

$$\sin\theta = 0 \text{ or } \cos\theta = -\tfrac{1}{\sqrt{2}}.$$

Find the tangent directions at the vertices. Sketch the curve.

**8.19.** Show that the limaçon with polar equation

$$r = \sqrt{2} - \cos\theta$$

has inflexions where $\cos\theta = \tfrac{2\sqrt{2}}{3}$. By checking that the appropriate determinant is zero, or otherwise, show also that the limaçon has vertices where $\sin\theta = 0$ or $\cos\theta = \tfrac{1}{\sqrt{2}}$. Find the tangent directions at the vertices. Sketch the curve.

$$\left[\kappa = \frac{-3\sqrt{2}\cos\theta + 4}{\left(2\sqrt{2}\cos\theta - 3\right)^{\frac{3}{2}}} : i, i \pm 1\right]$$

**8.20.** State the four vertex theorem. Find the vertices of the limaçon with polar equation

$$r = 1 + 2\cos\theta,$$

and make a careful drawing of this curve using polar graph paper. Why does this curve not provide a counterexample to the four vertex theorem?

$$[0, \pi; \text{ not simple}]$$

**8.21.** Show that the parametric curve $t \mapsto \mathbf{r}(t) = (t^2 + t^4, t^2 + t^5)$ has exactly one non-regular point and that this is a cusp but not an ordinary cusp. Write down the limiting tangent direction at the cusp. Show that $z(t)$ has zero curvature where $10t^3 + 15t - 8 = 0$. Use the intermediate value theorem and Rolle's theorem to deduce that $z(t)$ has one inflexion between $t = 0$ and $t = \frac{1}{2}$ and has no other inflexions. Show that $|\kappa(t)|$ tends to $\frac{1}{\sqrt{2}}$ as $t$ tends to 0. Draw the curve for $-1.1 \le t \le 1.1$ using 5 cm as 1 unit.

$$\left[\mathbf{r}'''(0) = \mathbf{0}; (1, 1); \text{ inflexion since } \frac{d}{dt}(10t^3 + 15t - 8) \ne 0\right]$$

**8.22.** Show that the parametric curve $t \mapsto \mathbf{r}(t) = (\frac{1}{4}t^4, t^2 + t^3)$ has exactly one non-regular point and that this is a cusp but not an ordinary cusp. Determine the limiting tangent direction at the cusp. Find the curvature $\kappa(t)$ at a general regular point, and show that $|\kappa(t)|$ tends to $\frac{1}{2}$ at the cusp. Show that there is exactly one point $P$ at which the curve has an inflexion. Determine the tangent direction at $P$. Draw the curve taking 10 cm as 1 unit. Indicate the point $P$ and plot about 9 other points. (Use the portrait mode with the origin near the left.)

**8.23.** Show that the parametric curve $t \mapsto \mathbf{r}(t) = (t^2, -t^4 + t^5)$ has exactly one non-regular point and that this is a cusp but not an ordinary cusp. Determine the limiting tangent direction at the cusp. Also determine the point $P$ on the curve, not at the cusp, at which the tangent is parallel to the $x$−axis. Find the curvature $\kappa(t)$ at a general regular point, and show that $|\kappa(t)|$ tends to 2 at the cusp. Show that there is exactly one point $Q$ at which the curve has an inflexion. Draw the curve for $-1.1 \le t \le 1.1$ using 5 cm as 1 unit. Indicate the points $P$ and $Q$ and plot about 9 other points. (Use the portrait mode with the origin near the top left: first calculate $\mathbf{r}(t)$ for $t = \pm 1.1$.)

$$\left[\mathbf{r}'''(0) = \mathbf{0}; (1, 0); \frac{4}{5}, \frac{8}{15}\right]$$

**8.24.** Show that the parametric curve $t \mapsto \mathbf{r}(t) = (t^2 - t^3, t^5)$ has exactly one non-regular point and that this is a cusp but not an ordinary cusp. Determine the cuspidal tangent line. Also determine the point $P$ on the curve, not at the cusp, at which the tangent is parallel to the $y$−axis. Find the curvature $\kappa(t)$ at a general regular point. Show that there is exactly one point $Q$ at which the curve has an inflexion. Draw the curve for $-1.1 \leq t \leq 1.1$ using 5 cm as 1 unit. Indicate the points $P$ and $Q$ and plot about 9 other points. (Use the landscape mode with the origin at the centre.)

# 9

# Circle of curvature

We consider now the radius of curvature, the circle of curvature, and the osculating circle at regular points of parametrised curves. The *radius of curvature* is the absolute value of the reciprocal of the curvature. The *circle of curvature* at a point on the curve is the circle passing through the point and having the same curvature and same tangent direction as the curve at the point. The radius of the circle of curvature is equal to the radius of curvature. The *osculating circle* at a point on the curve is the unique circle having $\geq$ 3–point contact with the curve at the point. We show that the circle of curvature coincides with the osculating circle.

## 9.1 Centre of curvature and circle of curvature

In general there is a unique circle having the same curvature and same tangent direction as the curve at a common point which is regular for the curve. The centre of this circle is called the centre of curvature at the point and the radius of the circle is called the radius of curvature at the point. The centre of curvature can also be described as the limiting position of the point of intersection of 'consecutive' normals. We now define the centre of curvature and the circle of curvature.

**Definition 9.1.** The *standard parametrisation* of a circle having curvature $\kappa$ $(\kappa \neq 0)$, and centre $z_0$, is

$$\boxed{z(t) = z_0 + \frac{1}{\kappa}e^{i\kappa t} \quad \text{or} \quad \mathbf{r}(t) = \mathbf{r}_0 + \frac{1}{\kappa}(\cos \kappa t, \sin \kappa t)} \quad \left(0 \leq t \leq \frac{2\pi}{|\kappa|}\right).$$

Notice that the standard parametrisation of a circle is a unit speed (arclength) parametrisation (i.e., $|z'| = 1$), and that $t$ is not the argument (polar angle) of $z - z_0$. This circle is positively oriented (counter-clockwise) if $\kappa > 0$, and is negatively oriented (clockwise) if $\kappa < 0$.

**Definition 9.2.** At a regular point $\mathbf{r}(t_0)$, respectively $z(t_0)$, of the curve $t \mapsto \mathbf{r}(t)$, respectively $t \mapsto z(t)$, the *centre of curvature* is

$$z_* = z_*(t_0) = z(t_0) + \frac{1}{\kappa} \frac{iz'(t_0)}{|z'(t_0)|}, \quad \text{or, equivalently,}$$

$$\mathbf{r}_* = \mathbf{r}_*(t_0) = \mathbf{r}(t_0) + \frac{1}{\kappa|\mathbf{r}'(t_0)|}(-y'(t_0), x'(t_0)) \, .$$

Thus the centre of curvature lies along the normal to the curve at $z(t_0)$, at an 'oriented distance' $\dfrac{1}{\kappa}$ to the 'left' of the curve. This means that the centre of curvature lies on the 'left' if $\kappa > 0$, and lies on the 'right' if $\kappa < 0$. 'Left' and 'right' here are determined by the direction of the parametrisation. We shall use the notations $t \mapsto \mathbf{r}(t)$ and $t \mapsto z(t)$ interchangeably for a curve. Many statements, proofs, and solutions are simpler using the curve $t \mapsto z(t)$ in the Argand diagram.

**Definition 9.3.** At a regular point $\mathbf{r}(t_0)$ of the curve $t \mapsto \mathbf{r}(t)$ the *radius of curvature* is $\dfrac{1}{|\kappa|}$ for $\kappa \neq 0$, and is $\infty$ for $\kappa = 0$.

The centre of curvature (see Figure 9.1) can be considered as the limiting position of the point of intersection of 'consecutive' normals to the curve. Notice that the radius of curvature of a circle of radius $R$, whatever its orientation, is $R$.

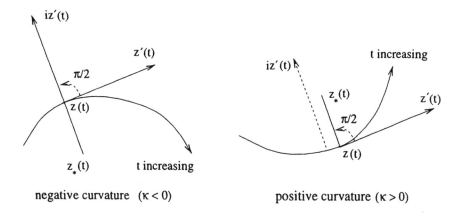

negative curvature $(\kappa < 0)$        positive curvature $(\kappa > 0)$

Figure 9.1 *Centre of curvature.*

**Definition 9.4.** The *circle of curvature* at a regular point $\mathbf{r}(t_0)$, at which the curvature is non-zero, is the circle centre $\mathbf{r}_*(t_0)$ and radius $\dfrac{1}{|\kappa|}$.

The equation of the circle of curvature in standard complex parametric form is

$$Z(s) = z_*(t_0) + \frac{1}{\kappa}e^{i\kappa s} \qquad \left(0 \le s \le \frac{2\pi}{|\kappa|}\right),$$

and the equation in standard real parametric form is

$$\mathbf{R}(s) = \mathbf{r}_*(t_0) + \frac{1}{\kappa}(\cos \kappa s, \sin \kappa s).$$

The algebraic equation is

$$(X - x_*)^2 + (Y - y_*)^2 = \frac{1}{\kappa^2}.$$

Other forms of the equation are

$$|Z - z_*| = \frac{1}{|\kappa|} \quad \text{or} \quad (Z - z_*)(\overline{Z} - \overline{z_*}) = \frac{1}{\kappa^2}.$$

Here we use $Z = X + iY$ and $\mathbf{R} = (X, Y)$ rather than $z = x + iy$ and $\mathbf{r} = (x, y)$ for the variable.

**Theorem 9.5.** *Let $\mathbf{r}(t_0)$ be a regular point of the curve $t \mapsto \mathbf{r}(t)$. Then there is a unique standardly parametrised circle, namely the circle of curvature, which passes through $\mathbf{r}(t_0)$, and which at $\mathbf{r}(t_0)$ has the same tangent line, the same direction of tangent vector, and the same curvature as the curve. Furthermore the sign of the orientation of this circle (counterclockwise is positive) coincides with the sign of the curvature.*

*Proof.* From the definition of the centre of curvature, we deduce that

$$|z_* - z(t_0)| = \frac{1}{|\kappa|}.$$

Thus the circle of curvature

$$|Z - z_*| = \frac{1}{|\kappa|}$$

passes through $z(t_0)$. Let

$$Z(s) = a + \frac{1}{\kappa}e^{i\kappa s}$$

be a standardly parametrised circle which passes through $z(t_0)$, where $\kappa$ is the curvature of the curve at $t = t_0$ (see Definition 9.1). Differentiating this equation gives $Z' = ie^{i\kappa s}$. This circle meets the curve at the point where $z(t_0) = Z(s)$. At this point we have

$$Z(s) - a = \frac{1}{\kappa}e^{i\kappa s} = -\frac{1}{\kappa}iZ'(s), \quad \text{and,}$$

$$z(t_0) - z_*(t_0) = -\frac{1}{\kappa}i\frac{z'(t_0)}{|z'(t_0)|}.$$

Therefore $a = z_*(t_0)$ if, and only if,

$$Z'(s) = \frac{z'(t_0)}{|z'(t_0)|},$$

that is if, and only if, the tangent vector $z'(t_0)$ to the curve and the tangent vector $Z'(s)$ to the circle are parallel and have the same direction (recall that $s \mapsto Z(s)$ is a unit speed curve). Since $a = z(t_0)$ for the circle of curvature, the existence, and uniqueness of the required circle follows.  □

The centre of curvature and the circle of curvature at a point lie on the 'concave side' of the curve at that point. Notice that the 'concave side' of the curve does not depend on the orientation of the curve, that is the direction of the parametrisation. The circle of curvature coincides with the osculating circle, and is often called the osculating circle. The osculating circle is the unique circle having $\geq$ 3–point contact with the curve at $\mathbf{r}(t_0)$ (see §9.2).

**Mnemonic**

> The circle of curvature has the same direction as the curve at their point of tangency.

**Worked example 9.6.** Find the centre of curvature and the circle of curvature of the curve $\mathbf{r} = (t^3 - t, t^2)$ at the origin.

*Solution.* The origin is given by the parameter value $t = 0$. By Example 7.6, we have that, at $t = 0$, $\kappa = -2$, and $\mathbf{r}' = (-1, 0)$. Thus the centre of curvature is

$$\mathbf{r}_* = \mathbf{r} + \frac{1}{\kappa |\mathbf{r}'|}(-y', x') = 0 - \frac{1}{2}(0, -1) = \frac{1}{2}(0, 1).$$

The circle of curvature is

$$(X - 0)^2 + (Y - \frac{1}{2})^2 = \left(\frac{1}{2}\right)^2,$$

that is

$$X^2 + Y^2 - Y = 0.$$

□

**Worked example 9.7.** Show that the circle of curvature of the curve

$$y = 2x + 3x^2 - 2xy + y^2 + 2y^3$$

at the origin is $3x^2 + 3y^2 - 5y + 10x = 0$.

*Solution.* Let $f = -y + 2x + 3x^2 - 2xy + y^2 + 2y^3$. We have

$$\frac{\partial f}{\partial y}(0, 0) = 1 \neq 0.$$

Therefore by Theorem 4.22, we can assume that $y$ is given as a function of $x$, $y = y(x)$, near the origin. Differentiating the defining equation implicitly with respect to $x$ we have

$$y' = 2 + 6x - 2xy' - 2y + 2yy' + 6y^2 y', \text{ and}$$

$$y'' = 6 - 4y' - 2xy'' + 2(y')^2 + 2yy'' + 6y^2 y'' + 12yy'y''.$$

Putting $x = y = 0$ in the first gives $y' = 2$. Substituting into the second then gives $y'' = 6$. Therefore

$$\kappa = \frac{6}{(1+4)^{\frac{3}{2}}} = \frac{6\sqrt{5}}{25},$$

and $(x_*, y_*) = \frac{5}{6}(-2, 1)$. The equation of the circle of curvature is therefore

$$(x + \tfrac{5}{3})^2 + (y - \tfrac{5}{6})^2 = \tfrac{125}{36},$$

which simplifies to give

$$3x^2 + 3y^2 - 5y + 10x = 0.$$

□

**Exercises**

**9.1.** Find the circle of curvature at $t = 0$ of the following curves.

a) the crunodal cubic $\mathbf{r}(t) = (t^2 - 1, t(t^2 - 1))$.

$$\left[ (x + \tfrac{1}{2})^2 + y^2 = \tfrac{1}{4} \right]$$

b) the acnodal cubic $\mathbf{r}(t) = (t^2 + 1, t(t^2 + 1))$.

**9.2.** Find the coordinates of the centre of curvature of the curve $y = \cosh x$ at the point $(0, 1)$. Find the circle of curvature.

(*Hint:* Recall that the derivatives of sinh and cosh are cosh and sinh respectively.)

**9.3.** Find a point on the curve $y = x^2 + x - 1$ for which the centre of curvature lies on the $y$-axis. $[x = -1]$

**9.4.** Write down the formula for the centre of curvature for a curve $y = f(x)$ parametrised by its $x$-coordinate.

$$\left[ \mathbf{r}(x) = (x, f(x)) + \frac{1 + (f'(x))^2}{f''(x)}(-f'(x), 1) \right]$$

**9.5.** A curve is parametrised by

$$t \mapsto \mathbf{r}(t) = \left( \frac{t^2}{1 + t^2}, \frac{t^3}{1 + t^2} \right).$$

Prove that the centre of curvature is $(-t^2 - \tfrac{1}{6}t^4, \tfrac{4}{3}t)$.

**9.6.** Show that the circles of curvature of the parabola $y^2 = 4ax$ for the ends of the latus rectum are $x^2 + y^2 - 10ax \pm 4ay - 3a^2 = 0$, and that they cut the curve again in the points $(9a, \mp 6a)$.

**9.7.** Prove that the curvature of the curve $y = \frac{1}{4}x^2 - \frac{1}{2}\log x$ $(x > 0)$ is

$$\frac{4x}{(1+x^2)^2}.$$

Find the point at which the curve is parallel to the $x$-axis, and prove that the circle of curvature at this point touches the $y$-axis.

$$[(1, \tfrac{3}{4})]$$

**9.8.** For the equi-angular spiral $r = ae^{m\theta}$, show that the centre of curvature is at the point where the line through the pole $r = 0$ and orthogonal to the radius vector meets the normal.

**9.9.** For what value of $\lambda$ does the parabola $y^2 = 4\lambda x$ have the same circle of curvature as the ellipse

$$\left(\frac{x-a}{a}\right)^2 + \frac{y^2}{b^2} = 1$$

at the origin?

$$\left[\frac{b^2}{16a}\right]$$

**9.10.** Show that the circle of curvature at the origin for the curve

$$x + y = ax^2 + by^2 + cx^3$$

is

$$(a + b)(x^2 + y^2) = 2x + 2y.$$

(*Hint:* Use Theorem 8.13. Recall that

$$\mathbf{r}_* = \mathbf{r} + \frac{1}{\kappa}\mathbf{n},$$

where $\mathbf{n}$ is the unit normal vector appropriate to the choice of the sign of the curvature. Alternatively use the method of Exercise 9.7.)

**9.11.** Calculate the curvature of the ellipse $\dfrac{x^2}{a^2} + \dfrac{y^2}{b^2} = 1$ at the point $(a\cot t, b\sin t)$. On an ellipse with $a = 8$ cm and $b = 4$ cm, draw accurately the circle of curvature at the point $\left(\dfrac{\sqrt{3}a}{2}, \dfrac{b}{2}\right)$. On a separate copy of the ellipse draw the circles of curvature at $(a, 0)$ and at $(0, b)$. Describe the relation of the points $z(t)$ of a curve to a circle tangent to the curve at $z(t_0)$ for $t$ sufficiently close to $t_0$, where the circle is not the circle of curvature at $z(t_0)$. Illustrate this by drawing two further circles tangent to the ellipse at $\left(-\dfrac{\sqrt{3}a}{2}, \dfrac{b}{2}\right)$ and noting the relevant behaviour on your diagram.

**9.12.** Let $y = f(x)$ be a curve whose centres of curvature all lie in the $x$-axis. Prove that $yy'' + (1 + (y')^2) = 0$. Hence show, by integrating this equation, that the points on the curve must satisfy an equation of the form $(x - a)^2 + y^2 = c^2$.

**9.13.** Show that the centre of curvature at a non-singular point $(a, b)$ of an algebraic curve $f(x, y) = 0$ is

$$\mathbf{r}_* = (a, b) - \frac{|\mathbf{grad}\, f|^2}{f_y^2 f_{xx} - 2 f_x f_y f_{xy} + f_x^2 f_{yy}}\, \mathbf{grad}\, f \,,$$

where all partial derivatives are evaluated at $(a, b)$.

## 9.2 Contact between curves and circles

We considered contact between an algebraic curve and a parametrised curve in Chapter 5. We consider here the special case where the algebraic curve is a circle that intersects a given parametric curve. We describe the significance of 1–point contact, 2–point contact and 3–point contact in this case. In general there is a unique circle having $\geq$ 3–point contact with a curve at a regular point. This circle is called the *osculating circle*. We show also that the osculating circle coincides with the circle of curvature at the point. This gives us an alternative way of defining the circle of curvature.

Let

$$|\mathbf{r} - (a, b)|^2 = k^2 \quad (k > 0)$$

be a fixed circle and let

$$t \mapsto \mathbf{r}(t)$$

be a parametrised curve. Define the function $\gamma$ by

$$\gamma(t) = |\mathbf{r}(t) - (a, b)|^2 - k^2 = (\mathbf{r}(t) - (a, b)) \cdot (\mathbf{r}(t) - (a, b)) - k^2$$
$$= (x(t) - a)^2 + (y(t) - b)^2 - k^2 .$$

The two curves intersect where $\gamma(t) = 0$. We assume that $t \mapsto \mathbf{r}(t)$ is smooth (or, more generally, suitably differentiable). In the special case where $x(t)$ and $y(t)$ are polynomials, $\gamma(t)$ is a polynomial in $t$.

We now consider a number of special cases.

*9.2.1 At least 1–point contact: $\gamma(t_0) = 0$*

This means that the curve intersects the circle at $\mathbf{r}(t_0)$. In the case of exactly 1–point contact, the circle and the curve will not have the same tangent line at the point of intersection.

*9.2.2 At least 2–point contact: $\gamma(t_0) = \gamma'(t_0) = 0$*

We assume that $t_0$ is a regular value. The second condition is

$$\gamma'(t_0) = 2\mathbf{r}' \cdot (\mathbf{r} - (a, b)) = 0 . \tag{9.1}$$

Thus the curve intersects the circle at $\mathbf{r}(t_0)$ and the tangent to the curve there is orthogonal to that radius of the circle which passes through $\mathbf{r}(t_0)$, that is the curve and the circle have the same tangent line at $\mathbf{r}(t_0)$. There are, through $\mathbf{r}(t_0)$, many such circles that have the same tangent line at $\mathbf{r}(t_0)$ as the curve; for each given radius there are two, one on each side of the curve.

*9.2.3 At least 3-point contact:* $\gamma(t_0) = \gamma'(t_0) = \gamma''(t_0) = 0$

We assume again that $t_0$ is a regular value. The third condition is that

$$\gamma''(t_0) = \mathbf{r}'' \cdot (\mathbf{r} - (a, b)) + |\mathbf{r}'|^2 = 0. \qquad (9.2)$$

From Equation 9.1 we have that $\mathbf{r} - (a, b)$ is orthogonal to $\mathbf{r}'$, and therefore $\mathbf{r} - (a, b) = \alpha(-y', x')$ where $\alpha$ is a real number. From the equation $\gamma(t_0) = 0$ we then have $\alpha^2((y')^2 + (x')^2) = k^2$, and therefore $\alpha \neq 0$. On substituting into the Equation 9.2, we obtain $\alpha \mathbf{r}'' \cdot (-y', x') + |\mathbf{r}'|^2 = 0$, and therefore the curvature of the curve is

$$\kappa = \frac{1}{|\mathbf{r}'|^3}(-y', x') \cdot \mathbf{r}'' = -\frac{1}{\alpha|\mathbf{r}'|} \ .$$

Thus there is at least 3-point contact if, and only if, the centre of the circle of curvature is

$$\mathbf{r}_* = \mathbf{r}(t_0) + \frac{1}{\kappa|\mathbf{r}'|}(-y', x') = \mathbf{r} - \alpha(-y', x') = \mathbf{r} - (\mathbf{r} - (a, b)) = (a, b),$$

that is if, and only if, the circle is the circle of curvature. The circle of curvature can therefore be regarded as the limiting position of circles passing through three distinct points on the curve as the three points move into coincidence. For this reason the circle of curvature is also called the osculating circle, the unique circle having the highest order of contact with the curve. All other circles have at most 2-point contact with the curve at the point. Recall that there is a unique circle passing through 3 non-collinear points of the plane. We have proved the following theorem.

**Theorem 9.8.** *At a regular point of a parametrised curve, the curve has at least 3-point (second-order) contact with its circle of curvature. Furthermore the circle of curvature is the unique circle having at least 3-point contact with the curve at the point.*

The osculating circle crosses the curve at the point of contact in case the circle of curvature has 3-point contact or more generally $(2n + 1)$-point contact $(n \geq 1)$. The osculating circle stays on one side of the curve near the point of contact in case the circle of curvature has 4-point contact or more generally $2n$-point contact $(n \geq 2)$.

**Remark 9.9.** For completeness we give a general definition of osculating curves. Let $P$ be a point on a curve $\Gamma$, and let $\mathbf{F}$ be a family of curves. The curve $\Gamma$ and a curve $\Omega$ in the family $\mathbf{F}$ *osculate* or are *osculating curves* if they have the highest possible point-contact at $P$ for all possible choices of

$\Omega$ in the family $\mathbf{F}$. For example, if $\mathbf{F}$ is the family of all lines and $P$ is a regular/non-singular point, the osculating line at $P$ is the tangent line to $\Gamma$ at $P$; and if $\mathbf{F}$ is the family of all circles and $P$ is a regular/non-singular point, the osculating circle at $P$ is the circle of curvature of $\Gamma$ at $P$. In each of these two cases there is a unique osculating curve.

## Exercises

**9.14.** Show that the curve $t \mapsto \mathbf{r}(t)$ has $\geq 4$–point contact with its osculating circle at a regular point $\mathbf{r}(t)$ if, and only if,

$$\frac{x'y'' - y'x''}{(x')^2 + (y')^2} \, (3\mathbf{r}'' \cdot \mathbf{r}') = \mathbf{r}''' \cdot (-y', x').$$

**9.15.** Find the points on the following curves at which the osculating circle has $\geq 4$–point contact:

    a) the ellipse $\mathbf{r}(\varphi) = (a \cos \varphi, b \sin \varphi)$    $(a \neq b)$,      $\left[ 0, \frac{1}{2}\pi, \pi, \frac{3}{2}\pi \right]$

    b) the cycloid $\mathbf{r}(\varphi) = (a\varphi - b \sin \varphi, a - b \cos \varphi)$    $(a \neq b)$.

(*Hint:* Use Exercise 9.14.)

**9.16.** Show that, if two curves have $\geq 3$–point contact at a common point, then they have the same osculating circle.

(*Hint:* Use Exercise 5.6.)

# 10

# Limaçons

We give an analysis of the various types of limaçons and their properties; in particular we find their non-regular points, inflexions, vertices, and undulations. As an application of curvature, we use these results to classify limaçons into five classes. Limaçons are algebraic curves of degree four. They are intimately connected with the orthotomics and reflection caustics of a circle which we study in §14.8. History and applications of limaçons are given in Chapter 13.

## 10.1 The equation

The polar equation of the limaçon is

$$t \mapsto z(t) = (a + b\cos t)e^{it}$$

where $a, b > 0$. The algebraic equation of this curve is

$$(x^2 + y^2 - bx)^2 = a^2(x^2 + y^2),$$

and its other equations are given in §3.7. The parameter $t$ in the polar equation is the polar angle of $z(t)$ if $a + b\cos t > 0$ and is the polar angle of $-z(t)$ if $a + b\cos t < 0$. There are three initial cases; in each of them, the limaçon intersects the initial line $y = 0$ at $z = a + b$ and at $z = b - a$ (see Figure 10.1).

Case i)   $a > b > 0$. The origin is within the curve. In this case $t$ is always the polar angle since

$$a + b > a - b > 0 \text{ and } a + b \geq a + b\cos t \geq a - b > 0$$

for each value of $t$. Such a limaçon consists of one loop.

Case ii)   $a = b > 0$. This is the cardioid with a cusp at the origin given by $t = \pi$. The cardioid consists of one loop.

Case iii)   $b > a > 0$. In addition to intersecting the initial line at

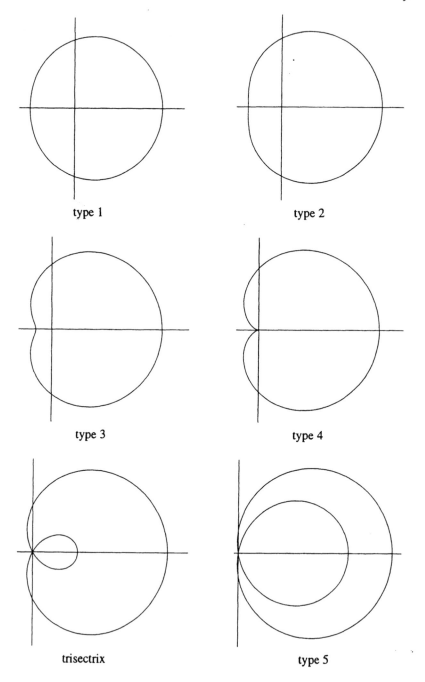

Figure 10.1 *The five classes of limaçons.*

$z = a + b$ and at $z = b - a$, the curve passes through the origin twice. The parameter values at which this happens are the solutions of the equation

$$\cos t = -\frac{a}{b}.$$

Under the given hypotheses we have

$$-1 < -\frac{a}{b} < 0,$$

so there are two values of the parameter $t$ satisfying the equation. The origin, in this case, is a nodal double point of the corresponding algebraic curve.

## 10.2 Curvature

We calculate the curvature of limaçons using complex number techniques; the use of real number techniques would make these calculations somewhat lengthier. We write $c = \cos t$ and $s = \sin t$. We have the following equations:

$$
\begin{aligned}
z(t) &= (a + b\cos t)e^{it}, \\
z' &= (-b\sin t)e^{it} + (a + b\cos t)ie^{it} = i\left(ae^{it} + be^{2it}\right), \\
z'' &= -(a + 2b\cos t)e^{it} + (-2b\sin t)ie^{it} = -\left(ae^{it} + 2be^{2it}\right), \\
z''' &= -i\left(ae^{it} + 4be^{2it}\right), \\
z'\,\overline{z'} &= (a + bc)^2 + b^2 s^2 = a^2 + b^2 + 2ab\cos t, \\
-\operatorname{im}(z'\,\overline{z''}) &= (a + 2bc)(a + bc) + 2b^2 s^2 = (a^2 + 2b^2) + 3ab\cos t, \\
\kappa &= \frac{-\operatorname{im}(z'\,\overline{z''})}{|z'|^3} = \frac{(a^2 + 2b^2) + 3ab\cos t}{(a^2 + b^2 + 2ab\cos t)^{\frac{3}{2}}}, \quad \text{and} \\
\kappa' &= \frac{3ab^2 \sin t\, (b + a\cos t)}{(a^2 + b^2 + 2ab\cos t)^{\frac{5}{2}}}, \quad \text{these last two for } |z'| \neq 0.
\end{aligned}
$$

The complex number techniques used here to calculate $z'\,\overline{z'} = |z'|^2$ and $x'y'' - y'x'' = -\operatorname{im}(z'\,\overline{z''})$ considerably shorten the calculations which would otherwise be needed. Replacing $\cos t + i\sin t$ by $e^{it}$ in the formula for $z'$ gives a further possible simplification in this specific case.

## 10.3 Non-regular points

For each fixed $t$, the (mutually orthogonal) unit vectors $e^{it}$ and $ie^{it}$ are linearly independent (over **R**); thus, we have above that

$$z' = (a + b\cos t)ie^{it} + (-b\sin t)e^{it} = 0$$

if, and only if, the coefficients (which are real) of $e^{it}$ and $ie^{it}$ are both zero. This is the case if, and only if, $a + b\cos t = 0$ and $\sin t = 0$, that is, $a = b$ and $t = \pi$. Thus the limaçons other than the cardioid are regular. The cardioid has precisely one non-regular point; this is at $t = \pi$ and, since $z''(\pi) \neq 0$, it is a cusp. The cusp is ordinary since $z''(\pi) = a - 2b$ and $z'''(\pi) = i(a - 4b)$ are linearly independent.

The reader should note well the above linear independence technique for complex numbers. Rather than writing $z'$ in terms of the standard (fixed) orthonormal vectors 1 and $i$, we write it in terms of the orthonormal vectors $e^{it}$ and $ie^{it}$. The set of axes given by the latter set of orthonormal vectors is 'moving'; that is it varies with $t$.

## 10.4 Inflexions

Now $\kappa = 0$ at a regular point if, and only if, $a^2 + 2b^2 = -3ab \cos t$. Since $|\cos t| \leq 1$, this can happen if, and only if, $3ab \geq a^2 + 2b^2$; that is, $0 \geq (a - 2b)(a - b)$, or, equivalently, $b \leq a \leq 2b$. Again $\kappa = \kappa' = 0$ at a regular point if, and only if, either

$$\cos t = -\frac{a^2 + 2b^2}{3ab} = -\frac{b}{a},$$

or

$$\cos t = -\frac{a^2 + 2b^2}{3ab} \quad \text{and} \quad \sin t = 0.$$

The first equations are satisfied if, and only if, $a^2 = b^2$; that is, $a = b$ and $t = \pi$. The second equations are satisfied if, and only if, $t = \pi$ and $a^2 + b^2 = 3ab$; that is, $t = \pi$, and either $a = b$ or $a = 2b$. Since $t = \pi$ gives a non-regular point in the case $a = b$, we have that $\kappa = \kappa' = 0$ if, and only if, $t = \pi$ and $a = 2b$. We deduce the following results.

Case i)   $a < b$ or $a > 2b$. There are no inflexions.

Case ii)   $a = b$. This is a cardioid. We have

$$- \mathrm{im}(z'\,\overline{z''}) \equiv 3a^2(1 + \cos t) = 0$$

if, and only if, $t = \pi$. However $t = \pi$ is a non-regular value, giving a simple cusp; therefore, $\kappa(\pi)$ is not defined. Thus there are no inflexions.

Case iii)   $b < a < 2b$. There are two points of simple inflexion where

$$\cos t = -\frac{a^2 + 2b^2}{3ab}.$$

Case iv)   $a = 2b$. There is one point satisfying $- \mathrm{im}(z'\,\overline{z''}) = 0$; this is where $t = \pi$. However this is a point of undulation (see §10.6).

## 10.5 Vertices

The number of vertices also depends on the relation between $a$ and $b$.

Case i)  $b < a$. There are four vertices: two are given by $t = 0, \pi$, and two are given by

$$\cos t = -\frac{b}{a}.$$

Case ii)  $b = a$. There is one vertex where $t = 0$. As $a$ approaches $b$ $(a > b)$, the three vertices at $t = \pi$ and

$$\cos t = -\frac{b}{a}$$

coalesce, and the resulting point is a cusp. The two inflexions also coalesce into this cusp.

Case iii)  $b > a$. There are two vertices where $t = 0, \pi$.

**Exercise**

**10.1.** Explain these results, in view of the four-vertex theorem.

## 10.6 Undulations

In the case where $a = 2b$ there is a point of simple undulation where $t = \pi$. Otherwise there are no undulations. Where $a = 2b$ and $t = \pi$, we have proved above that $\kappa = \kappa' = 0$. Also, by Lemma 8.4, in this case we have

$$\kappa''(\pi) = \frac{3ab^2 \cos \pi \, (b + a \cos \pi)}{(a^2 + b^2 + 2ab \cos \pi)^{\frac{5}{2}}} \neq 0.$$

## 10.7 The five classes of limaçons

The above results, on non-regularity, inflexions, vertices and undulations, divide the limaçons into five classes, including two limiting cases, depending on the relations between $a$ and $b$. We assume $a, b > 0$. These five classes are given as follows.

1)  $a > 2b$.  Since $\kappa > 0$ the centre of curvature lies to the left of the curve. The closed curve is therefore convex and simple. It has four vertices, no inflexions, and no undulations.

2)  $a = 2b$.  Again $\kappa > 0$, except that $\kappa(\pi) = 0$, so that the centre of curvature lies to the left of the curve. The closed curve is therefore convex and simple. It has four vertices, no inflexions, and one simple undulation.

3)  $2b > a > b$.  From §10.4 we have that

$$0 < \frac{a^2 + 2b^2}{3ab} < 1$$

in this case. Therefore $\kappa < 0$ at the points where

$$\cos t < -\frac{a^2 + 2b^2}{3ab}.$$

The centre of curvature lies to the right of the curve at such points. The closed curve is simple, and has four vertices, two simple inflexions and no undulations.

4) $a = b$. Again $\kappa > 0$, except that $\kappa$ is not defined at the cusp given by $t = \pi$. The centre of curvature lies to the left of the curve except at the cusp, where the centre of curvature is not defined. The closed curve is simple, and has one ordinary cusp, one vertex, no inflexions, and no undulations.

5) $b > a$. Again $\kappa > 0$ everywhere. The closed curve is not simple; it has a double point given by the two values of $t$ which satisfy

$$\cos t = -\frac{a}{b}.$$

Also there are two vertices, no inflexions, and no undulations.

Examples of limaçons in each of the five classes are shown in Figure 10.1. The limiting cases given by $a = 0$ or $b = 0$ are both circles. The parameter range needed for the limiting case $a = 0$ differs from that of all the other cases. In case $a = 0$ the parametrisation $t \mapsto b \cos t\, e^{it}$ $(0 \le t \le 2\pi)$ covers the circle twice, so we restrict the parameter to the range $0 \le t \le \pi$.

## 10.8 An alternative equation

In Chapter 13 we obtain the equation of the limaçon, with respect to a different coordinate origin as

$$\varphi \mapsto z(\varphi) = 2ae^{i\varphi} + (c - 2a)^{2i\varphi}.$$

The limaçon is obtained there as a roulette and the origin is at the centre of the fixed circle. Again, in this equation, the parameter $\varphi$ is not the polar angle of $z(\varphi)$. It is the polar angle of the centre of the rolling circle and also of the point of contact of the two circles. The algebraic form of this equation can, by a translation of origin, be transformed into the algebraic equation given in §10.1.

# 11

# Evolutes

---

The centre of curvature at a point of a curve was defined in Chapter 9. The evolute of the curve is the locus of the centre of curvature at a point on the curve as the point moves along the curve.

## 11.1 Definition and special points

We parametrise the evolute in the following definition using the same parameter as that used for the curve, thus the centre of curvature of the curve at the point with parameter value $t$ is the point on the evolute having the same parameter value.

**Definition 11.1.** The *evolute* of a regular smooth curve $t \mapsto \mathbf{r}(t)$ is the locus of the centre of curvature; that is, it is the curve $t \mapsto \mathbf{r}_*(t)$, respectively $t \mapsto z_*(t)$, where

$$z_* = z_*(t) = z(t) + \frac{1}{\kappa} \frac{iz'(t)}{|z'(t)|}, \quad \text{or, equivalently,}$$

$$\mathbf{r}_* = \mathbf{r}_*(t) = \mathbf{r}(t) + \frac{1}{\kappa|\mathbf{r}'(t)|}(-y'(t), x'(t)) .$$

The reader should note carefully the formulae

$$\frac{1}{\kappa|\mathbf{r}'(t)|} = \frac{(x')^2 + (y')^2}{x'\,y'' - y'\,x''}, \quad \text{and,}$$

$$x'\,y'' - y'\,x'' = -\operatorname{im}\left(z'\,\overline{z''}\right),$$

which are used later. Note the minus sign in the last formula.

There is a simple formula for the derivative of $t \mapsto \mathbf{r}_*(t)$ given by the following proposition.

**Proposition 11.2.** *For a smooth curve* $t \mapsto z(t)$, *we have*

$$z'_*(t) = -\frac{\kappa'}{\kappa^2} \, i \, \frac{z'}{|z'|}, \quad \text{or}$$

$$r'_*(t) = -\frac{\kappa'}{\kappa^2} \frac{1}{|r'|} (-y', x') .$$

*Proof.* We have, by Theorem 7.15,

$$z'_*(t) = z'(t) + \frac{d}{dt} \left( \frac{1}{\kappa(t)} \frac{iz'(t)}{|z'(t)|} \right)$$

$$= z'(t) + \frac{1}{\kappa(t)} \frac{d}{dt} \left( \frac{iz'(t)}{|z'(t)|} \right) + \frac{iz'(t)}{|z'(t)|} \frac{d}{dt} \left( \frac{1}{\kappa(t)} \right)$$

$$= z'(t) + (-z'(t)) + \frac{iz'(t)}{|z'(t)|} \frac{(-\kappa'(t))}{(\kappa(t))^2}$$

$$= -\frac{\kappa'}{\kappa^2} \, i \, \frac{z'}{|z'|}$$

as required.                                                    □

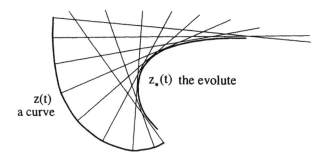

Figure 11.1 *The evolute of a curve.*

Using this proposition, we have the following theorem.

**Theorem 11.3.** *For a smooth curve* $t \mapsto r(t)$, *we have* $r'_*(t_0)$ *is orthogonal to* $r'(t_0)$ *for points which are defined and regular for both curves. Thus at values* $t_0$ *which are regular for both curves we have that the normal to* $t \mapsto r(t)$ *at* $r(t_0)$ *is the tangent, at the centre of curvature* $r_*(t_0)$, *to* $t \mapsto r_*(t)$, *and vice versa.*

This theorem tells us that the evolute $t \mapsto r_*(t)$ is the envelope of the normals to the curve $t \mapsto r(t)$; the normals to the curve are tangents to the evolute (see Figure 11.1). We consider envelopes in Chapter 14.

## Mnemonic

> The normals to a curve are tangents to the evolute of the curve.
> The evolute is the envelope of the normals to a curve.

Next we consider special points on $\mathbf{r}(t)$ and $\mathbf{r}_*(t)$, especially points where one of the curves is not regular.

### 11.1.1 Linear points of $t \mapsto \mathbf{r}(t)$

The evolute $\mathbf{r}_*(t)$ is clearly not defined at a linear point of $t \mapsto \mathbf{r}(t)$ (i.e., a point where $\kappa = 0$, or, equivalently, where $\mathbf{r}'$ and $\mathbf{r}''$ are parallel). Indeed as such a point is approached along the curve, $|\mathbf{r}_*(t)|$ becomes arbitrarily large.

### 11.1.2 Non-regular points of $t \mapsto \mathbf{r}(t)$

We define $\mathbf{r}_*(t_0) = \mathbf{r}(t_0)$, at a non-regular point $t_0$ of $t \mapsto \mathbf{r}(t)$ provided that

$$\lim_{t \to t_0} \frac{(x')^2 + (y')^2}{x'\,y'' - y'\,x''} = -\lim_{t \to t_0} \frac{(\mathbf{r}')^2}{\mathbf{r}' \cdot (-y'', x'')}$$

exists. With this definition, the curve and its evolute meet at $\mathbf{r}(t_0)$, and the evolute $t \mapsto \mathbf{r}_*(t)$ is continuous at $t_0$. At a regular point $t_0$ which is not a linear point, $\mathbf{r}_*(t_0)$ and $\mathbf{r}(t_0)$ are always distinct.

### 11.1.3 Ordinary cusps of $t \mapsto \mathbf{r}(t)$

The condition of §11.1.2 is satisfied at an ordinary cusp of $t \mapsto \mathbf{r}(t)$.

**Theorem 11.4.** *Let $t_0$ give an ordinary cusp of $t \mapsto \mathbf{r}(t)$, then $\mathbf{r}_*$ is continuous at $t_0$, where $\mathbf{r}_*(t_0) = \mathbf{r}(t_0)$.*

*Proof\**. Let $f = \mathbf{r}' \cdot \mathbf{r}'$ and $g = \mathbf{r}' \cdot (-y'', x'')$. Then $f$, $f' = 2\mathbf{r}' \cdot \mathbf{r}''$, $g$, and $g' = \mathbf{r}' \cdot (-y''', x''')$ are all zero at the cusp. Also $f''(t_0) = 2\mathbf{r}'' \cdot \mathbf{r}'' \neq 0$, from the definition of cusp; and $g''(t_0) = \mathbf{r}'' \cdot (-y''', x''') \neq 0$, from the definition of simple cusp (recall that $\mathbf{r}''$ and $\mathbf{r}'''$ are 'not parallel' if, and only if, $\mathbf{r}'' \cdot (-y''', x''') \neq 0$). Therefore by L'Hôpital's theorem (see §4.7.5)

$$\lim_{t \to t_0} \frac{|\mathbf{r}'|^2}{\mathbf{r}' \cdot (-y'', x'')} = \lim_{t \to t_0} \frac{f}{g}$$

$$= \lim_{t \to t_0} \frac{f'}{g'} = \lim_{t \to t_0} \frac{2\mathbf{r}' \cdot \mathbf{r}''}{\mathbf{r}' \cdot (-y''', x''')}$$

$$= \lim_{t \to t_0} \frac{f''}{g''} = \lim_{t \to t_0} \frac{2\mathbf{r}' \cdot \mathbf{r}''' + 2\mathbf{r}'' \cdot \mathbf{r}''}{\mathbf{r}'' \cdot (-y''', x''') + \mathbf{r}' \cdot (-y^4, x^4)}$$

$$= \frac{2\,\mathbf{r}'' \cdot \mathbf{r}''}{\mathbf{r}'' \cdot (-y''', x''')}\Bigg|_{t=t_0},$$

which is finite and non-zero. The result now follows from §11.1.2.     ☐

### 11.1.4 Vertices of $t \mapsto r(t)$

By Proposition 11.2, we have immediately the following theorem.

**Theorem 11.5.** *At a regular value $t_0$ of $t \mapsto r(t)$, at which $\kappa \neq 0$, we have that $r_*(t_0)$ is a non-regular point of $t \mapsto r_*(t)$ if, and only if, $r(t_0)$ is a vertex of $t \mapsto r(t)$.*

**Worked example 11.6.** Find the evolute of the curve given by
$$r(t) = (\tfrac{1}{2}t^2, \tfrac{1}{3}t^3).$$
Investigate this evolute at the non-regular points of $t \mapsto r(t)$.

*Solution.* We have
$$r' = (t, t^2) = t(1, t),$$
$$r'' = (1, 2t),$$
$$|r'| = |t|\sqrt{1 + t^2},$$
$$\frac{(x')^2 + (y')^2}{x'y'' - y'x''} = \frac{|t|^2(1 + t^2)}{t^2} = 1 + t^2 \text{ for } t \neq 0, \text{ and}$$
$$r_* = (\tfrac{1}{2}t^2, \tfrac{1}{3}t^3) + (1 + t^2)(-t^2, t)$$
$$= (-\tfrac{1}{2}t^2 - t^4, t + \tfrac{2}{3}t^3) \text{ for } t \neq 0.$$

The curve $r$ has an ordinary cusp at $t = 0$. Since
$$\lim_{t \to t_0} \frac{(x')^2 + (y')^2}{x'y'' - y'x''} = 1$$
exists, we define $r_*(0) = r(0)$. Thus we have
$$r_* = (-\tfrac{1}{2}t^2 - t^4, t + \tfrac{2}{3}t^3) \text{ for all } t.$$

In this case, since the parametrisation of the evolute is by polynomials, it is smooth; indeed, the parametrisation is analytic for all values of the parameter.     ☐

## 11.2 A matrix method for calculating evolutes

We recall that the inverse of a matrix $A = \begin{pmatrix} a & b \\ c & d \end{pmatrix}$ is
$$A^{-1} = \frac{1}{|A|} \begin{pmatrix} d & -b \\ -c & a \end{pmatrix} = \frac{1}{ad - bc} \begin{pmatrix} d & -b \\ -c & a \end{pmatrix}.$$

---

\* For the meaning of \* see the preface.

**Theorem 11.7.** *The evolute of* $t \mapsto r(t)$ *(at non-linear points) is* $t \mapsto r_*(t)$ *given by*

$$(\mathbf{r}_* - \mathbf{r})^t = \begin{pmatrix} \mathbf{r}' \\ \mathbf{r}'' \end{pmatrix}^{-1} \begin{pmatrix} 0 \\ \mathbf{r}' \cdot \mathbf{r}' \end{pmatrix},$$

*where* $\begin{pmatrix} \mathbf{r}' \\ \mathbf{r}'' \end{pmatrix} = \begin{pmatrix} x' & y' \\ x'' & y'' \end{pmatrix}$ *and* $()^t$ *denotes the transpose matrix.*

*Proof.* We have

$$\begin{pmatrix} \mathbf{r}' \\ \mathbf{r}'' \end{pmatrix}^{-1} \begin{pmatrix} 0 \\ \mathbf{r}' \cdot \mathbf{r}' \end{pmatrix} = \frac{1}{x'\,y'' - y'\,x''} \begin{pmatrix} y'' & -y' \\ -x'' & x' \end{pmatrix} \begin{pmatrix} 0 \\ \mathbf{r}' \cdot \mathbf{r}' \end{pmatrix}$$

$$= \frac{(\mathbf{r}')^2}{x'\,y'' - y'\,x''} \begin{pmatrix} -y' \\ x' \end{pmatrix}.$$

The result follows from the definition of evolute. $\qquad\qquad\square$

**Worked example 11.8.** Find the evolute of $\mathbf{r} = (t^2, t^3)$.

*Solution.* We have

$$\begin{pmatrix} \mathbf{r}' \\ \mathbf{r}'' \end{pmatrix}^{-1} \begin{pmatrix} 0 \\ \mathbf{r}' \cdot \mathbf{r}' \end{pmatrix} = \begin{pmatrix} 2t & 3t^2 \\ 2 & 6t \end{pmatrix}^{-1} \begin{pmatrix} 0 \\ 4t^2 + 9t^4 \end{pmatrix}$$

$$= \frac{1}{6t^2 - 2} \begin{pmatrix} 6t & -3t^2 \\ -2 & 2t \end{pmatrix} \begin{pmatrix} 0 \\ 4t^2 + 9t^4 \end{pmatrix}$$

$$= \frac{(4t^2 + 9t^4)}{6t^2 - 2} \begin{pmatrix} -3t^2 \\ 2t \end{pmatrix}$$

and therefore

$$\mathbf{r}_* = (t^2, t^3) + \frac{(4t^2 + 9t^4)}{6t^2 - 2}(-3t^2, 2t).$$

$$\square$$

## 11.3 Evolutes of the cycloid and the cardioid

As further examples we obtain the evolutes of the cycloid and the cardioid. Note in each case the relation between the cusps, vertices, and points where the curves meet. For the purpose of comparison, we calculate the evolute in each case using both the real and the complex methods.

**Worked example 11.9.** Find the evolute of the cycloid

$$t \mapsto \mathbf{r}(t) = (at - a\sin t, a - a\cos t).$$

*Solution.* We have

$$\mathbf{r}' = (a - a\cos t, \, a\sin t), \quad \mathbf{r}'' = (a\sin t, a\cos t),$$

$$\kappa = \frac{(a - a\cos t)(a\cos t) - (a\sin t)(a\sin t)}{a^3((1 - \cos t)^2 + \sin^2 t)^{\frac{3}{2}}} = -\frac{1}{2^{\frac{3}{2}}a\sqrt{1 - \cos t}}, \quad \text{and}$$

$$
\begin{aligned}
\mathbf{r}_* &= \mathbf{r} + \frac{1}{\kappa|\mathbf{r}'|}(-y', x') \\
&= a(t - \sin t, \, 1 - \cos t) - 2a(-\sin t, \, 1 - \cos t) \\
&= a(t + \sin t, \, \cos t - 1) \\
&= a((t - \pi) + \sin t + \pi, \, (1 + \cos t) - 2) \\
&= \mathbf{r}(t - \pi) + a(\pi, -2).
\end{aligned}
$$

Thus, for the cycloid, the evolute is the same curve moved to a different position. See Figure 11.2.

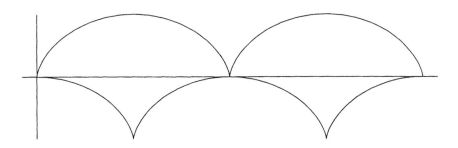

Figure 11.2 *The cycloid and its evolute.*

The corresponding complex equations are

$$z = a(t + i) - aie^{-it}, \quad z' = a - ae^{-it}, \quad \text{and} \quad z'' = aie^{-it}.$$

Thus

$$z'\,\overline{z''} = -a^2ie^{it} + a^2i, \quad \text{and} \quad z'\,\overline{z'} = a^2(2 - e^{it} - e^{-it}),$$

and we have

$$-\mathrm{im}(z'\,\overline{z''}) = -a^2(1 - \cos t) \quad \text{and} \quad |z'| = a\sqrt{2(1 - \cos t)}.$$

Thus the curvature is

$$\kappa = -\frac{1}{2^{\frac{3}{2}}a\sqrt{1 - \cos t}},$$

and, as in the real case, the evolute is

$$z_* = a(t+i) - aie^{-it} - 2ai(1 - e^{-it})$$
$$= a(t-i) + aie^{-it}$$
$$= a(t - \pi + i + (\pi - 2i) - ie^{-i(t-\pi)})$$
$$= z(t - \pi) + a(\pi - 2i).$$

□

Notice that, on using complex numbers, the calculation of $\kappa$ for this curve becomes shorter and simpler. It is important, when using complex numbers in these examples, not to replace the unit vector $e^{it}$ by $\cos t + i \sin t$ in order to determine real and imaginary parts until after simplifications have been made.

For curves not given parametrisations involving trigonometric functions, it is usually simpler to use the real form for the calculation of curvature.

For each of the cycloid, cardioid, astroid, epicycloid, and hypocycloid the evolute is a similar curve to itself, but may be a different size (e.g., magnified) and rotated to a different position. These curves are described in more detail as roulettes in Chapter 13.

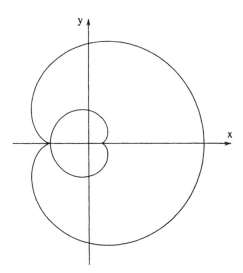

Figure 11.3 *The cardioid and its evolute.*

**Worked example 11.10.** Find the evolute of the cardioid

$$t \mapsto \mathbf{r}(t) = (2a \cos t + a \cos 2t, \ 2a \sin t + a \sin 2t).$$

*Solution.* We have

$$\mathbf{r}' = a(-2\sin t - 2\sin 2t,\ 2\cos t + 2\cos 2t),$$
$$\mathbf{r}'' = a(-2\cos t - 4\cos 2t,\ -2\sin t - 4\sin 2t),\quad \text{and}$$
$$\kappa = \frac{12a^2(1 + \cos t)}{2^3 2^{\frac{3}{2}} a^3 (1 + \cos t)^{\frac{3}{2}}} = \frac{3}{a2^{\frac{5}{2}}\sqrt{1 + \cos t}},$$

on using $\cos t = \cos(2t - t) = \cos 2t \cos t + \sin 2t \sin t$. Therefore

$$\mathbf{r}_* = \mathbf{r}(t) + \frac{2}{3}(-y'(t),\ x'(t))$$
$$= \left(\frac{2a}{3}\cos t - \frac{a}{3}\cos 2t,\ \frac{2a}{3}\sin t - \frac{a}{3}\sin 2t\right)$$
$$= \left(-\frac{2a}{3}\cos(t + \pi) - \frac{a}{3}\cos 2(t + \pi),\ -\frac{2a}{3}\sin(t + \pi) - \frac{a}{3}\sin 2(t + \pi)\right).$$

On comparing this with with the equation of the original cardioid, we see that the evolute is a cardioid which is $\frac{1}{3}$ the size of the original cardioid and rotated through an angle $\pi$ about the origin. See Figure 11.3.

The corresponding complex equations are

$$z = 2ae^{it} + ae^{2it},\quad z' = 2aie^{it} + 2aie^{2it},\quad \text{and}\quad z'' = -2ae^{it} - 4ae^{2it}.$$

Thus

$$z'\,\overline{z''} = -4ia^2(3 + e^{it} + 2e^{-it}),\quad \text{and}$$
$$z'\,\overline{z'} = (2ai)(-2ai)(e^{it} + e^{2it})(e^{-it} + e^{-2it})$$
$$= 4a^2(2 + e^{it} + e^{-it})$$
$$= 8a^2(1 + \cos t),$$

and we have $-\text{im}\,(z'\,\overline{z''}) = 12\,a^2(1 + \cos t)$ and $|z'| = a2^{\frac{3}{2}}\sqrt{1 + \cos t}$. The formula for the curvature is now immediate. Notice that the unit tangent vector to the cardioid is $ie^{\frac{3}{2}it}$.                                                    □

For the cardioid, $t = \pi$ is a non-regular value, and is a cusp. The evolute passes through this cusp (see 11.1.2 above). The second equation, $t \mapsto \mathbf{r}_*(t)$, is given as the locus of $P$ as indicated on the right where the parameter measures the angular position of the centre of the rolling circle of radius $\frac{a}{3}$ measured from the initial half-line. The first equation $t \mapsto \mathbf{r}(t)$ is the locus of $P'$, where each circle has radius $a$.

## Exercises

**11.1.** A curve is given by

$$x = 3\sin t - \sin^3 t,\quad y = 3\cos t - \cos^3 t.$$

Prove that the evolute is

$$x^{\frac{2}{3}} + y^{\frac{2}{3}} = 3^{\frac{2}{3}}.$$

**11.2.** A curve is given by

$$x = 3\sin t - 2\sin^3 t, \quad y = 3\cos t - 2\cos^3 t.$$

Prove that $\mathbf{r}' = 3(\cos^2 t - \sin^2 t)(\cos t, \sin t)$, and calculate the curvature. Prove that the evolute is

$$x^{\frac{2}{3}} + y^{\frac{2}{3}} = 2^{\frac{4}{3}}.$$

**11.3.** The curve $x^3 + xy^2 = y^2$ is parametrised by

$$x = \frac{t^2}{1+t^2}, \quad y = \frac{t^3}{1+t^2}.$$

Show that the equation of its evolute is $512x + 288y^2 + 27y^4 = 0$.

**11.4.** The parabola $y^2 = 4ax$, where $a > 0$, is parametrised by

$$\mathbf{r}(t) = (at^2, 2at).$$

Prove that the curvature $\kappa(t)$ at the point with parameter $t$ is given by

$$\kappa(t) = \frac{-1}{2a\left(1+t^2\right)^{\frac{3}{2}}}.$$

Hence prove that the evolute of the parabola is the curve

$$\mathbf{r}_*(t) = \left(2a + 3at^2, \ -2at^3\right).$$

Find the algebraic equation of this evolute. Calculate the arc-length of the evolute from $(2a, 0)$ to $(5a, -2a)$.

**11.5.** On a drawn parabola $y^2 = 4x$, where 1 unit = 1 cm, mark, in colour, the centre of curvature at the point at which the parabola meets its axis. Draw normals as follows to this parabola. Consider a circle with centre at the focus $(1, 0)$; do not draw the whole circle but only the intersections needed. The normal to the parabola at a point at which the circle meets the parabola is the line given by joining that point to the appropriate one of the two points in which the circle meets the $x$−axis. Choose the sequence of such circles of radii 1.2, 1.6, 2.0, 2.5, 3.0, 4.0, 5.0, and 6.0. Each of these will meet the parabola in two points (14 points in all). Outline the part of the evolute of the parabola as the curve touched by these normals.

**11.6.** Let $\mathbf{r}_*(t) = \left(2a + 3at^2, \ -2at^3\right)$ be the evolute of the parabola. Calculate the curvature $\kappa_*$ of $\mathbf{r}_*$, and the evolute $\mathbf{r}_{**}$ of $\mathbf{r}_*$. Verify that the condition for $\mathbf{r}_{**}$ to be defined at the non-regular point of $\mathbf{r}_*$ is satisfied.

In the case where $a = \frac{1}{4}$, plot and draw the curves $\mathbf{r}_*$ for $-2 \le t \le 2$, and $\mathbf{r}_{**}$ for $-1 \le t \le 1$ on the same diagram at a scale of 2 cm = 1 unit. (Use the portrait position with the origin at the centre of the graph paper.)

$$\left[\kappa_* = \frac{-1}{6at(1+t^2)^{3/2}}, \mathbf{r}_{**} = (-6at^4 - 3at^2 + 2a, -6at - 8at^3)\right]$$

**11.7.** The ellipse

$$\frac{x^2}{a^2} + \frac{y^2}{b^2} = 1,$$

where $a, b > 0$, is parametrised by $\mathbf{r}(t) = (ac, bs)$, where $c = \cos\varphi$ and $s = \sin\varphi$. Prove that the curvature $\kappa(\varphi)$ at the point with parameter $\varphi$ is given by

$$\kappa(\varphi) = \frac{ab}{(a^2s^2 + b^2c^2)^{\frac{3}{2}}}.$$

Calculate the radius of curvature at those points at which the ellipse meets its major and its minor axes. Hence prove that the evolute of the ellipse is the curve

$$\mathbf{r}_*(t) = \left(\frac{a^2 - b^2}{a}c^3, \ -\frac{a^2 - b^2}{b}s^3\right).$$

Show that the evolute is regular except at four points. Write down the Cartesian equation of the evolute. Prove that the length of the whole evolute is

$$4\left(\frac{a^2}{b} - \frac{b^2}{a}\right).$$

**11.8.** On a drawn ellipse $\frac{x^2}{6^2} + \frac{y^2}{4^2} = 1$, where 1 unit = 1 cm, mark, in colour, the centres of curvature at each of the four points at which the ellipse meets its major and minor axes. Draw normals as follows to this ellipse. Consider a circle with centre on the minor axis and which passes through the two foci. The normal to the ellipse at a point at which the circle meets the ellipse is the line given by joining that point to the appropriate one of the two points in which the circle meets the minor axis. Choose the sequence of such circles whose centres lie at distances 0, 1, 2, 3, 4, 6, and 12 cm from the origin in one direction (7 points in all). Each of these will meet the ellipse in two points. Outline the part of evolute of the ellipse as the curve touched by these normals.

**11.9.** Show that the cycloid $\mathbf{z}(t) = (t - \sin t, 1 - \cos t)$ has speed

$$2\left|\sin\frac{t}{2}\right|.$$

Find the non-regular points of $\mathbf{z}$. Find the evolute $\mathbf{z}_*$ of $\mathbf{z}$, and show that $\mathbf{z}_*(t) = \mathbf{z}(t - \pi) + (\pi, 2)$. Hence show that $\mathbf{z}_*$ is congruent to a reparametrisation of $\mathbf{z}$. Show that the arc length $s(t)$ of $\mathbf{z}$ from the point parametrised by 0 to the point parametrised by $t$ is

$$s(t) = 4\left(1 - \cos\frac{t}{2}\right) \text{ for } 0 \le t \le 2\pi.$$

Write down a unit speed reparametrisation of the restriction of $\mathbf{z}$ to the open interval $(0, 2\pi)$. Sketch $\mathbf{z}$ and $\mathbf{z}_*$ on the same diagram for

$$-2\pi \le t \le 4\pi.$$

$$\left[ 2n\pi, -\frac{1}{2^{\frac{3}{2}}\sqrt{1 - \cos t}}, t = 2\cos^{-1}\left(1 - \frac{s}{4}\right) \right]$$

**11.10.** Given that the evolute $\mathbf{e}$ of a parametric curve $\mathbf{r}$, without inflexions, is defined everywhere by

$$\mathbf{r}' \cdot \mathbf{e} = \frac{1}{2}\frac{d}{dt}(\mathbf{r} \cdot \mathbf{r}) \quad \text{and} \quad \mathbf{r}'' \cdot \mathbf{e} = \frac{1}{2}\frac{d^2}{dt^2}(\mathbf{r} \cdot \mathbf{r}),$$

deduce that $\mathbf{e} = \mathbf{r} + \rho\mathbf{n}$ where $\rho$ is a number and $\mathbf{n}$ is a unit vector, with

$$\mathbf{r}' + \rho\mathbf{n}' = 0 \quad \text{and} \quad \mathbf{e}' = \rho'\mathbf{n}$$

everywhere. The curve $\mathbf{r}$ is said to have an ordinary vertex at $t = a$ if

$$\mathbf{r}''' \cdot \mathbf{e} = \frac{1}{2}\frac{d^3}{dt^3}(\mathbf{r} \cdot \mathbf{r}) \quad \text{but} \quad \mathbf{r}^4 \cdot \mathbf{e} \neq \frac{1}{2}\frac{d^4}{dt^4}(\mathbf{r} \cdot \mathbf{r})$$

at $t = a$. Prove that $\mathbf{r}$ has an ordinary vertex at $t = a$ if, and only if, $\rho' = 0$ but $\rho'' \neq 0$.

**11.11.** Show that the curve $\mathbf{z}(t) = (\cos t, \sin t) + t(\sin t, -\cos t)$ is regular except where $t = 0$. Prove that its evolute $t \mapsto \mathbf{z}_*(t)$ is the unit circle. Let $s_*(t)$ $(t \ge 0)$ be the arc length along $\mathbf{z}_*$ from $\mathbf{z}_*(0)$ to $\mathbf{z}_*(t)$. Show by direct calculation that $s_*(t) = |\mathbf{z}_*(t) - \mathbf{z}(t)|$ for $t > 0$.

**11.12.** a) Show that the curvature of the evolute $t \mapsto \mathbf{z}_*(t)$ of the curve $t \mapsto \mathbf{z}(t)$ is

$$\kappa_* = -\frac{\kappa^3 |z'|}{|\kappa'|}.$$

b) Let $\psi$ be the angle measured counter-clockwise from the positive $x$-axis to the tangent of $t \mapsto \mathbf{z}(t)$, and $s$ the arc-length of $t \mapsto \mathbf{z}(t)$. Using the formula

$$\frac{d^2 s}{d\psi^2} = \frac{d}{dt}\left(\frac{ds}{d\psi}\right)\frac{dt}{d\psi},$$

prove that $\rho_*$, the radius of curvature of the evolute, satisfies

$$\rho_* = \left|\frac{d^2 s}{d\psi^2}\right|.$$

$$\left(\text{Hint: Recall that } \kappa = \frac{d\psi}{ds}.\right)$$

**11.13.** Find the Cartesian equation of the evolute of the hyperbola

$$x = a\cosh t, \quad y = b\sinh t.$$

**11.14.** Find the evolute of the cardioid $z = (1 + \cos\theta)e^{i\theta}$. Show that

i) $-\operatorname{im}\left(z' \overline{z''}\right) = 3(1 + \cos\theta),$

ii) $\kappa = \dfrac{3}{2\sqrt{2}\sqrt{1 + \cos\theta}}$ , and

iii) $z_* = \frac{1}{3}(1 + \cos\theta)e^{i\theta} - \frac{2}{3}\sin\theta ie^{i\theta}$.

**11.15.** Show that the evolute of an equi-angular spiral is an equi-angular spiral.

**11.16.** Show that the evolute of an astroid is an astroid.

**11.17.** Show that the evolute of an epicycloid is an epicycloid.

**11.18.** Show that the evolute of a hypocycloid is a hypocycloid.

# 12

# Parallels, involutes

We now consider parallels and involutes of a curve. A parallel is the locus of points lying a fixed distance along the normals to the curve (as in parallel lines), and an involute of a curve $t \mapsto \mathbf{w}(t)$ is a curve of which $t \mapsto \mathbf{w}(t)$ is the evolute. Involutes are not unique. Any parallel to an involute is an involute, and any two involutes are parallel to one another. An involute is the locus of a point on a piece of string as the string is wrapped along the curve or unwrapped from the curve.

## 12.1 Parallels of a curve

Let $t \mapsto \mathbf{z}(t) = (x(t), y(t))$ be a regular smooth curve in $\mathbf{R}^2$. Given such a curve $\mathbf{z}$ in the real plane $\mathbf{R}^2$, we denote the corresponding curve in the Argand diagram $\mathbf{C}$ by $t \mapsto z(t) = x(t) + iy(t)$, and *vice versa*. The reader should be familiar with these two interchangeable forms of the curve. We give proofs using either of the two forms. Along each normal vector (oriented from the tangent vector $\mathbf{z}'(t)$ by rotation through $\dfrac{\pi}{2}$ counterclockwise) we measure a fixed oriented-distance $c$, where $c$ may be positive, negative, or zero. The locus of the point obtained in this way, as $t$ varies, is called a *parallel* to the curve.

**Definition 12.1.** The *parallel* at an oriented distance $c$ to the left of a regular curve $t \mapsto \mathbf{z}(t) = (x(t), y(t))$ is the parametrised curve

$$t \mapsto w(t) = z(t) + ci\frac{z'(t)}{|z'(t)|}, \quad \text{or}$$

$$t \mapsto \mathbf{w}(t) = \mathbf{z}(t) + c\frac{1}{|\mathbf{z}'(t)|}(-y'(t), x'(t)) .$$

Recall that

$$\frac{1}{|\mathbf{z}'(t)|}(-y'(t), x'(t)),$$

respectively

$$i\,\frac{z'(t)}{|z'(t)|}\,,$$

is the unit normal vector to the left of the oriented curve.

**Example 12.2.** The parallels of a circle are concentric circles, and the parallels of a line are parallel lines (see Figure 12.1).

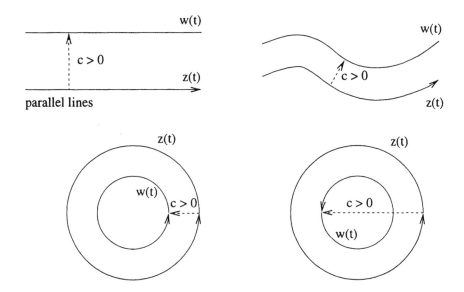

Figure 12.1 *Parallel curves.*

### 12.1.1 *Non-regular values of* $t \mapsto \mathbf{z}(t)$

The parallel may also be defined at a parameter value $t_0$, at which $t \mapsto z(t)$ is not regular, provided that the unit tangent $\dfrac{z'(t)}{|z'(t)|}$ tends to a limit $\alpha$, as $t$ tends to $t_0$. We define $w(t_0) = z(t_0) + ci\alpha$. This will give a parametrisation of the parallel which is continuous at $t_0$. Differentiability of the parallel at points corresponding to such non-regular values needs further examination. Where $t_0$ gives a cusp of $t \mapsto \mathbf{z}(t)$, separate parallels are defined in an analogous way for $t \geq t_0$ and for $t \leq t_0$. We show in §12.1.4 that for the ellipse there are two parallels whose non-regular points are not cusps. Therefore the ellipse itself is a parallel of a curve which has non-regular points.

### 12.1.2 Tangents and centres of curvature

We now consider the relation between tangents of the curve and a parallel. Also we show that in general a curve and a parallel have the same centres of curvature and therefore the same evolute.

**Theorem 12.3.** *For a smooth regular function $t \mapsto z(t)$, the parallel $t \mapsto w(t)$ is smooth and satisfies*

$$\mathbf{w}' = (1 - \kappa c)\, \mathbf{z}'.$$

*Proof.* For the unit vector $\dfrac{z'(t)}{|z'(t)|}$, we have

$$\frac{d}{dt} \frac{z'(t)}{|z'(t)|} = \kappa i z'(t),$$

by Proposition 11.2. Differentiating $w = z + ci\, \dfrac{z'}{|z'|}$ gives

$$w' = z' - c\kappa z'.$$

Since $z$ and $\kappa$ are smooth, it follows that $w$ is also smooth. The higher derivatives are given by formally differentiating $w' = z' - c\kappa z'$. $\square$

This result has the following immediate consequences.

**Corollary 12.4.** *Let $t \mapsto z(t)$ be regular at a parameter value $t$. Then $t$ is a non-regular value for the parallel at (oriented) distance $c$ if, and only if,*

$$1 - \kappa c = 0.$$

**Corollary 12.5.** *Let $t \mapsto z(t)$ and the parallel $t \mapsto w(t)$ at (oriented) distance $c$ be regular at a parameter value $t$. Then the tangents vectors to the two curves at the parameter value $t$ are parallel. They have the same direction if $1 - \kappa c > 0$, and opposite directions if $1 - \kappa c < 0$.*

The normal lines to the curve and a parallel are also the same at a value $t$ which is regular for each curve, though, in a similar way, the standard normal vectors, determined by the parametrisation, may have opposite directions.

**Theorem 12.6.** *The centres of curvature of a curve and a parallel are identical at a value of $t$, which, for both, is regular and not a point of inflexion. Thus parallel curves have the same evolute at values of $t$ which, for both, are regular and not a point of inflexion.*

*Proof.* Let the curvature of the parallel be $L = L(t)$. Then, differentiating

$$\frac{z'(t)}{|z'(t)|} = + \frac{w'(t)}{|w'(t)|} \quad \text{in case } 1 - \kappa c > 0, \text{ and}$$

$$\frac{z'(t)}{|z'(t)|} = - \frac{w'(t)}{|w'(t)|} \quad \text{in case } 1 - \kappa c < 0,$$

we have, since

$$\frac{d}{dt}\frac{z'(t)}{|z'(t)|} = \kappa i z'(t),$$

by Proposition 11.2,

$\kappa z' = Lw' = L(1 - \kappa c)z'$, and thus $\dfrac{1}{L} = \dfrac{1}{\kappa} - c$ if $1 - \kappa c > 0$, and

$\kappa z' = -Lw' = -L(1 - \kappa c)z'$, and thus $-\dfrac{1}{L} = \dfrac{1}{\kappa} - c$ if $1 - \kappa c < 0$.

Therefore the centre of curvature of the parallel $w = (u, v)$ in either case is

$$\ddot{w}_* = w + \frac{1}{L}\frac{1}{|w'|}(-v', u') =$$

$$z + c\frac{1}{|z'|}(-y', x') + \left(\frac{1}{\kappa} - c\right)\frac{1}{|z'|}(-y', x') = z_*.$$

Note that

$$\frac{1}{|w'|}(-v', u') = -\frac{1}{|z'|}(-y', x')$$

if $1 - \kappa c < 0$.       □

### 12.1.3 Non-regular points and inflexions

A parallel and the evolute meet at the non-regular points of the parallel. We give a condition that such points are cusps. Also we show that in general simple inflexions of a curve give rise to simple inflexions of the parallels.

**Theorem 12.7.** *A parallel and the evolute of a smooth curve* $t \mapsto z(t)$ *meet at a point with parameter value* $t_0$ *(i.e.,* $z_*(t_0) = w(t_0)$*), which is regular for* $t \mapsto z(t)$*, if, and only if,* $t_0$ *is non-regular for the parallel.*[†]

*Proof.* The parallel and evolute are

$$w(t) = z(t) + \frac{c}{|z'|}(-y', x'), \text{ and}$$

$$z_*(t) = z(t) + \frac{1}{\kappa|z'|}(-y', x').$$

The result follows from Corollary 12.4 on comparing these equations.     □

On differentiating

$$w'(t) = (1 - c\kappa(t))z'(t)$$

for general $t$ we have

$$w'' = (1 - c\kappa)z'' - c\kappa'z', \text{ and} \tag{12.1}$$
$$w''' = (1 - c\kappa)z''' - 2c\kappa'z'' - c\kappa''z'.$$

[†] Compare this result with §11.1.2.

**Theorem 12.8.** *Let $t_0$ be a non-regular value of the parallel $t \mapsto \mathbf{w}(t)$ and a regular value of $t \mapsto \mathbf{z}(t)$, and let the curvature of $t \mapsto \mathbf{z}(t)$ satisfy $\kappa'(t_0) \neq 0$. Then $\mathbf{w}(t_0)$ is a cusp of $t \mapsto \mathbf{w}(t)$. It is an ordinary cusp if, and only if, $\mathbf{z}(t_0)$ is not a point of inflexion of $t \mapsto \mathbf{z}(t)$.*

*Proof.* Since $\mathbf{w}'(t_0) = 0$, we have $1 - c\kappa = 0$. Thus

$$\mathbf{w}''(t_0) = -c\kappa'\mathbf{z}' \neq 0,$$

and therefore $\mathbf{w}(t_0)$ is a cusp of $t \mapsto \mathbf{w}(t)$. Now

$$\mathbf{w}'''(t_0) = -2c\kappa'\mathbf{z}'' - c\kappa''\mathbf{z}'.$$

Since $c\kappa' \neq 0$ at $t_0$, we have that $\mathbf{z}'$ and $\mathbf{z}''$ are linearly independent if, and only if, $\mathbf{w}''$ and $\mathbf{w}'''$ are linearly independent. Therefore the cusp is ordinary if, and only if, $\mathbf{z}(t_0)$ is not a point of inflexion by Criterion 5.18. □

**Theorem 12.9.** *Let $t_0$ give a regular point of simple inflexion of $t \mapsto \mathbf{z}(t)$ and a regular point of the parallel $t \mapsto \mathbf{w}(t)$, then $t_0$ gives a point of simple inflexion of the parallel.*

*Proof.* Let $\mathbf{z}'$ be parallel to $\mathbf{z}''$. Then $\mathbf{w}' = (1 - \kappa c)\mathbf{z}'$ is parallel to

$$\mathbf{w}'' = (1 - \kappa c)\mathbf{z}'' - \kappa'c\mathbf{z}'.$$

Given further that $\mathbf{z}'''$ is not parallel to $\mathbf{z}'$, then

$$\mathbf{w}''' = (1 - c\kappa)\mathbf{z}''' + \text{ a multiple of } \mathbf{z}'$$

is not parallel to $\mathbf{z}'$, since $1 - c\kappa \neq 0$ and $\mathbf{z}''' \neq \mathbf{0}$. □

### 12.1.4 Parallels of the ellipse

As an example we consider parallels to the ellipse. Possible parallels to an ellipse include the ones indicated in Figure 12.2. The ellipse is given by $\mathbf{z}(t) = (a\cos t, b\sin t)$ where $a > b > 0$. We have

$$\mathbf{z}'(t) = (-a\sin t, b\cos t), \quad \mathbf{z}''(t) = (-a\cos t, -b\sin t),$$

$$\kappa = \frac{ab}{\left(a^2 \sin^2 t + b^2 \cos^2 t\right)^{3/2}} \quad \text{and} \quad \kappa' = \frac{-\frac{3}{2}ab.2\sin t\cos t\left(a^2 - b^2\right)}{\left(a^2 \sin^2 t + b^2 \cos^2 t\right)^{5/2}}.$$

The parametrisation of the ellipse is regular and has four vertices given by $t = 0$, $\frac{\pi}{2}$, $\pi$, and $\frac{3\pi}{2}$. Since $a > b$ we have $\frac{a}{b^2} > \frac{b}{a^2}$. The maximum and minimum values of the curvature, which occur at the vertices, are these two numbers $\frac{a}{b^2}$ and $\frac{b}{a^2}$ respectively. For $0 < t < \frac{\pi}{2}$, we have $\kappa' < 0$ and therefore $\kappa$ decreases from $\frac{a}{b^2}$ to $\frac{b}{a^2}$ in the interval $\left[0, \frac{\pi}{2}\right]$. Let the parallel

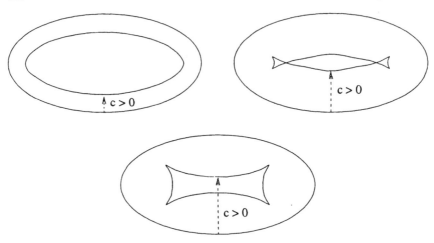

Figure 12.2 *Parallels to an ellipse.*

at oriented-distance $c$ (which may be negative) have curvature $L$. Then, as above, we have

$$\pm\frac{1}{L} = \frac{1}{\kappa} - c.$$

For the case where

$$\frac{b^2}{a} < c < \frac{a^2}{b},$$

the parallel has precisely one ordinary cusp for $0 < t < \dfrac{\pi}{2}$ by Theorem 12.8 and the whole curve has four ordinary cusps by symmetry. In the cases where $c > \dfrac{a^2}{b}$ or $c < \dfrac{b^2}{a}$ the parallel is regular; the latter case includes all negative values of $c$. We make a special note of the case where $c = \dfrac{b^2}{a}$. In this case the parallel is regular except at the values $t = 0$ and $t = \pi$. By Equations 12.1 we have $\mathbf{w}'(0) = \mathbf{w}''(0) = 0$, but $\mathbf{w}'''(0) = -c\kappa''\mathbf{z}' \neq 0$, and $\kappa''(0) < 0$ using Lemma 8.4. Now $t = 0$ gives a non-regular point of the parallel which is not a cusp, but, as in §6.4, there is a tangent line to the parallel curve at $t = 0$, and this is parallel to the tangent line to the ellipse at the corresponding point. A similar result holds at $t = \pi$. For the parallel given by $c = \dfrac{a^2}{b}$ similar results hold at $t = \dfrac{\pi}{2}$ and $t = \dfrac{3\pi}{2}$.

**Remark 12.10.** Compare the argument above with the situation at a cusp. At a cusp the limiting tangent line exists by L'Hôpital's theorem (see §4.7.5), but a limiting tangent vector does not exist since the limiting tangent vector directions on the two branches are opposite. In the case of the parallels considered here, the unit tangent vector

$$\frac{1}{|\mathbf{w}'(t)|}\,\mathbf{w}'(t)$$

tends to the limit

$$\frac{1}{|\mathbf{z}'(0)|}\,\mathbf{z}'(0)$$

as $t$ tends to zero.

### Exercises

**12.1.** Find the order of contact between the parallel to the ellipse

$$\mathbf{z}(t) = (a\cos t, b\sin t) \quad (a > b > 0)$$

and the tangent line to the ellipse at $t = 0$ in the case where

$$c = \frac{b^2}{a}.$$

[4]

**12.2.** Find the non-regular points of the parallel to the ellipse

$$\mathbf{z}(t) = (a\cos t, b\sin t) \quad (a > b > 0)$$

in the case where $c = \dfrac{a^2}{b}$, and show that they are not cusps. Find the tangent line to the parallel at $t = \dfrac{\pi}{2}$.

**12.3.** On a drawn ellipse $\dfrac{x^2}{6^2} + \dfrac{y^2}{4^2} = 1$, where 1 unit = 1 cm, mark, in colour, the centres of curvature at each of the four points at which the ellipse meets its major and minor axes. Draw normals as follows to this ellipse. Consider a circle with centre on the minor axis and which passes through the two foci. The normal to the ellipse at a point at which the circle meets the ellipse is the line given by joining that point to the appropriate one of the two points in which the circle meets the minor axis. Choose the sequence of such circles whose centres lie at distances 0, 1, 2, 3, 4, 6, and 12 cm from the origin in one direction (7 points in all). Each of these will meet the ellipse in two points. Outline the part of evolute of the ellipse as the curve touched by these normals. Draw the corresponding part of the parallel to the ellipse at a distance 4 cm measured along the inward normals.

(*Hint:* Use the results of Exercise 11.7.)

**12.4.** On a drawn parabola $y^2 = 4x$, where 1 unit = 1 cm, mark, in colour, the centre of curvature at the point at which the parabola meets its axis. Draw normals as follows to this parabola. Consider a circle with centre at the focus $(1,0)$; do not draw the whole circle but only the intersections

needed. The normal to the parabola at a point at which the circle meets
the parabola is the line given by joining that point to the appropriate one
of the two points in which the circle meets the $x$−axis. Choose the sequence
of such circles of radii 1.2, 1.6, 2.0, 2.5, 3.0, 4.0, 5.0, and 6.0. Each of these
will meet the parabola in two points (14 points in all). Outline the part of
the evolute of the parabola as the curve touched by these normals. Draw
the corresponding part of the parallel to the parabola at a distance 4 cm
measured along the 'inward' normals.
(*Hint:* Use the results of Exercise 11.4.)

**12.5.** The cycloid-arch $\Gamma$ is given by

$$z(t) = t + i - ie^{-it}, \ 0 < t < 2\pi.$$

Prove that the parametrisation is regular. Show that the unit tangent vector
at $z(t)$ is $ie^{-i\frac{t}{2}}$ and find the unit normal vector obtained by rotating this
through an angle $\dfrac{\pi}{2}$ in a positive direction. Find a parametrisation $t \mapsto z_\lambda(t)$
for the parallel curve at oriented distance $\lambda$ from $\Gamma$.

For each $\lambda$, show that $t$ is a non-regular value of $z_\lambda$ if and only if

$$\sin \frac{t}{2} = -\frac{\lambda}{4}.$$

Prove that $z'_\lambda(t) = z''_\lambda(t) = 0$ can only occur for $\lambda = -4$, and that $z_\lambda$ has
two cusps for each fixed $\lambda$ satisfying $0 > \lambda > -4$.

Using the envelope technique draw the parallel to the cycloid supplied,
(scale 1 unit = 4cm), in the case where $\lambda = -2.5$. Outline the envelope
clearly.

$$\left[ -e^{-i\frac{t}{2}}, t + i - ie^{-it} - \lambda e^{-i\frac{t}{2}} \right]$$

**12.6.** Find the evolute $z_*$ of the ellipse $z = (\cos t, 2\sin t)$. Sketch $z$ and $z_*$
and show in your sketch two parallels of $t \mapsto z(t)$ which have cusps lying
on the evolute.

**12.7.** Show that the unit tangent vector to the curve

$$r(t) = (t + \sin t, 1 - \cos t) \qquad -\pi < t < \pi$$

is

$$\left( \cos \frac{t}{2}, \ \sin \frac{t}{2} \right),$$

and write down the unit normal vector.

Write down a parametric equation for the parallel at distance $\lambda$ and
verify that its velocity is zero for values of $t$ such that $\lambda = 4\cos \dfrac{t}{2}$ (or
$\lambda = -4\cos \dfrac{t}{2}$ if the opposite normal has been chosen). Classify these non-
regular points for different parallels. (You may assume that at any cusp
the second of the derivatives to be non-zero is independent of the first to
be non-zero.)

Illustrate your account with careful sketches.

$$\left[\left(-\sin\frac{t}{2}, \cos\frac{t}{2}\right), (t+\sin t, 1-\cos t) + \lambda\left(-\sin\frac{t}{2}, \cos\frac{t}{2}\right)\right]$$

**12.8.** Let $z$ be the ellipse given by $z(t) = (a\cos t, b\sin t)$. Given that the curvature $\kappa$ satisfies

$$(a^2 - b^2)\cos^2 t = a^2 - \left(\frac{ab}{\kappa}\right)^{\frac{2}{3}},$$

find the points with $\kappa = \frac{1}{5}$ on the given ellipse, where $a = 6$ cm and $b = 4$ cm, and mark them on a diagram.

Construct the parallels to the ellipse with $d = \pm 5$ as an envelope. Mark on your drawing the apparent non-regular points of the envelope.

**12.9.** Let $w = z + ci\dfrac{z'}{|z'|}$ be a parallel of a regular curve $t \mapsto z(t)$. Show that $t \mapsto z(t)$ is a parallel of $t \mapsto w(t)$ where $t \mapsto w(t)$ is regular.

## 12.2 Involutes

The relation of evolute to involute is comparable to the relation of differentiation to indefinite integration. Integration is inverse to differentiation up to an arbitrary constant, that is adding a constant to an integral of a function $f$ gives another integral of $f$ and any two integrals of $f$ differ by a constant. Taking involute is inverse to taking evolute up to a constant shift (to a parallel) along the normals. In general a parallel to an involute of a curve is also an involute and any two involutes of a curve are parallels of one another.

**Definition 12.11.** An *involute* of a curve $t \mapsto \mathbf{w}(t)$ is a curve $t \mapsto \mathbf{z}(t)$ whose evolute is $t \mapsto \mathbf{w}(t)$.

Involutes are not unique; indeed, we have the following theorem.

**Theorem 12.12.** *Any parallel of an involute of a given curve, near a parameter value which is regular and not a point of inflexion for both the involute and the parallel, is also an involute of the curve. Conversely any two involutes of the curve, near a parameter value which is regular and not a point of inflexion for both, are parallels of one another.*

*Proof.* Parallel curves have the same evolute by Theorem 12.6. Conversely suppose that $t \mapsto z(t)$ and $t \mapsto w(t)$ are two curves having the same evolute. Thus

$$z_* = z + \frac{1}{\kappa}i\frac{z'}{|z'|} = w_* = w + \frac{1}{L}i\frac{w'}{|w'|}.$$

Differentiating this gives

$$-\frac{\kappa'}{\kappa^2}i\frac{z'}{|z'|} = -\frac{L'}{L^2}i\frac{w'}{|w'|},$$

by Proposition 11.2. Thus, since $\dfrac{z'}{|z'|}$ and $\dfrac{w'}{|w'|}$ are unit vectors,

$$\frac{\kappa'}{\kappa^2} = \frac{L'}{L^2} \quad \text{if } z' \text{ has the same direction as } w', \text{ and}$$

$$\frac{\kappa'}{\kappa^2} = -\frac{L'}{L^2} \quad \text{if } z' \text{ has the opposite direction to } w'.$$

Integrating gives $\dfrac{1}{\kappa} = \dfrac{1}{L} + c$ or, respectively, $\dfrac{1}{\kappa} = -\dfrac{1}{L} + c$. In either case we have

$$w = z + \frac{1}{\kappa} i \frac{z'}{|z'|} - \frac{1}{L} i \frac{w'}{|w'|} = z + ci \frac{z'}{|z'|},$$

and therefore $w$ is a parallel of $z$. $\qquad\square$

## Mnemonic

> A parallel of an involute is an involute.
> Any two involutes are parallel.

### 12.2.1  Geometrical interpretation of the involute

An involute $t \mapsto \mathbf{z}(t)$ of a curve $t \mapsto \mathbf{w}(t)$ is the locus of a point on a piece of string as the string is either unwrapped from $t \mapsto \mathbf{w}(t)$ or wrapped along $t \mapsto \mathbf{w}(t)$ (see Figure 12.3).

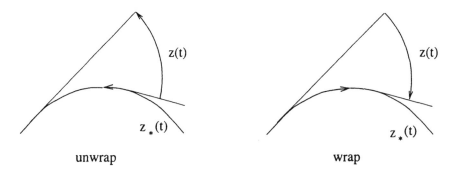

unwrap                                    wrap

Figure 12.3 *Involutes by wrapping or unwrapping.*

Let $t \mapsto \mathbf{z}(t)$ be a curve with curvature $\kappa(t)$ and with evolute $t \mapsto \mathbf{z}_*(t)$, and let $s_*(t)$ be the arc-length of the evolute from a given point $\mathbf{z}_*(t_0)$. (The arc length is positive for $t > t_0$ and negative for $t < t_0$.) By Theorem 11.5 the vertices of $t \mapsto \mathbf{z}(t)$ correspond to non-regular points of the evolute. Since

$$\rho' = \pm \frac{\kappa'}{\kappa},$$

vertices are given by $\rho' = 0$, except where $\kappa = 0$.

**Theorem 12.13.** *Let $t \mapsto z(t)$ be smooth, regular, and have no inflexion on the interval $[t_0, t]$ (or $[t, t_0]$ if $t < t_0$) and let either $\rho' > 0$ on the interval or $\rho' < 0$ on the interval, except that $\rho'$ can be 0 at the end points. Then we have either a) $s_* = \rho - \rho_0$ or b) $s_* = \rho_0 - \rho$, where $\rho(t) = \dfrac{1}{|\kappa(t)|}$ is the radius of curvature and $\rho_0 = \rho(t_0)$ is constant.*

*Proof.* We have

$$s_*(t) = \int_{t_0}^t |\mathbf{z}'_*(t)| \, dt,$$

and

$$\frac{d}{dt} s_*(t) = |\mathbf{z}'_*(t)| = \frac{|\kappa'|}{\kappa^2}$$

by Proposition 11.2. Now $\dfrac{|\kappa'|}{\kappa^2}$ is $-\dfrac{d}{dt}\left(\dfrac{1}{\kappa}\right)$ for $\kappa' \geq 0$, and is $\dfrac{d}{dt}\left(\dfrac{1}{\kappa}\right)$ for $\kappa' < 0$. Therefore

$$\frac{d}{dt}\left(s_* + \frac{1}{\kappa}\right) = 0 \quad \text{for} \quad \kappa' \geq 0, \quad \text{and}$$

$$\frac{d}{dt}\left(s_* - \frac{1}{\kappa}\right) = 0 \quad \text{for} \quad \kappa' < 0.$$

Hence either $s_* - \rho$ is constant or $s_* + \rho$ is constant. The result now follows. Note that $\rho_0 > 0$ and $\rho > 0$, but $s_*$ may be positive or negative. $\qquad\square$

### Mnemonic

> The arc-length of the portion of the evolute corresponding to an arc of the original curve on which $\rho$ constantly increases or constantly decreases is equal to the difference of the radii of curvature at the extremities.

### 12.2.2 Existence of involutes

We now show that involutes of regular curves exist in general, and give a formula for describing them.

**Theorem 12.14.** *Let $t \mapsto w(t)$ be a smooth curve which is regular on the interval $[t_0, T]$ (or, where appropriate, the interval $[T, t_0]$). Then*

$$z(t) = w(t) - \sigma(t) \frac{w'(t)}{|w'(t)|}, \quad \text{or}$$

$$\mathbf{z}(t) = \mathbf{w}(t) - \frac{\sigma(t)}{|\mathbf{w}'(t)|} \mathbf{w}'(t),$$

where $\sigma(t)$ *is the arc-length of* $t \mapsto \mathbf{w}(t)$ *measured from* $\mathbf{w}(t_0)$, *is the invo-lute of* $t \mapsto \mathbf{w}(t)$ *which begins at* $\mathbf{w}(t_0)$. *Different choices of* $t_0$ *give different involutes.*

*Proof.* We have

$$(\mathbf{z}(t) - \mathbf{w}(t)) \cdot (\mathbf{z}(t) - \mathbf{w}(t)) = (\sigma(t))^2.$$

Differentiating gives

$$(\mathbf{z}(t) - \mathbf{w}(t)) \cdot (\mathbf{z}'(t) - \mathbf{w}'(t)) = \sigma(t) \cdot \sigma'(t).$$

Therefore

$$(\mathbf{z}(t) - \mathbf{w}(t)) \cdot \mathbf{z}'(t) = (\mathbf{z} - \mathbf{w}) \cdot \mathbf{w}' + \sigma \cdot \sigma' \qquad (12.2)$$

$$= -\frac{\sigma}{|\mathbf{w}'|} \mathbf{w}' \cdot \mathbf{w}' + \sigma \cdot \sigma' = 0,$$

since $|\mathbf{w}'| = \sigma'$. Hence $\mathbf{z}(t) - \mathbf{w}(t)$ is parallel to the normal to $t \mapsto \mathbf{z}(t)$ at parameter value $t$. Thus

$$0 = (\mathbf{z} - \mathbf{w}) \cdot \mathbf{z}' = -\frac{\sigma}{|\mathbf{w}'|} \mathbf{w}' \cdot \mathbf{z}'$$

and therefore $\mathbf{w}' \cdot \mathbf{z}' = 0$. Hence the tangent to $t \mapsto \mathbf{z}(t)$ is perpendicular to the tangent to $t \mapsto \mathbf{w}(t)$. Collecting these results we have i) $\mathbf{z}(t) - \mathbf{w}(t)$ is normal to $t \mapsto \mathbf{z}(t)$, and ii) $\mathbf{z}(t) - \mathbf{w}(t)$ is tangent to $t \mapsto \mathbf{w}(t)$ (see Figure 12.4). Thus $\mathbf{z} - \mathbf{w} = \theta \cdot (-y', x')$ where $\theta$ is some scalar function of $t$. Differentiating Equation 12.2 above gives $(\mathbf{z} - \mathbf{w}) \cdot \mathbf{z}'' + \mathbf{z}' \cdot \mathbf{z}' = 0$, since $\mathbf{w}' \cdot \mathbf{z}' = 0$. Therefore

$$-|\mathbf{z}'|^2 = (\mathbf{z} - \mathbf{w}) \cdot \mathbf{z}'' = \theta \cdot \kappa \cdot |\mathbf{z}'|^3.$$

Thus $\theta = -\dfrac{1}{\kappa|\mathbf{z}'|}$ where $\kappa$ is the curvature of $t \mapsto \mathbf{z}(t)$, and hence

$$\mathbf{w} = \mathbf{z} + \frac{1}{\kappa|\mathbf{z}'|}(-y', x').$$

Therefore $t \mapsto \mathbf{w}(t)$ is the evolute of $t \mapsto \mathbf{z}(t)$.  ☐

Whether an involute is obtained by wrapping a piece of string along a curve or unwrapping a piece of string from a curve depends on whether the chord from the curve to its involute lies in front of, respectively behind, the point $\mathbf{z}(t)$ with respect to the forward direction of the tangent vector, where the direction of the tangent vector is determined by the given direction of parametrisation, or orientation of the curve. At any given point on a curve there are involutes in front of the point $\mathbf{z}(t)$ (along the tangent vector) and other involutes behind $\mathbf{z}(t)$ (subject to the hypotheses of Theorem 12.14 on the existence of involutes). At each point of a curve at which the curvature

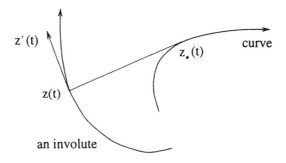

Figure 12.4 *Evolute and involute are orthogonal.*

is not zero, there is an involute meeting the curve at the point and having a cusp there.

**Worked example 12.15.** Find the involute of the circle $w = ae^{i\theta}$ which begins at $z = 1$.

*Solution.* The arc-length of the circle from $z = 1$ is

$$\sigma(\theta) = \int_0^\theta a \, dt = a\theta.$$

Thus

$$z(\theta) = w(\theta) - \sigma(\theta)\frac{w'(\theta)}{|w'(\theta)|} = ae^{i\theta} - a\theta i e^{i\theta} = a(1 - \theta i)e^{i\theta}.$$

□

**Exercises**

**12.10.** The spiral curve $\Gamma$ is given by $t \mapsto z(t) = e^{it} - ite^{it}$. Prove that $\Gamma$ is regular except for a simple cusp at $t = 0$. Find the curvature $\kappa(t)$ of $\Gamma$ and show that its evolute is the circle $t \mapsto e^{it}$.

Draw the curve $\Gamma$ as an involute using the equation $z(t) = e^{it} - ite^{it}$. Use polar graph paper in the portrait mode with 1 unit = 4 cm, and increments of $t$ in radians corresponding to 10 degrees in the range $-\frac{\pi}{6} \le t \le \frac{5\pi}{6}$. At each relevant point $e^{it}$ on the circle draw the tangent to the circle using the angle bisection method with compasses. Hence plot the points $z(t) = e^{it} - ite^{it}$ and draw $\Gamma$.

**12.11.** Show that the involute of the catenary $y = c\cosh\frac{x}{c}$, which begins at the point $(0, c)$, is the tractrix

$$x = c\cosh^{-1}\frac{c}{y} - \sqrt{c^2 - y^2}.$$

# 13

# Roulettes

---

In Chapter 12 we showed that where a line $\ell$ is rolled, without slipping, on a given curve $\Gamma$, a fixed point on the line traces out an involute to the curve $\Gamma$; moreover, all involutes to the curve $\Gamma$ are obtained in this way. We now consider the more general situation where a general curve $\Gamma_1$ rolls without slipping on a fixed curve $\Gamma_2$, and determine the curve traced out by a point which is rigidly attached (by rods for example) to $\Gamma_1$. Such a curve is called a *roulette*. Examples of roulettes are cycloids, limaçons, the nephroid, rose curves, the deltoid, and the astroid; we consider these and other specific roulettes. As the curve $\Gamma_1$ rolls on $\Gamma_2$, the plane in which $\Gamma_1$ is fixed performs a rigid motion. We show conversely, and perhaps surprisingly, that any rigid motion of the plane is given by rolling some curve in the plane on a second curve in a fixed plane. Because of this the general theory of roulettes has importance in geometry and kinematics.

## 13.1 General roulettes

In this chapter we generally regard the plane as the Argand diagram $\mathbf{C}$, rather than $\mathbf{R}^2$, since this enables us to determine the equations of the roulettes far more easily.

**Definition 13.1.** A *roulette* is the locus of a point (fixed) in a moving plane, where the motion is obtained by rolling, without slipping, a curve (fixed) in the moving plane on a curve in the fixed plane.

The moving plane $\Pi_1$ contains the curve $\Gamma_1$, and the fixed plane $\Pi_2$ contains the curve $\Gamma_2$. The plane $\Pi_1$ lies on $\Pi_2$ and the motion of $\Pi_1$ is given by rolling $\Gamma_1$ on $\Gamma_2$. At each instant of the motion the two curves are tangent to one another at the point of contact, and there is no slipping (see Figure 13.1).

We assume that the moving curve $\Gamma_1$ is parametrised by $t \mapsto z(t)$ and that the fixed curve $\Gamma_2$ is parametrised by $t \mapsto w(t)$, where, $z(t)$ and $w(t)$ are the position vectors referred to the fixed plane $\Pi_2$ before motion commences.

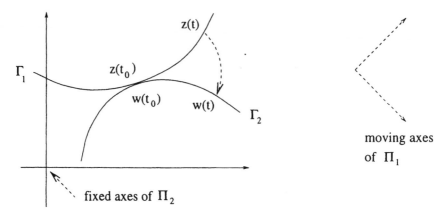

Figure 13.1 *Rolling curves.*

We assume that the parametrisations are such that, as rolling takes place, the point of $\Gamma_2$ which was initially $z(t)$ coincides with $w(t)$ as the curves come into contact at $w(t)$. Thus before rolling takes place, the two curves are in contact at the points with parameter value $t = t_0$; that is $z(t_0) = w(t_0)$, and the tangent lines to the two curves are the same at this point. Furthermore the arc-length from $z(t_0)$ to $z(t)$ along $\Gamma_1$ will be the same as the arc-length from $w(t_0)$ to $w(t)$ along $\Gamma_2$. Notice that when $z(t)$ coincides with $w(t)$, the point $z(t)$ is instantaneously at rest and therefore the normals at the corresponding points on each roulette pass through $w(t)$.

**Definition 13.2.** Parametrisations of two curves $\Gamma_1$ and $\Gamma_2$ are *comparable*, if, at each stage during rolling, the two points in contact have the same parameter values.

Our construction of roulettes depends on the two curves $\Gamma_2$ and $\Gamma_2$ being given comparable parametrisations. The next results are useful when constructing comparable parametrisations.

**Criterion 13.3.** *Regular smooth parametrisations* $t \mapsto w(t)$ *and* $t \mapsto z(t)$ *of curves* $\Gamma_1$ *and* $\Gamma_2$ *are comparable provided that*
    *i) their initial points coincide, that is*

$$w(t_0) = z(t_0),$$

*ii) at $t_0$ their tangent vectors are equal, that is*

$$w'(t_0) = z'(t_0), \quad and$$

*iii)*                          $|w'(t)| = |z'(t)| \neq 0$ *for each* $t.$

*Proof.* Conditions i and ii tell us that the two curves have the same initial position, and have the same tangent lines and same orientation (direction of the curve) there. Condition iii tells us that the tangent vectors of the two curves are well defined and non-zero at each value of the parameter, and that the two curves have the same arc-length since

$$\int_{t_0}^t |w'(u)|\, du = \int_{t_0}^t |z'(u)|\, du.$$

Thus the arc-length of $\Gamma_2$ from $w(t_0)$ to $w(t)$ is equal to the arc-length of $\Gamma_1$ from $z(t_0)$ to $z(t)$, and therefore $w(t)$ rolls into the position of $z(t)$. The result now follows. □

In case the fixed curve is a line, the rolling curve has a comparable parametrisation given by the following corollary.

**Corollary 13.4.** *Given a regular curve* $t \mapsto z(t)$, *the comparable parametrisation of the tangent line at* $z(t_0)$ *is*

$$t \mapsto z(t_0) + s(t)\frac{z'(t_0)}{|z'(t_0)|} = w(t).$$

*Proof.* We have

$$w'(t) = s'(t)\frac{z'(t_0)}{|z'(t_0)|}$$

and therefore

i) $w(t_0) = z(t_0)$,

ii) $w'(t_0) = s'(t_0)\dfrac{z'(t_0)}{|z'(t_0)|} = z'(t_0)$, and

iii) $|w'(t)| = s'(t) = |z'(t)|$ for each value of $t$. □

In case both curves are regular, suitable arc-length parametrisation are comparable as in the following theorem.

**Theorem 13.5.** *Two regular smooth curves which have* $\geq 2$–*point contact at an initial point can be given comparable parametrisations by arc-length.*

*Proof.* We can choose the arc length parametrisation of each curve which satisfies

i) $z(0) = w(0) = $ initial point, and

ii) $z'(0) = w'(0)$.

The second of these conditions is that the curves are given the same direction at the initial point. Arc-length parametrisations satisfy

$$|w'(s)| = |z'(s)| = 1 \neq 0$$

for each $s$. The result follows from Criterion 13.3. □

For comparable parametrisations we have $|w'(t)| = |z'(t)|$ for each $t$, but $|w'(t)| = |z'(t)|$ alone does not imply that the parametrisations are comparable. In practice, using arc-length parametrisations can often considerable shorten the determination of comparable parametrisations, particularly where the two curves are circles, or a circle and a line. However the arc-length of many curves is given by a complicated formula and it is then usually preferable to use other parametrisations. Other parametrisations may be more appropriate for example where the two curves are congruent.

We next give an extension of Criterion 13.3 which covers the case where the initial points of the fixed and rolling curves may be cusps. Its proof is similar to the proof of Criterion 13.3.

**Criterion 13.6.** *Let $t \mapsto w(t)$ and $t \mapsto z(t)$ $(t \geq t_0)$ be smooth parametrisations, which are regular except possibly at $t_0$, of curves $\Gamma_1$ and $\Gamma_2$. The parametrisations are comparable provided that*
   *i) their initial points coincide, that is,*

$$w(t_0) = z(t_0),$$

   *ii) at $t_0$ their limiting unit tangents exist and are equal, that is,*

$$\lim_{t \to t_0} \frac{w'(t)}{|w'(t)|} = \lim_{t \to t_0} \frac{z'(t)}{|z'(t)|}, \quad and$$

   *iii)*        $|w'(t)| = |z'(t)| \neq 0$ *for each $t > t_0$.*

We now derive the equation of the roulette. We roll $\Gamma_1$ until $z(t)$ on $\Gamma_1$ reaches the position of $w(t)$ on the fixed curve $\Gamma_2$; the curve $\Gamma_1$ in this new position will be tangent to the curve $\Gamma_2$ at the point $w(t)$. We say that this position of the rolling occurs '*at time $t$*'. We wish to determine the equation of the curve traced out by a fixed point $c$ of $\Pi_1$ as the rolling takes place.

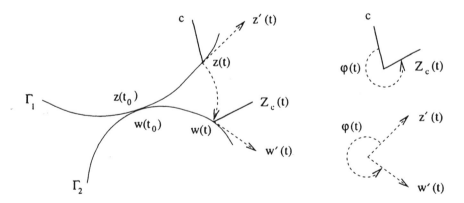

Figure 13.2 *The general roulette.*

**Theorem 13.7.** *The equation of the curve traced out by the point c in the moving* $\Pi_1$ *is*

$$\boxed{Z_c(t) = w(t) + \frac{w'(t)}{z'(t)}(c - z(t)).}$$ (13.1)

*The curve* $t \mapsto Z_c(t)$ *lies in the fixed plane* $\Pi_2$.

*Proof.* The movement of the plane $\Pi_1$ from time $t_0$ to time $t$ corresponds to a translation of $\Pi_1$ which carries the vector $z(t)$ to $w(t)$ and a rotation of $\Pi_1$ which carries $z'(t)$ into $w'(t)$ (recall that $z'(t)$ and $w'(t)$ are directed vectors of equal lengths). The point of $\Pi_1$, which describes the roulette, corresponds, to a point

$$c = z(t) + (c - z(t))$$

of $\Pi_2$ before rolling commences. The position vectors we consider here are all with respect to the fixed plane $\Pi_2$. When $\Gamma_1$ has been rolled so that the point $z(t)$ of $\Pi_1$ (referred to $\Pi_2$ before rolling commences) comes into tangency with the fixed curve at $w(t)$, the vector $c - z(t)$ has been rotated to $a(t)(c - z(t))$ and the point $c$ has moved to

$$Z_c(t) = w(t) + a(t)(c - z(t))$$ (13.2)

in the fixed plane, where $a(t)z'(t) = w'(t)$ and $|a'(t)| = 1$: $a(t)$ is instantaneously of the form $e^{i\varphi(t)}$. Notice that the motion at time $t$ corresponds to a rotation of the plane through an angle $\varphi(t)$ (counter-clockwise) together with a translation of the origin (see Figure 13.2). □

**Remark* 13.8.** The real version of Equation 13.2 is

$$\mathbf{Z_c}(t) = \mathbf{w}(t) + \begin{pmatrix} \cos\varphi(t) & -\sin\varphi(t) \\ \sin\varphi(t) & \cos\varphi(t) \end{pmatrix}(\mathbf{c} - \mathbf{z}(t))$$

where the vectors here are regarded as column vectors, and the multiplication in the last term is matrix multiplication. Note that in Equation 13.1 $\dfrac{w'(t)}{z'(t)}$ denotes division of complex numbers and the resulting complex number is then multiplied by $c - z(t)$. The real (in $\mathbf{R}^2$) version of the equation of a roulette is far more complicated, for example we replace

$$\frac{w'(t)}{z'(t)} = \frac{w'(t)\,\overline{z'(t)}}{|z'(t)|^2}$$

by

$$\frac{1}{|\mathbf{z'}|^2}\left((u'\,x' + v'\,y')\,,\,(-u'\,y' + v'\,x')\right) = (\cos\varphi(t), \sin\varphi(t)).$$

* For the meaning of * see the preface.

where $z = (x, y)$ and $w = (u, v)$. We then need to 'multiply' this by $(c_1 - x(t), c_2 - y(t))$ in a manner corresponding to complex multiplication. Complex multiplication does not, of course, correspond to the scalar product, but corresponds to matrix multiplication. The resulting formula is more complicated than is the formula of Theorem 13.7. The reader is recommended to use the complex form.

We have shown above that rolling one curve upon another determines a motion, parametrised by $t$ of the plane of the moving curve. The point $c$ in the moving plane undergoes the motion $t \mapsto Z_c(t)$. We show in §13.5 conversely that a motion, parametrised by $t$, of a moving plane on a fixed plane is given by the rolling of a curve in the moving plane on a curve in the fixed plane.

We next give examples of roulettes obtained by rolling a circle on a line, and by rolling a circle on, or in, a second circle. The reader carefully should learn the techniques used to find comparable parametristions.

## 13.2 Parametrisation of circles

The standard unit-speed parametrisations (see Definition 9.1) of a circle having centre $z_0$ and radius $\rho$ are

$$\text{positively described: } z(t) = z_0 + \rho e^{\frac{it}{\rho}} \quad (0 \le t \le 2\pi\rho), \text{ and}$$

$$\text{negatively described: } z(t) = z_0 + \rho e^{-\frac{it}{\rho}} \quad (0 \le t \le 2\pi\rho).$$

The initial and final points of these parametrisations are

$$z(0) = z(1) = z_0 + \rho.$$

In order to determine comparable parametrisations in the examples that follow, we need to consider unit-speed parametrisations of the circle for which the initial and final point is at a general point on the circumference. We obtain such parametrisations from the standard parametrisations by rotating the circle about its centre. We multiply $\rho e^{\frac{it}{\rho}}$, respectively $\rho e^{-\frac{it}{\rho}}$, by $e^{i\alpha}$, which gives the parametrisations

$$\boxed{\begin{aligned} &\text{positively described: } z(t) = z_0 + \rho e^{\frac{it}{\rho}} e^{i\alpha} \quad (0 \le t \le 2\pi\rho), \text{ and} \\ &\text{negatively described: } z(t) = z_0 + \rho e^{-\frac{it}{\rho}} e^{i\alpha} \quad (0 \le t \le 2\pi\rho). \end{aligned}}$$

Each of these parametrisations has initial and final point at

$$\boxed{z_0 + \rho e^{i\alpha}.}$$

Special cases are indicated in Figure 13.3.

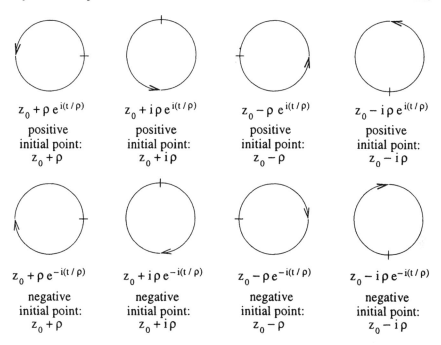

Figure 13.3 *Unit-speed parametrisations of the circle.*

The reader should be familiar with the following formulae involving the exponential function:

$$e^0 = 1, \ e^{\frac{i\pi}{2}} = i, \ e^{i\pi} = -1, \ e^{\frac{3i\pi}{2}} = e^{-\frac{i\pi}{2}} = -i.$$

### 13.3 Cycloids: rolling a circle on a line

We first consider the example where a circle is rolled on a line. The opposite of this is where a line is rolled on a circle; points on the line trace out the involute of the circle. We choose the line to be the real axis in the Argand diagram, and the rolling circle to have radius $a$ and centre at $w = ia$. In the initial position the two curves are tangent at the origin. The standard arc-length parametrisation of the real axis (or $u$–axis) is

$$t \mapsto w(t) = t$$

in the Argand diagram (or $t \mapsto \mathbf{w}(t) = (t, 0)$ in $\mathbf{R}^2$). The standard parametrisation (by arc-length) of the rolling circle, in its initial position, is $t \mapsto ia + ae^{\frac{it}{a}}$, but this satisfies $0 \mapsto ia + a$. We want $t = 0$ to correspond to

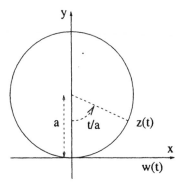

Figure 13.4 *Rolling a circle on a line.*

the origin (see Figure 13.4), so we must rotate the circle about its centre through $-\dfrac{\pi}{2}$. We do this by multiplying $ae^{\frac{it}{a}}$ by $-i$ and obtain the arc-length parametrisation

$$t \mapsto z(t) = ia + a(-i)e^{\frac{it}{a}} = ia(1 - e^{\frac{it}{a}})$$

where now $0 \mapsto 0$. (Recall that for an arc-length parametrisation $t \mapsto w(t)$, we have

$$|w'(t)| = s'(t) = 1$$

where $s$ denotes the arc-length of $t \mapsto w(t)$.) Applying this rotation is an essential technique for finding the comparable parametrisation in the case where the second curve is a circle. It remains to check that Condition ii of Criterion 13.3 is satisfied, which is easily done. Thus the equation of the roulette is

$$Z_c(t) = t + \frac{1}{e^{\frac{it}{a}}}\left(c - ai\left(1 - e^{\frac{it}{a}}\right)\right)$$

$$= t + ai + e^{-\frac{it}{a}}(c - ai).$$

Usually we choose $c = qi$, so that at time $t$ the point $c$ is on the vertical line through the centre of the circle.

The form of these roulettes is given in Figure 13.5. In the case where the point $c$ is on the rolling circle, the roulette has cusps. In the case where the point $c$ is outside the rolling circle, the roulette has nodal double points. In the case where the point $c$ is inside the rolling circle, but not at its centre, the roulette has inflexions.

The real form of the equation of this roulette is

$$Z_c(t) = \left(t - (a - q)\sin\frac{t}{a}, \quad a - (a - q)\cos\frac{t}{a}\right)$$

where $c = iq$ and $q$ is real. The point $c$ is on the $v$-axis.

We now give a standard parametrisation of the roulettes given by rolling a circle on a line, which is not, in general, by arc-length. This is obtained by replacing $t$ by $a\varphi$ in the above parametrisation.

**Definition 13.9.** The *standard parametrisation* of the roulettes given by rolling a circle on a line is

$$\varphi \mapsto a\varphi + ai + e^{-i\varphi}(c - ai).$$

The range of the parameter $\varphi$ is all the real numbers.

The parameter $\varphi$ used here is the angle of rotation, measured in the clockwise (not counter-clockwise) direction, of a radius of the circle. It is not the polar angle of the point of the roulette. Angles here are measured in radians; thus a full rotation increases $\varphi$ by $2\pi$ and moves the point of contact a (linear) distance $2\pi a$. The reader should note that $c$ is in general a complex number.

We now give a classification of the roulettes given by rolling a circle on a line. It is convenient to take $c$ on the $v$-axis, thus $c = iq$ where $q$ is real. This does not result in a loss of generality, since the point $c$ in the moving plane would be in such a position at some stage during the motion. The roulettes given by rolling a circle on a line are classified as follows (see Figure 13.5).

**Cycloid.** The cycloid is given by $c = 0$ (or $c = 2a$). It has cusps at $\varphi = 2n\pi$, for all integers $n$.

**Curtate cycloid.** A curtate cycloid is given by $c = iq$ where $0 < q < a$ (or $a < q < 2a$); curtate cycloids have simple inflexions at the points given by

$$\cos\varphi = \frac{a - q}{a}.$$

**Prolate cycloid.** A prolate cycloid is given by $c = iq$ where $q < 0$ (or $q > 2a$); prolate cycloids have nodal double points.

Finally the case $c = ia$ gives a straight line.

These curves are transcendental curves, not algebraic curves. From

$$(x - a\varphi)^2 + (y - a)^2 = (a - q)^2$$

we have

$$a\varphi = x \pm \sqrt{(q - y)(q + y - a)}.$$

Eliminating $\varphi$ from this equation and $y = a - (a - q)\cos\varphi$, we obtain the Cartesian equation.

### 13.3.1 History and applications of cycloids

These are the curves described, for example, by points on the wheel of a moving railway train. The cycloid, known to the ancient Greeks was studied

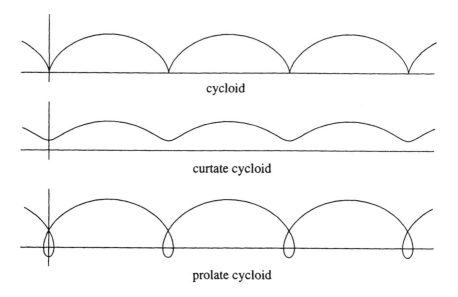

Figure 13.5  *Cycloids.*

(c. 1599) by Galileo. In 1658 Pascal, under a *nom de plume,* offered prizes of 20 and 40 Spanish doubloons for solutions to three outstanding problems about the curve; the problems, however, remained unresolved. The inverted cycloid is a tautochrone; the time taken for a particle to fall down the curve to the lowest point is independent of the initial point (Huygens 1673). It is also a  brachistochrone (Johann Bernoulli); a particle constrained to fall along a curve from an upper point to a lower point reaches the lower point earliest if it falls from the cusp to the vertex of an inverted cycloid. Huygens (1656) made the first pendulum clock; its pendulum described an inverted cycloidal arc. The inverted cycloid is an isochrone.

## 13.4  Trochoids: rolling a circle on or in a circle

**Definition 13.10.** A *trochoid* is a roulette given by rolling a circle on or in a second circle.

There are two cases. The first case is where neither circle is within the other; such curves are called epitrochoids. The second case is where one circle is within the other; such curves are called hypotrochoids. In this latter case either the larger circle is fixed or the smaller circle is fixed; the radii of the two circles cannot be equal.

## 13.4.1 Epitrochoids

These trochoids are given where a circle rolls '*outside*' a fixed circle, in the sense that neither circle is contained in the other.

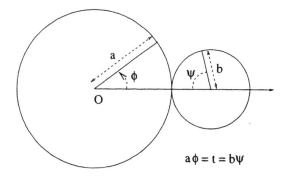

$$a\phi = t = b\psi$$

Figure 13.6 *Rolling a circle outside a circle.*

We parametrise the fixed circle by $w(t) = ae^{\frac{it}{a}}$. This is the standard positively-oriented unit-speed parametrisation for a circle with centre at the origin and radius $a$. The rolling circle has radius $b$ and we can choose it to have centre at $z = a+b$. In this case of one circle rolling outside another, the comparable parametrisation of the rolling circle must be negatively-oriented (see Figure 13.6). The standard such unit-speed circle is

$$t \mapsto a + b + be^{-\frac{it}{b}};$$

but for this circle we have $0 \mapsto a + 2b$ rather than $a$. The parametrisation we require will therefore be of the form

$$t \mapsto a + b + b\omega e^{-\frac{it}{b}},$$

where $\omega$ is a complex number of unit modulus. The number $\omega$ will determine the initial point (given by $t = 0$) of the parametrisation of the rolling circle, and multiplication by $\omega$ corresponds to rotating the circle about its centre. The number $\omega$ can be seen to be $-1$ from geometrical considerations, and the second circle must be rotated through $\pi$ in order that the points given by $t = 0$ are in the same position (see Figure 13.6). Alternatively $\omega$ can be found by solving $z(0) = w(0)$ algebraically. The comparable parametrisation of the rolling circle is therefore

$$z(t) = a + b - be^{-\frac{it}{b}}.$$

Recall that we must have $w(0) = z(0)$ and the same value of the parameter $t$ must give the pair of points on the two curves which come into contact at time $t$. So we must check that $|w'(t)| = |z'(t)|$ for each $t$ and that $w'(t)$ and $z'(t)$ are oriented in the correct directions along the tangent lines; that is, that the parametrisations are comparable. The parametrisations given here are by arc-length. The reader should check that Criterion 13.3 is satisfied. Therefore the equation of the roulette is

$$Z_c(t) = (a+b)e^{\frac{it}{a}} + (c-(a+b))e^{\frac{it(a+b)}{ab}}.$$

It is convenient to take $c = p$ where $p$ is real. In the case where $\dfrac{a}{b}$ is not rational, the image of the curve appears to fill in much of the annulus

$$\{a \le |z| \le a+2b\}.$$

We assume that $\dfrac{a}{b}$ is a rational number; indeed, in the examples which we consider, $\dfrac{a}{b}$ is an integer.

We now give a standard parametrisation of the roulettes given by rolling a circle on a circle, which is not, in general, by arc-length. This is obtained by replacing $t$ by $a\varphi$ and $n$ by $\dfrac{a+b}{b}$ in the above parametrisation.

**Definition 13.11.** The *standard parametrisation* of the roulettes given by rolling a circle on a circle is

$$Z(\varphi) = nbe^{i\varphi} + (c-nb)e^{in\varphi}.$$

The range of the parameter $\varphi$ is the interval $[0, 2\pi]$.

The parameter $\varphi$ used here is the polar angle, measured in the counter-clockwise direction, of the rotating point of contact, or, equivalently, of the centre of the rotating circle. It is not the polar angle of the point of the roulette, nor is this new parametrisation in general by arc-length.

We next consider some special epitrochoids (see Figure 13.7 and Figure 13.8).

**Epicycloids.** These are the epitrochoids obtained in the case where the point $c$ is on the circumference of the rolling circle. This occurs if $c = a+2b$ or $c = a$, for example. In the former case the curve is

$$Z(\varphi) = b\left(ne^{i\varphi} + e^{ni\varphi}\right)$$

where $nb = a+b$ and $a\varphi = t$. The epicycloid, where $n$ is an integer, has $n-1 = \dfrac{a}{b}$ ordinary cusps.

**Limaçons.** These are epitrochoids for which the fixed and rolling circles have the same radius $a$. Their equation is

$$Z(\varphi) = 2ae^{i\varphi} + (c-2a)e^{2i\varphi}.$$

A polar equation using a different origin of coordinates is given in Chapter 10. The shapes of the various types of limaçons are indicated in Figure 10.1. Limaçons are quartic curves. Their algebraic equation corresponding to the above parametric equation can be obtained by eliminating $\cos\varphi$

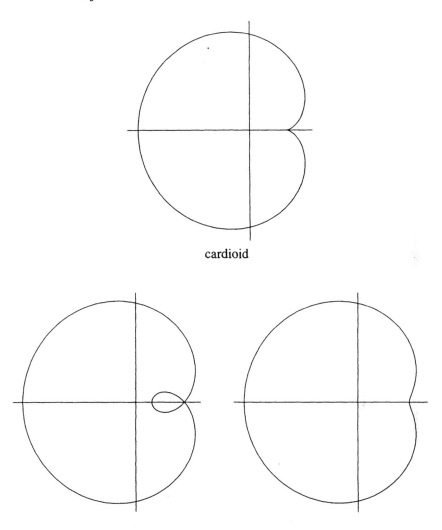

cardioid

Figure 13.7 *Epitrochoids: limaçons.*

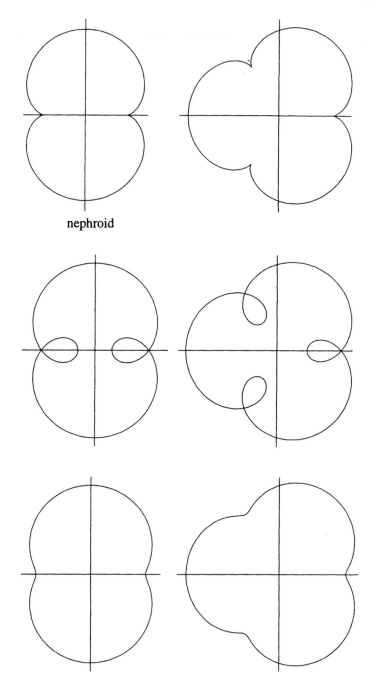

nephroid

Figure 13.8 *Epitrochoids II.*

from the equations

$$x = 2a \cos \varphi + (c - 2a)(2 \cos^2 \varphi - 1), \text{ and}$$
$$x^2 + y^2 = 8a^2 - 4ac + c^2 + 4a(c - 2a) \cos \varphi.$$

A simpler algebraic equation with the different coordinate origin is given in Chapter 10.

**Cardioids.** These are limaçons where the point $c$ is on the circumference of the rolling circle. Equivalently they are the epicycloids for which the fixed and rolling circles have the same radii. Thus we have $c = 3a$ or $c = a$; in the former case the equation is

$$Z(\varphi) = a \left(2e^{i\varphi} + e^{2i\varphi}\right).$$

**Trisectrix.** This is a limaçon for which $c = 4a$. It passes through the centre of the fixed circle. It can be used for trisecting angles.

**Nephroid.** This is an epicycloid for which $a = 2b$, that is $n = 3$. It has two cusps. It has the parametric equation

$$Z(\varphi) = b \left(3e^{i\varphi} + e^{3i\varphi}\right).$$

It is an algebraic curve of degree 6 with equation

$$(x^2 + y^2 - 4b^2)^3 = 108\, b^4 x^2.$$

### 13.4.2 Hypotrochoids

These trochoids are given where a circle of radius $b$ rolls in a fixed circle of radius $a$ with one circle inside the other; we must have $b \neq a$. There are two possibilities, first that the rolling circle is contained within the fixed circle (that is $b < a$), and second that the fixed circle is contained within the rolling circle (that is $b > a$). We parametrise the fixed circle by

$$w(t) = ae^{\frac{it}{a}}$$

and the rolling circle by

$$z(t) = (a - b) + be^{\frac{it}{b}}.$$

These parametrisations are by arc-length. The argument giving the comparable parametrisation of the rolling circle is similar to that given for epitrochoids above. In this case however both circles have positively-oriented parametrisations; this follows since, in the initial position, their tangent vectors have the same direction (see Figure 13.9). Notice that the same formula for $z(t)$ applies both in the case where the rolling circle is contained in the fixed circle and in the case where the fixed circle is contained

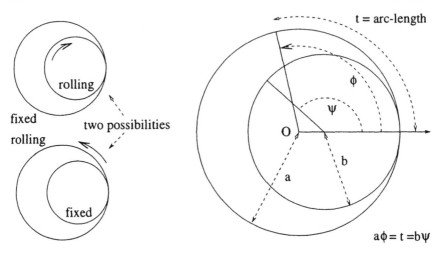

Figure 13.9 *Rolling a circle 'inside' a circle.*

in the rolling circle. The reader should check that Criterion 13.3 is satisfied. The the equation of the roulette is

$$Z_c = (a - b)e^{\frac{it}{a}} + (c - (a - b))e^{\frac{it(b-a)}{ab}}.$$

Again it is convenient to take $c = p$ where $p$ is real. We assume $\frac{a}{b}$ is a rational number; indeed, in the examples we consider, $\frac{a}{b}$ is an integer.

We now give a standard parametrisation of the roulettes given by rolling a circle in a circle, which is not, in general, by arc-length. This is obtained by replacing $t$ by $a\varphi$ and writing $m = \dfrac{a - b}{b}$ in the above parametrisation.

**Definition 13.12.** The *standard parametrisation* of the roulettes given by rolling a circle in a circle is

$$Z(\varphi) = mbe^{i\varphi} + (c - mb)e^{-im\varphi}.$$

The range of the parameter $\varphi$ is the interval $[0, 2\pi]$.

The parameter $\varphi$ used here is the polar angle, measured in the counterclockwise direction, of the rotating point of contact. If $b < a$, $\varphi$ is also the polar angle of the centre of the rolling circle; if $b > a$, the polar angle of the centre of the rolling circle is $\varphi + \pi$. It is not the polar angle of the point of the roulette, nor is this new parametrisation in general by arc-length.

We next consider some special cases of hypotrochoids (see Figure 13.10 and Figure 13.11).

**Hypocycloids.** These are hypotrochoids for which $c$ is on the circumference of the rolling circle. This occurs if $c = a$ or $c = a - 2b$, for example.

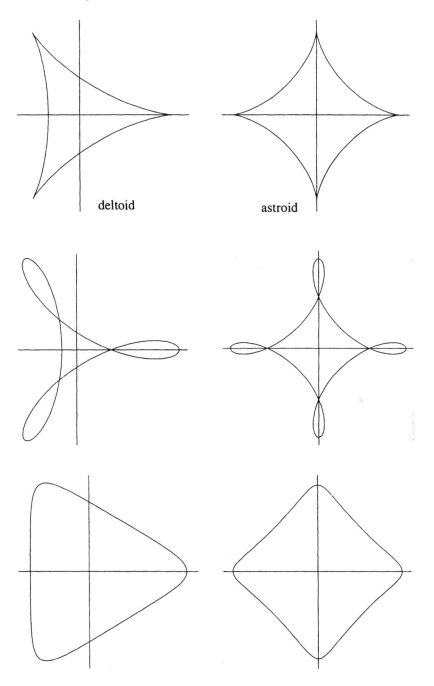

deltoid  astroid

Figure 13.10 *Hypotrochoids.*

In the former case the curve is

$$Z(\varphi) = b\left(me^{i\varphi} + e^{-im\varphi}\right).$$

The hypocycloid, where $m = \dfrac{a-b}{b} > 1$ is an integer, has $m+1 = \dfrac{a}{b}$ cusps.

**Deltoid.** This is a hypocycloid for which $m = 2$. It has three cusps. Its parametric equation is

$$Z(\varphi) = b\left(2e^{i\varphi} + e^{-2i\varphi}\right).$$

It is a quartic algebraic curve with algebraic equation given by eliminating $\cos\varphi$ from the two equations:

$$x = b(2\cos^2\varphi + 2\cos\varphi - 1), \text{ and}$$
$$x^2 + y^2 - 5b^2 = 4b^2\cos\varphi.$$

**Astroid.** This is a hypocycloid for which $a = 4b$ and $m = 3$. It has four cusps. Its parametric equation is

$$Z(\varphi) = b\left(3e^{i\varphi} + e^{-3i\varphi}\right)$$
$$= 4b(\cos^3 + i\sin^3).$$

The Cartesian equation is $x^{\frac{2}{3}} + y^{\frac{2}{3}} = a^{\frac{2}{3}}$, and the algebraic equation of this sextic curve, given by repeatedly cubing the Cartesian equation, is

$$(a^2 - x^2 - y^2)^3 = 27\,a^2x^2y^2.$$

**Rose curves** or **Rhodonea.** The rhodonea include the hypotrochoids for which $c = 0$ (see Figure 13.11). These hypotrochoids pass through the origin. Their equation is

$$Z = mb\left(e^{i\varphi} - e^{-im\varphi}\right)$$
$$= 2mb\cos\left(\frac{m+1}{m-1}\theta\right)e^{i\theta},$$

where

$$\theta = -\frac{m-1}{2}\varphi.$$

This equation gives a rose curve with $m+1$ petals ($m \geq 1$ an integer). For $m = 2$ the curve is a trifolium and has three petals; it is a quartic algebraic curve with equation

$$(x^2 + y^2)^2 = 4b(4x^3 - 3x(x^2 + y^2)).$$

For $m = 3$ the curve is a quadrifolium and has four petals; it is a sextic algebraic curve with equation

$$(x^2 + y^2)^3 = 36\,b^2(x^2 - y^2)^2.$$

trifolium

quadrifolium

Figure 13.11 *Hypotrochoids: rhodonea.*

For $m = 4$ the curve has five petals, and is an algebraic curve of degree 8.. More generally a rhodonea is a curve of the form

$$Z = A \cos (K\theta)\, e^{i\theta}.$$

For $K > 0$ an integer, the curve has $K$ non-intersecting petals.

**Worked example 13.13.** A nephroid is the locus of a point on the circumference of a circle of radius $\frac{1}{2}$ which rolls on the outside of the circle $t \mapsto e^{it}$ of radius 1. Given that the two circles are initially in contact at the point $z = 1$, show, using the general equation of rolling motion, that the equation of the nephroid is

$$z(t) = \tfrac{3}{2}e^{it} - \tfrac{1}{2}e^{3it}.$$

*Solution.* (Method 1 – the standard method.) We first find a comparable parametrisation for the rolling circle. By inspection we have

$$w(t) = e^{it}, \quad z(t) = \tfrac{3}{2} + \tfrac{1}{2}e^{i\pi}e^{2it}.$$

The two circles are oriented counter-clockwise. To check that the parametrisations are comparable, we have i) $w(0) = 1 = z(0)$, ii) $w'(0) = i = z'(0)$, and iii) $|w'(t)| = 1 = |z'(t)|$ for each $t$. Applying the standard formula for rolling curves we have

$$R(t) = w + \frac{w'}{z'}(1-z) = e^{it} + \frac{ie^{it}}{ie^{-2it}}\left(-\tfrac{1}{2} + \tfrac{1}{2}e^{-2it}\right) = \tfrac{3}{2}e^{it} - \tfrac{1}{2}e^{3it}.$$

□

*Solution.* (Method 2 – an *ad hoc* method.) At parameter value $t$ the centre of the rolling circle has position vector $\tfrac{3}{2}e^{it}$. The vector from the initial centre of the rolling circle to the centre of the fixed circle has increased its *angle* (measured counter-clockwise from the initial line) from $\pi$ to $\pi + 3t$, and thus the point of the rolling circle initially in contact with the fixed circle has moved to

$$\tfrac{3}{2}e^{it} + \tfrac{1}{2}e^{i\pi+3it} = \tfrac{3}{2}e^{it} - \tfrac{1}{2}e^{3it}.$$

(Note that it is $e^{3it}$ and not $e^{2it}$.)                                                           □

### 13.4.3 History and applications of trochoids

These roulettes are the loci describes for example by points on a railway wheel rolling around a circle. The cardioid and limaçons were drawn by the famous etcher and painter Dürer (1525). The name limaçon, the French biological word for snail, was introduced by Roberval (c. 1630). The arc-length of the cardioid was determined by La Hire in 1708. Gear teeth cut in epicycloid and hypocycloid shapes were used for constant velocity transfer from one mechanical drive to another; in automobiles for example it is important that the ratio between the angular velocities of the drives on either side of the gear unit does not fluctuate, since this would result in vibration of the vehicle and possible subsequent failure of the gear unit. Certain types of microphone and aerial systems have directional sensitivity curves of cardioid shape. The cardioid is the caustic of a circle for a light source on its circumference (Johann and Jakob Bernoulli 1672), and the nephroid is the caustic (teacup caustic) for parallel rays of light (Huygens 1678) – see Figure 14.11. The general limaçon is the caustic of a circle for light source at other points (St. Laurent 1826). If light from a narrow slit between the curtains of a partially darkened room is allowed to fall onto a cup of tea, the reflection from the inside of the cup onto the surface of the liquid shows the shape one half of the nephroid. The trisectrix is so named because it can be used for trisecting angles. A famous classical problem was to trisect angles using ruler and compass techniques; this is now known to be impossible. The limaçon is the conchoid of a circle with respect to a point ‾on the circumference. Where the ratios of the radii of the circles determining the roulettes are rational, the epicycloid and hypocycloid are algebraic curves.

**Exercises**

**13.1.** A circle of radius $b$ rolls on the outside of a fixed circle. Given that the fixed circle is parametrised by $\varphi \mapsto ae^{i\varphi}$, where $\varphi$ is the polar angle of the centre of the moving circle, write down the comparable parametrisation for the moving circle. Check your answer by substituting $t = a\varphi$ in the above results.

**13.2.** Show that the parametrisations of the epicycloid

$$Z(\varphi) = b\left(ne^{i\varphi} + e^{ni\varphi}\right)$$

and the line

$$w(\varphi) = b(n+1) + 4ib\frac{n}{n-1}\sin\left(\frac{n-1}{2}\varphi\right)$$

are comparable.

**13.3.** A circle of radius 1 rolls without slipping on the outside of a fixed circle also of radius 1; the centre of the fixed circle is at the origin. The circles are initially in contact at the point $z = 1$. Let the fixed circle be parametrised by $\varphi \mapsto e^{i\varphi}$. Find a comparable parametrisation for the rolling circle, and write down the conditions which ensure that it is comparable. Prove that the point, which is linked to the rolling circle and which is initially at $z = 4$, describes the curve

$$\varphi \mapsto 2e^{i\varphi}\left(1 + e^{i\varphi}\right) = Z(\varphi).$$

Prove further that $Z(\varphi) = 4e^{\frac{3i\varphi}{2}}\cos\dfrac{\varphi}{2}$.

$$[2 - e^{-i\varphi}]$$

**13.4.** Plot and draw the roulette

$$\varphi \mapsto Z(\varphi) = 4e^{\frac{3i\varphi}{2}}\cos\frac{\varphi}{2}$$

using polar graph paper. Take 1 unit $= 2\frac{1}{2}$ cm, and plot about 18 points.

$\left(\text{Note. The polar angle of } Z(\varphi) \text{ is } \dfrac{3i\varphi}{2} \text{ or } \dfrac{3i\varphi}{2} + \pi \text{ depending on } \cos\dfrac{\varphi}{2}.\right)$

**13.5.** A circle of radius 1 rolls without slipping on the outside of a fixed circle of radius 2; the centre of the fixed circle is at the origin. The circles are initially in contact at the point $z = 2$. Let the fixed circle be parametrised by $\varphi \mapsto 2e^{\frac{i\varphi}{2}}$. Find a comparable parametrisation for the rolling circle, and write down the conditions which ensure that it is comparable. Prove that the point, which is linked to the rolling circle and which is initially at $z = 6$, describes the curve

$$\varphi \mapsto 3e^{\frac{i\varphi}{2}} + 3e^{\frac{3i\varphi}{2}} = Z(\varphi).$$

Prove further that $Z(\varphi) = 6e^{i\varphi}\cos\frac{\varphi}{2}$.

**13.6.** Plot and draw the roulette

$$\varphi \mapsto Z(\varphi) = 6e^{i\varphi}\cos\frac{\varphi}{2}$$

using polar graph paper. Take 1 unit $= \frac{5}{3}$ cm, and plot about 18 points in the range $0 \leq \varphi \leq 2\pi$.

$\left(\text{The polar angle of } Z(\varphi) \text{ is } \varphi \text{ or } \varphi + \pi \text{ depending on the sign of } \cos\frac{\varphi}{2}.\right)$

**13.7.** A circle of radius $a$ rolls without slipping on the outside of a fixed circle of radius $2a$; the centre of the fixed circle is at the origin. The circles are initially in contact at the point $z = 2a$. Let the fixed circle be parametrised by $\varphi \mapsto 2ae^{i\varphi}$. Find a comparable parametrisation for the rolling circle, and write down the conditions which ensure that it is comparable. Prove that the point, which is linked to the rolling circle and which is initially at $z = 2a$, describes the curve

$$\varphi \mapsto 3ae^{i\varphi} - ae^{3i\varphi} = Z(\varphi).$$

Prove that $\varphi \mapsto Z(\varphi)$ has precisely two non-regular points, and that its curvature is

$$\kappa = \frac{\sqrt{2}}{3a}\frac{1}{\sqrt{1 - \cos 2\varphi}}.$$

Draw a rough sketch of the curve $\varphi \mapsto Z(\varphi)$ indicating carefully the non-regular points.

$$[3a - ae - 2i\varphi; 0, \pi]$$

**13.8.** A circle of radius 1 rolls without slipping on the inside of a fixed circle of radius 3; the centre of the fixed circle is at the origin. The circles are initially in contact at the point $z = 3$. Let the fixed circle be parametrised by $t \mapsto 3e^{\frac{it}{3}}$. Find a comparable parametrisation for the rolling circle, and write down the conditions which ensure that it is comparable. Prove that the point, which is linked to the rolling circle and which is initially at $z = 4$, describes the curve

$$t \mapsto 2e^{\frac{it}{3}} + 2e^{-\frac{2it}{3}} = Z(t).$$

Prove further that this curve is the rose curve

$$Z(t) = 4e^{-\frac{it}{6}}\cos\frac{t}{2} = 4e^{i\theta}\cos 3\theta,$$

where $\theta = -\dfrac{t}{6}$ .

By multiplying the polar equation $r = 4\cos 3\theta$ of the curve by $r^3$ and using $\cos 3\theta = 4\cos^3 \theta - 3\cos \theta$, or otherwise, prove that the points of the curve satisfy the algebraic quartic equation $(x^2 + y^2)^2 = 4x(x^2 - 3y^2)$.

$$[2 + e^{it}]$$

**13.9.** Plot and draw the roulette $r = 4\cos 3\theta$ using polar graph paper. Take 1 unit = 2 cm, and plot in the range $0 \le \theta \le \pi$ in increments of 10 degrees.

(*Note.* The polar angle of $4e^{i\theta}\cos 3\theta$ is $\theta$ or $\theta + \pi$ depending on $\cos 3\theta$.)

**13.10.** The ellipse $x^2 + \dfrac{y^2}{4} = 1$ is parametrised by $t \mapsto z(t) = c + 2is$, where $c = \cos t$ and $s = \sin t$. A second ellipse $(x-2)^2 + \dfrac{y^2}{4} = 1$ is parametrised by $t \mapsto z(t) = 2 - c + 2is$. The second ellipse rolls on the first with the two being in contact at time zero at the point $z = 1$. Show that the parametrisations are comparable. Prove that the centre of the rolling curve which is initially at the point $z = 2$ describes the roulette given by

$$t \mapsto Z(t) = X(t) + iY(t) = \frac{4i}{s + 2ic} = \frac{4(2c + is)}{s^2 + 4c^2} .$$

Show that this hippopede of Proclus satisfies the algebraic equation

$$\left(x^2 + y^2\right)^2 = 16\left(\frac{x^2}{4} + y^2\right)$$

and deduce that its polar equation is

$$r^2 = 4 + 12\sin^2 \theta .$$

**13.11.** Plot and draw the roulette $r = 2\sqrt{1 + 3\sin^2 \theta}$ in the range $-\pi \le \theta \le \pi$. (Use the scale 1 unit = 2 cm, with the polar graph paper in the portrait mode.)

**13.12.** The parabola $\Gamma : 4y = x^2$ is parametrised by $t \mapsto 2t + t^2 i$. A second parabola $\Omega : 4y = -x^2$ is parametrised by $t \mapsto 2t - t^2 i$.

The parabola $\Omega$ rolls on the parabola $\Gamma$ with their vertices being initially in contact. Prove that the parametrisations are comparable. Using the standard formula for rolling motion, or otherwise, show that the motion described by the vertex of $\Omega$ is given by

$$t \mapsto Z(t) = \frac{2t^2(t - i)}{t^2 + 1} .$$

Show that the points $Z(t) = X(t) + iY(t)$ on this roulette satisfy the algebraic equation $Y(X^2 + Y^2) = -2X^2$. Show further that the algebraic curve given by this latter equation has polar equation $R = -2\dfrac{\cos^2 \theta}{\sin \theta}$.

Prove similarly that $\Omega$ satisfies the polar equation $r = -4\,\dfrac{\sin\theta}{\cos^2\theta}$. Deduce that each line through the origin which meets the two curves meets them at points which satisfy $Rr = 8$. Determine the values of $R$ for $\theta = 14°$ and for $\theta = 70°$.

**13.13.** Using a drawn parabola at a scale of 1 unit $= 1$ cm, regarded as $\Omega$ in its initial position, draw the roulette of Exercise 13.12 by using the formula $Rr = 8$ connecting $r$ and $R$ applied to a suitable collection of lines through the origin. Determine about 10 points using this method and plot the points of the roulette corresponding to $\theta = 14°, 70°, 110°$, and $176°$ using a protractor.

**13.14.** The branch $\Gamma : t \mapsto w(t) = 3t^2 - 2it^3$ $(t \geq 0)$ of a semi-cubical parabola rolls without slipping on the branch

$$\Omega : t \mapsto w(t) = 3t^2 + 2it^3 \quad (t \geq 0)$$

of the same semi-cubical parabola. The curves are initially in contact at the origin. Given that the parametrisations are comparable, prove that the point of the rolling curve, which is initially at the origin, describes the curve $\Sigma$ given by

$$t \mapsto Z(t) = \frac{2(t^4 - it^3)}{1 + t^2}.$$

Show that the points $Z(t) = X(t) + iY(t)$ on this roulette satisfy the algebraic equation $Y^2(X^2 + Y^2) = 2X^3$. Show further that the algebraic curve given by this latter equation has the polar equation

$$R = \frac{2\cos^3\theta}{\sin^2\theta}.$$

Prove similarly that $\Gamma$ and $\Omega$ both satisfy the polar equation

$$r = \frac{27\sin^2\theta}{4\cos^3\theta}.$$

Deduce that each line through the origin, which meets the two curves, meets them at points which satisfy $Rr = \frac{27}{2}$. Determine the value of $R$ for $\theta = 60°$ and $\theta = 70°$.

**13.15.** Using a semi-cubical parabola drawn to a scale of 1 unit $= 1$ cm, draw the roulette $\Sigma$ of Exercise 13.14 by using the formula connecting $r$ and $R$ applied to a suitable collection of lines through the origin. Determine about 8 points using this method, and plot the points of $\Sigma$ corresponding to $\theta = 60°, 70°$, and $90°$ using your protractor.

**13.16.** The branch $\Gamma : t \mapsto z(t) = t + i - ie^{-it}$ $(0 \leq t \leq 2\pi)$ of a cycloid rolls without slipping on the branch

$$\Omega : t \mapsto w(t) = -t + i - ie^{it} \quad (0 \leq t \leq 2\pi)$$

of the same cycloid. The curves are initially in contact at the origin. Given that the parametrisations are comparable, prove that the point of the rolling curve, which is initially at the origin, describes the curve $\Sigma$ given by

$$t \mapsto Z(t) = t\left(1 + e^{-it}\right) + 2i\left(1 - e^{-it}\right).$$

Show that the points $Z(t) = X(t) + iY(t)$ on this roulette satisfy the polar equation

$$R = 4(\sin\theta - \theta\cos\theta).$$

**13.17.** Plot and draw the roulette

$$R = 4(\sin\theta - \theta\cos\theta)$$

for $0 \le \theta \le \dfrac{\pi}{2}$. (Use the scale 1 unit = 2.5 cm, with the polar graph paper in the portrait mode. Be aware that the angle $\theta$ used in the given polar equation is measured in radians.)

**13.18.** The parabola $y^2 = x$ rolls on the line $x = 0$. Show that the locus of the focus is the catenary $x = \frac{1}{4}\cosh 4y$.
(*Hint:* Parametrise the parabola by $t \mapsto (t^2, t)$. The arc length is

$$s(t) = \int_0^t \sqrt{1 + 4u^2}\,du = \tfrac{1}{4}\sinh^{-1}(2t) + \tfrac{1}{2}t\sqrt{1 + 4t^2}.$$

Then $t \mapsto is(t)$ gives a comparable parametrisation of the line.)

**13.19.** The epicycloid $Z(\varphi) = b\left(ne^{i\varphi} + e^{ni\varphi}\right)$ is rolled on the line

$$w(\varphi) = b(n+1) + 4ib\frac{n}{n-1}\sin\left(\frac{n-1}{2}\varphi\right).$$

(See Exercise 13.2.) Show that the locus of 0 is an ellipse.

**13.20.** Show that the hypocycloid where $m = 1$ (the fixed circle has radius 2 and the rolling circle has radius 1) is a line segment.

**13.21.** Show that the hypotrochoid, where $m = 1$ (the fixed circle has radius 2 and the rolling circle has radius 1) and where $c$ lies on the $x$–axis and $c \ne 0, 2$, is an ellipse. $\hfill [\cos\varphi + i(2 - c)\sin\varphi]$

**13.22.** Show that the curvature of the roulette

$$Z(t) = w(t) + \frac{w'(t)}{z'(t)}(c - z(t))$$

is    $\kappa = \dfrac{\varphi'\,|c - z(t)|^2 + \operatorname{im}\left[(c - z(t))\,\overline{z'(t)}\right]}{|c - z(t)|^3\,|\varphi'|}$    where $e^{i\varphi(t)} = \dfrac{w'(t)}{z'(t)}$.

**13.23.** The line $x = a$ rolls on the circle $|z| = a$. Write down comparable arc-length parametrisations for the two curves. Find the equation of the locus of a point $c$ initially on the real axis. Show that the roulette has an ordinary cusp if, and only if, $c = a$. Show also that the roulette has two inflexions in case $a < c < 2a$, and that the tangent line to the roulette at the point $z = 2a$ has 4–point contact with the curve in case $c = 2a$.

$$[(c - it)e^{\frac{it}{a}}]$$

**13.24.** Show that the cycloid has cusps at $\varphi = 2n\pi$, and that they are ordinary cusps.

**13.25.** Show that the curtate cycloids have inflexions at the points given by $\cos \varphi = \dfrac{a - q}{a}$ and that they are simple inflexions.

**13.26.** Show that the point of the limaçon with parameter $\varphi$ is non-regular if, and only if, $a + (c - 2a)e^{i\varphi} = 0$. Deduce that the cardioid is the only limaçon having a non-regular point, and prove that this point is an ordinary cusp.

**13.27.** Show that the epicycloid, where $n = \dfrac{a + b}{b}$ is an integer, has $n - 1$ ordinary cusps.

**13.28.** Show that a limaçon with $\frac{3}{2}a > c > a$ has two simple inflexions.

$$\left( \text{Hint: Show that inflexions occur where } \cos \varphi = \frac{a^2 + 2(2a - c)^2}{3a(2a - c)} \cdot \right)$$

**13.29.** Show that the hypocycloid, where $m = \dfrac{a - b}{b} > 1$ is an integer, has $m + 1 = \dfrac{a}{b}$ cusps.

**13.30.** Show that the hypotrochoid has $2m + 2$ simple inflexions provided that

$$0 < |m^3(c - mb)^2 - m^2 b^2| < m^2(m - 1)b.$$

## 13.5 Rigid motions

Rolling one curve on another determines a rigid motion of the plane. We now prove the perhaps suprising converse, namely that general rigid motions can be obtained by rolling one curve on another.

First we recall some of the properties of rigid self-maps of the plane.

**Definition 13.14.** A *rigid self-map* of $\mathbf{R}^2$ or of $\mathbf{C}$ is a self-map which preserves distances (and therefore also preserves non-oriented angles).

A reflexion of the plane in any line is a rigid self-map; it changes the orientation of angles from counter-clockwise to clockwise.

**Lemma 13.15.** *A rigid self-map of* $\mathbf{R}^2$ *or of* $\mathbf{C}$, *which is not a reflexion, is a composition of a rotation and a translation; thus it is of the form*

$$z \mapsto e^{i\varphi}z + b = az + b$$

*where* $a$ *and* $b$ *are a complex numbers and* $|a| = 1$. *A rigid self-map, which is not a reflexion, preserves oriented angles.*

**Definition 13.16.** A *rigid motion* is a family of rigid self-maps, which are not reflexions, parametrised by a parameter $t$; that is

$$(z, t) \mapsto a(t)z + b(t) = F(z, t)$$

where $a(t)$ and $b(t)$ are a complex-valued functions and $|a(t)| = 0$. Usually we assume that $F(z, t_0) = z$, that is the rigid self-map corresponding to $t = t_0$ is the identity map $z \mapsto z$. A *smooth rigid motion* is a rigid motion for which $a(t)$ and $b(t)$ are smooth.

Given that $|a(t)| = 1$, we have $a(t) = e^{i\varphi(t)}$ for some angle $\varphi(t)$. The angle $\varphi(t)$ is not unique since $e^{i(\varphi(t)+2n\pi)} = e^{i\varphi(t)}$. However given that $a(t)$ is smooth on an interval containing $t_0$, it can be shown using the inverse function theorem that there is a unique smooth function $t \mapsto \varphi(t)$, determined by a specific choice of $\varphi(t_0)$, such that $a(t) = e^{i\varphi(t)}$ for each value of $t$. Notice that $\varphi(t)$ is the angle of rotation, measured in the counter-clockwise direction, of the plane.

**Theorem 13.17.** *The motion of the plane $\Pi_1$ (over the plane $\Pi_2$), obtained by rolling a regular curve in $\Pi_1$ on a regular curve in $\Pi_2$, is a rigid motion. Conversely, in any interval of $t$ in which $a'(t) \neq 0$ and $a'(t)b''(t) - a''(t)b'(t) \neq 0$, any rigid motion arises by rolling a regular curve $\Gamma_1$ in $\Pi_1$ on a regular curve $\Gamma_2$ in $\Pi_2$.*

*Proof.* For a rolling motion we have shown above that the locus of a point initially at $c$ is given by

$$\begin{aligned}
t \mapsto Z_c(t) &= w(t) + \frac{w'}{z'}(c - z) \\
&= w(t) + a(t)(c - z(t)) \\
&= w(t) + a(t)c + (w(t) - a(t)z(t)) \\
&= a(t)c + b(t),
\end{aligned}$$

where $|a(t)| = 1$, and the result is immediate. Conversely let

$$F(z, t) = a(t)z + b(t)$$

where $a(t) = e^{i\varphi(t)}$, $a(t_0) = 1$ and $b(t_0) = 0$. For each fixed $t$ this determines the rigid map $F_t$ of the Argand diagram given by

$$F_t(z) = a(t)z + b(t).$$

Let the rolling curve $t \mapsto z(t)$ be the solution $z = z(t)$ of $\dfrac{\partial F}{\partial t} = 0$, that is $a'z + b' = 0$; therefore, we have

$$z(t) = -\frac{b'(t)}{a'(t)}.$$

Define the fixed curve $t \mapsto w(t)$ by letting $w(t)$ be the image of $z(t)$ at parameter value $t$. Thus we have

$$w(t) = F_t(z(t)) = F(z(t), t) = a(t)z(t) + b(t).$$

Then $w' = a'z + az' + b' = az'$ since $a'z + b' = 0$ as above, and therefore

$$z' = \frac{d}{dt}\left(-\frac{b'(t)}{a'(t)}\right)$$
$$= -\frac{a'(t)b''(t) - a''(t)b'(t)}{(a')^2} \neq 0$$

and

$$w' = az' \neq 0.$$

In particular the curves are regular by the hypotheses. For any point $c$ in the Argand diagram we now have

$$F(c, t) = a(t)c + b(t) \qquad\qquad (13.3)$$
$$= w(t) + a(t)(c - z(t))$$
$$= w(t) + \frac{w'(t)}{z'(t)}(c - z(t)).$$

Notice that

$$\left|\frac{w'(t)}{z'(t)}\right| = |a(t)| = 1$$

and it is rolling which is taking place. From $w = az + b$ and $w' = az'$, we have $w(t_0) = z(t_0)$ and $w'(t_0) = z'(t_0)$. Therefore the parametrisations $w(t)$ and $z(t)$ are comparable. Equation 13.3 shows that the motion of any point $c$, under the rigid motion, is given by rolling $\Gamma_1$ ($t \mapsto z(t)$) on $\Gamma_2$ ($t \mapsto w(t)$). □

The condition $a'(t)b''(t) - a''(t)b'(t) \neq 0$ is needed. For example suppose that $b(t)$ is always zero, that is the motion is a rotation around the origin. In this case $z(t)$ and $w(t)$ as given by the formulae are both everywhere zero.

**Mnemonic**

> The rolling curve $z(t)$ is given by solving $\dfrac{\partial F}{\partial t} = 0$.
> The fixed curve is $w(t) = F(z(t), t)$.

**Worked example 13.18.** A rigid motion of the plane is given by

$$(z, t) \mapsto F(z, t) = t + ai + e^{-it/a}(z - ai).$$

Show that the motion is given by rolling a circle on a line.

*Solution.* The rolling curve is the solution (for $z$) of

$$\frac{\partial F}{\partial t} = 0,$$

that is of

$$1 - \frac{i}{a} e^{-it/a}(z - ai) = 0,$$

and therefore

$$z(t) = ai - aie^{it/a},$$

which is a circle centre $ai$ and of radius $a$. The fixed curve is

$$w(t) = F(z(t), t) = t + ai + e^{-it/a}(-aie^{it/a})$$
$$= t,$$

which is the real axis. $\qquad\qquad\square$

## 13.6 Non-regular points and inflexions of roulettes

Finally we investigate some of the properties of the locus described by a fixed point of the plane under a rigid motion.

### 13.6.1 Non-regular points

Given a rigid motion $(z, t) \mapsto a(t)z + b(t)$ we have that a point $c$ describes the curve $Z_c$ where $Z_c(t) = a(t)c + b(t)$.

**Theorem 13.19.** *In any interval in which $a'(t) \neq 0$, the parameter value $t$ is non-regular for the curve $t \mapsto Z_c(t)$ if, and only if,*

$$c = -\frac{b'(t)}{a'(t)} = z(t).$$

*Thus, for each $t$, there is precisely one $c$ for which $t \mapsto Z_c(t)$ is non-regular at $t$. Given further that $a'(t)b''(t) - a''(t)b'(t) \neq 0$, this point is a cusp.*

*Proof.* We have $Z'_c(t) = a'(t)c + b'(t)$, and the first result is immediate. At a non-regular point we then have

$$Z''_c(t) = a''(t)c + b''(t) = a''(t)\left(-\frac{b'(t)}{a'(t)}\right) + b''(t)$$
$$= \frac{a'(t)b''(t) - a''(t)b'(t)}{a'(t)} \neq 0,$$

and the second result follows. $\qquad\qquad\square$

The curve $t \mapsto z(t)$ in the theorem is the rolling curve in case the rigid motion is given by rolling (see Theorem 13.17). The roulette $Z_c$ in this case has a non-regular point if $c$ lies on the rolling curve. The non-regular point of $Z_c$ lies on the fixed curve $t \mapsto w(t)$ at

$$Z_c(t) = a(t)c + b(t) = a(t)z(t) + b(t) = w(t),$$

since $c = z(t)$.

**Mnemonic**

> $Z_c$ has a non-regular point if $c$ lies on the rolling curve.
> The non-regular point lies on the fixed curve at $ac + b$.

### 13.6.2 Inflexions*

We now consider the inflexions of the curve $t \mapsto Z_c(t)$ for a fixed value of $c$. We have $Z_c'(t) = a'(t)c + b'(t)$ and $Z_c''(t) = a''(t)c + b''(t)$. Thus $Z_c'(t) = 0$ if, and only if, $c = z(t)$. Recall that if there is a point of inflexion at a regular point $Z_c(t)$, then $Z_c'(t)$ is parallel to $Z_c''(t)$. Since $Z_c'(t) \neq 0$, this is equivalent to $Z_c''(t)$ being a real multiple of $Z_c'(t)$. Thus

$$\theta(a'c + b') = (a''c + b'')$$

where $\theta$ is real and therefore

$$\theta \left| \frac{a'}{a''} \right| e^{i\psi} \left( c + \frac{b'}{a'} \right) = \left( c + \frac{b''}{a''} \right), \tag{13.4}$$

where

$$\frac{a'}{a''} = \left| \frac{a'}{a''} \right| e^{i\psi}.$$

Thus, for fixed $t$, rotating the vector

$$c + \frac{b'}{a'}$$

through the (fixed) angle $\psi$ gives a vector which is parallel to

$$c + \frac{b''}{a''}.$$

The locus of points $c$ which satisfy Equation 13.4 for a fixed value of $t$ (and therefore a fixed value of $\psi$), but varying values of $\theta$, is a circle, Circle 1 (see Figure 13.12), on the chord $PQ$, where

$$P = -\frac{b''}{a''} \quad \text{and} \quad Q = -\frac{b'}{a'} = z(t).$$

This circle is parametrised by $\theta$. The points $c$ lying on this circle are the points for which $t \mapsto Z_c(t)$ has a point of inflexion at the fixed value $t$. The associated points $Z_c(t)$ where the points of inflexion occur on the curves

---

\* For the meaning of \* see the preface.

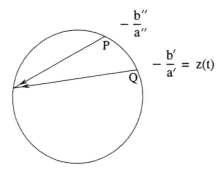

Figure 13.12 *Circle 1*.

$t \mapsto Z_c(t)$ lie on a congruent circle, Circle 2, through $w(t)$. This latter is the image of Circle 1 under $c \mapsto a(t)c + b(t)$ (fixed $t$). Now the tangent at $z(t)$ to Circle 1 is parallel to $\overrightarrow{QP}e^{-i\psi}$ (angle subtended by chord = angle between chord and tangent) and therefore to

$$\left(\frac{b'}{a'} - \frac{b''}{a''}\right)\frac{a''}{a'} = \frac{b'a'' - a'b''}{(a')^2}.$$

Also the tangent at $z(t) = -\dfrac{b'}{a'}$ to $t \mapsto z(t)$ is parallel to

$$z'(t) = \frac{b'a'' - a'b''}{(a')^2}.$$

Thus the curve $t \mapsto z(t)$ is tangent to Circle 1 at $z(t)$ (where $t = t_0$ is fixed). Therefore $t \mapsto w(t)$ is tangent to Circle 2 at $w(t) = a(t)z(t) + b(t)$.

**Definition 13.20.** The *inflexion circle* at $t_0$ is the set of points at which the curves $t \mapsto Z_c(t)$, as $c$ varies, have an inflexion at $t_0$.

The inflexion circle at $t$ is Circle 2. Any point of the moving plane which lies on the inflexion circle is at a point of inflexion of its path.

**Mnemonic**

> $Z_c$ has a point of inflexion if $c$ lies on Circle 1.
> The point of inflexion lies on the inflexion circle.

### 13.6.3 A method of finding the inflexion circle at $t$ *

It is sufficient to find one point of inflexion, that is one value of $c$ for which $t \mapsto Z_c(t)$ has a point of inflexion at $t$. For then the circle can be determined

---

* For the meaning of * see the preface.

since it passes through $w(t)$ and is tangent there to $t \mapsto w(t)$. The centre of the inflexion circle is

$$d = w + is \frac{w'}{|w'|},$$

where $|d - c|^2 = |s|^2$. The equation

$$\left| w + is\frac{w'}{|w'|} - c \right|^2 = |s|^2$$

has a unique (real) solution $s$; this determines $d$. The radius of the inflexion circle is $|s|$. (See Figure 13.13.)

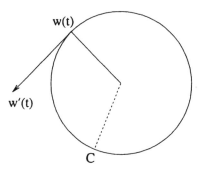

Figure 13.13 *The inflexion circle.*

Typically (for fixed $t$) there is precisely one $c$ among the inflexions of $Z_c(t)$ for which $Z_c$ has $\kappa = \kappa' = 0$ at $t$.

## Exercises

**13.31.** A plane slides over a fixed plane so that at time $t$ the point $(1,0)$ in the moving plane is at $(2\cos t - 1,\, 0)$ in the fixed plane, and the point $(-1, 0)$ in the moving plane is at $(-1,\, 2\sin t)$ in the fixed plane. Find the equation $F(c, t) = a(t)c + b(t)$ for the motion, and hence show that the motion is given by rolling a circle of radius 1 around the inside of a circle of radius 2.

Prove that the point $c = 1$ describes the real axis between $-3$ and $1$ under the motion, and that the point $c = \frac{1}{2}$ describes the ellipse

$$4\,(x + 1)^2 + 36\,y^2 = 9.$$

Draw the circles in the initial position. On the same diagram sketch the locus of the point $c = \frac{1}{2}$, carefully marking the lengths of the semi-axes.
$$\left[ e^{-it}c + e^{it} - 1;\, e^{2it}, 2e^{it} - 1 \right]$$

**13.32.** A rigid motion is given by

$$F(z, t) = t + ia + e^{-i\frac{t}{a}}(z - ia).$$

Prove that the rolling and fixed curves are respectively a circle and a line.

Draw a rough sketch indicating the rolling and fixed curves and the curves traced out by the point with initial position $3ai$ for $0 \le t \le 2\pi$.

**13.33.** The parabola $4y = x^2$ is parametrised by $t \mapsto 2t + t^2 i$ where

$$t = \tan\theta \qquad \left(-\frac{\pi}{2} < \theta < \frac{\pi}{2}\right)$$

is the slope of the tangent at $2t + t^2 i$. A second parabola $4y = -x^2$ is parametrised by $t \mapsto 2t - t^2 i$.

The second parabola rolls on the first with their vertices being initially in contact. Prove that the parametrisations are comparable. Using the standard formula for rolling motion, or otherwise, show that the motion described by a point $c$ is given by

$$t \mapsto Z_c(t) = -2i \sin\theta \tan\theta e^{i\theta} + ce^{2i\theta}.$$

By using the chain rule, or otherwise, prove that

$$\frac{dZ_c}{dt} = 2e^{i\theta}(\sin^2\theta \cos\theta) - 2ie^{i\theta}\sin\theta(1 + \cos^2\theta) + 2ice^{2i\theta}\cos^2\theta,$$

$$Z_c'(0) = 2ic,$$

$$Z_c''(0) = -4i - 4c.$$

$$\left(\textit{Hint: Use the formula } \frac{d}{d\theta}(f(\theta)g(\theta)) = f'(\theta)g(\theta), \text{ where } f(\theta) = 0.\right)$$

Hence find the inflexion circle at $t = 0$ for the motion.

**13.34.** Draw a rough sketch indicating the rolling and fixed curves of Example 13.18 in the initial position $t = 0$ and the curve traced out, for $0 \le t \le 2\pi$, by the point with initial position $3ai$.

# 14

# Envelopes

We have seen in Chapter 11 that the normal lines to a plane curve are tangents to its evolute. If we draw all these normal lines they will 'envelop' the evolute. We consider now the more general case where a 'family' of general curves envelops a curve. As well as families of lines we will give examples of families of other curves, including families of circles, parabolae, or ellipses. A geometrical envelope of a family of curves is a curve which at each of its points is tangent to a curve of the family. We consider three types of envelopes, namely singular-set envelopes, discriminant envelopes, and limiting positions envelopes. In individual cases these may give rise to different sets and may differ from the geometrical envelope depending, for example, on the existence of singularities. As examples of envelopes, we consider evolutes and parallel curves as envelopes, and orthotomics and caustics as envelopes, including, especially, the orthotomics and caustics of the circle. We also prove that a curve is the envelope of its circles of curvature. The approach we take in this chapter is that of problem solving, rather than of proving the technical theorems which occur in the general theory of envelopes.

**Mnemonic**

> At each point the geometrical envelope is tangent to a curve of the family.

As a pictorial example we consider the case where lines are drawn in the plane according to a smooth formula. These lines will perhaps appear to 'concentrate' along some apparent curve. This 'apparent curve of concentration' is the geometrical envelope.

**Example 14.1.** Draw lines $z_\lambda$ through $z(\lambda) = e^{i\lambda} + e^{i(\frac{\pi}{2}+\lambda)}$, making an angle $\frac{\pi}{2} + \lambda$ with the initial half-line for each $\lambda$. The point $z(\lambda)$ lies on the circle of centre 0 and radius $\sqrt{2}$, and the 'apparent curve of concentration' is the circle of centre 0 and radius 1 (see Figure 14.1).

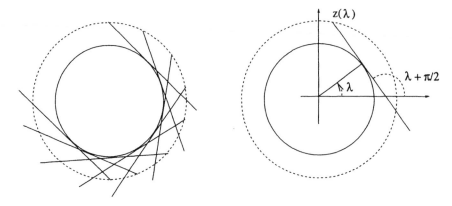

Figure 14.1 *The evolute as an envelope.*

## 14.1 Evolutes as a model

An important example of envelopes is given by the family of normals to a given curve. In this case the 'apparent curve of concentration' is the evolute (see Figure 11.1). We give a careful analysis of this example as a first step to finding a general method for determining envelopes.

Let $t \mapsto z(t)$ be a regular smooth curve in $\mathbf{C}$. The normal at $z(\lambda)$ is

$$z_\lambda(t) = z(\lambda) + t \frac{iz'(\lambda)}{|z'(\lambda)|}.$$

As $\lambda$ varies these normals form a family $\{z_\lambda\}$ of lines; thus, at each point $z(\lambda)$ of the curve $t \mapsto z(t)$, we have the normal line $t \mapsto z_\lambda(t)$. We now establish a criterion for determining the envelope of this family of lines; this will serve as a model for the definition of the singular-set envelope of a general family of parametrised curves. Let $\lambda \mapsto z_*(\lambda)$ be a smooth curve. We consider the following criterion.

**Criterion 14.2.**

*i) For each $\lambda$, $z_*(\lambda)$ lies on the normal line $t \mapsto z_\lambda(t)$ at $z(\lambda)$ to $t \mapsto z(t)$, indeed*

$$z_*(\lambda) = z_\lambda(t(\lambda))$$

*where $\lambda \mapsto t(\lambda)$ is a smooth function, and*

*ii) the tangent vector $z'_*(\lambda)$ to $\lambda \mapsto z_*(\lambda)$ at $z_*(\lambda)$ is parallel to the normal vector to $t \mapsto z(t)$ at $z(\lambda)$, indeed*

$$z'_*(\lambda) \quad \text{is a multiple of} \quad \frac{iz'(\lambda)}{|z'(\lambda)|} = z'_\lambda(t(\lambda)) = \frac{dz_\lambda}{dt}(t(\lambda))$$

*by a suitably differentiable scalar (i.e., real) function.*

**Proposition 14.3.** *Let* $t \mapsto z(t)$ *be a regular smooth curve whose curvature is not zero. Then the evolute* $\lambda \mapsto z_*(\lambda)$ *satisfies Criterion 14.2, and conversely any regular smooth curve which satisfies Criterion 14.2 is the evolute.*

*Proof.* The evolute satisfies Criterion 14.2 by definition and by the results of §11.1 (see Definition 11.1). In this case

$$t(\lambda) = \frac{1}{\kappa(\lambda)}$$

which is smooth, at points which are not points of inflexion, since $t \mapsto z(t)$ is smooth. We now consider the converse. Suppose that Criterion 14.2 is satisfied by $\lambda \mapsto z_*(\lambda)$. We must prove that $z_*$ is the evolute. Let

$$Z(\lambda, t) = z_\lambda(t),$$

then i of Criterion 14.2 implies (using Theorem 7.15) that

$$
\begin{aligned}
z_*' &= \frac{\partial}{\partial \lambda} Z(\lambda, t(\lambda)) \\
&= \frac{\partial Z}{\partial \lambda}(\lambda, t(\lambda)) + \frac{\partial Z}{\partial t}(\lambda, t(\lambda)) \frac{dt}{d\lambda} \\
&= z' + it(\lambda) \frac{d}{d\lambda}\left(\frac{z'}{|z'|}\right) + i\frac{z'}{|z'|}\frac{dt}{d\lambda} \\
&= (1 - t(\lambda)\kappa(\lambda))\, z' + \frac{1}{|z'|}\frac{dt}{d\lambda} iz'.
\end{aligned}
$$

Now $z'$ and $iz'$ are orthogonal vectors. Therefore ii is satisfied if, and only if, $t(\lambda)\kappa(\lambda) = 1$, and therefore if, and only if,

$$z_*(\lambda) = z_\lambda(t(\lambda)) = z(\lambda) + \frac{1}{\kappa(\lambda)}\frac{iz'(\lambda)}{|z'(\lambda)|},$$

that is, if, and only if, $z_*$ is the evolute. $\qquad\square$

The converse tells us that a curve which at each of its points touches a normal to a given curve $\Gamma$ is the evolute of $\Gamma$ (or part of the evolute).

## 14.2 Singular-set envelopes

We model the definition of the singular-set envelope of a parametrised family of parametrised curves on the criterion in the previous section. The plane curves, which we consider here, are regarded as curves in $\mathbf{C}$ or in $\mathbf{R}^2$ as appropriate. We adopt a notation which will cover each of these cases, although, in some examples, it is best to consider the curve in $\mathbf{C}$. In certain examples it will be convenient to use $\mathbf{z}(t)$, $\mathbf{z}_\lambda(t)$, $\mathbf{z}_*(\lambda)$, etc. in place of $z(t)$, $z_\lambda(t)$, $z_*(\lambda)$, etc.

Consider a family of curves $t \mapsto z_\lambda(t)$ where $t$ is in $J$ and $\lambda$ is in $K$ ($J$ and

$K$ are intervals). Define $Z : K \times J \to \mathbf{R}^2$, or $Z : K \times J \to \mathbf{C}$ as appropriate, by $Z(\lambda, t) = z_\lambda(t)$. We assume that $Z = (X, Y)$ or $Z = X + iY$ is smooth, that is that the partial derivatives of all orders of its components $X(\lambda, t)$ and $Y(\lambda, t)$ are continuous.

**Definition 14.4.** A *smooth family* of parametrised curves is a family $\{z_\lambda\}$ for which $Z(\lambda, t)$ is a smooth function.

We defined a geometrical envelope of a family of curves to be a curve which, at each of its points, is tangent to a curve of the family. We now give a precise definition of envelope which is somewhat more general since it can include 'singular' envelopes.

**Definition 14.5.** A *singular-set envelope*, or simply just an *envelope*, of a smooth family $\{z_\lambda\}$ of parametrised curves is a curve $\epsilon : M \to \mathbf{R}^2$ (or $\epsilon : M \to \mathbf{C}$) which satisfies

    i) for each $u$, $\epsilon(u)$ lies on the curve $t \mapsto z_\lambda(t)$, indeed

$$\epsilon(u) = z_{\lambda(u)}(t(u)), \text{ and,}$$

    ii) at $\epsilon(u)$, the tangent vectors

$$\epsilon'(u) \quad \text{and} \quad z'_{\lambda(u)}(t(u))$$

are parallel,

where $M$ is an interval, and $u \mapsto t(u) : M \to J$ and $u \mapsto \lambda(u) : M \to K$ are smooth functions for which $\dfrac{d\lambda}{du}$ has only isolated zeros.

The vectors $\epsilon'(u)$ and $z'_{\lambda(u)}(t(u))$ are the tangent vectors at $\epsilon(u)$ to the curves $u \mapsto \epsilon(u)$ and $t \mapsto r_{\lambda(u)}(t)$.

The two conditions of Definition 14.5 are that the curve $t \mapsto z_{\lambda(u)}(t)$ meets the envelope at the point determined by the parameter value $u$, and that the two curves at this point have the same tangent lines and therefore the same normal lines. Notice that Criterion 14.2 is a special case of this definition where $\lambda(u) = u$.

**Definition 14.6.** The *singular set* of $Z(\lambda, t)$ is the set defined by

$$\left| \frac{\partial(X, Y)}{\partial(\lambda, t)} \right| = \begin{vmatrix} \dfrac{\partial X}{\partial \lambda} & \dfrac{\partial X}{\partial t} \\[2ex] \dfrac{\partial Y}{\partial \lambda} & \dfrac{\partial Y}{\partial t} \end{vmatrix} = 0.$$

Here $Z(\lambda, t) = X(\lambda, t) + iY(\lambda, t)$ where $X$ and $Y$ are real valued functions.

Recall that

$$\begin{vmatrix} \dfrac{\partial X}{\partial \lambda} & \dfrac{\partial X}{\partial t} \\[2ex] \dfrac{\partial Y}{\partial \lambda} & \dfrac{\partial Y}{\partial t} \end{vmatrix} = 0$$

if, and only if, the vectors $\dfrac{\partial Z}{\partial \lambda}$ and $\dfrac{\partial Z}{\partial t}$ are parallel, and that two vectors

are parallel if, and only if, one is a multiple of the other by a real number.

**Mnemonic**

> The singular set is given by $\dfrac{\partial Z}{\partial \lambda}$ and $\dfrac{\partial Z}{\partial t}$ are parallel.

**Theorem 14.7.** *The curve $\epsilon : M \to \mathbf{R}^2$, given by $\epsilon(u) = z_{\lambda(u)}(t(u))$, is a singular-set envelope of the smooth family $\{z_\lambda\}$ of curves if, and only if, $(\lambda(u), t(u))$ belongs to the singular set of $Z(\lambda, t)$ for each $u$.*

*Proof.* From Definition 14.5 i) we have

$$\epsilon'(u) = \frac{d}{d\lambda} Z(\lambda(u), t(u))$$
$$= \frac{\partial Z}{\partial \lambda}(\lambda(u), t(u)) \frac{d\lambda}{du} + \frac{\partial Z}{\partial t}(\lambda(u), t(u)) \frac{dt}{du}.$$

Also

$$z'_{\lambda(u)}(t(u)) = \frac{\partial Z}{\partial t}(\lambda(u), t(u)).$$

Thus, for $\dfrac{d\lambda}{du} \neq 0$, $\dfrac{\partial Z}{\partial \lambda}$ and $\dfrac{\partial Z}{\partial t}$ are parallel if, and only if, $\epsilon'(u)$ and $z'_{\lambda(u)}(t(u))$ are parallel. Since $\dfrac{d\lambda}{du}$ has only isolated zeros, the result follows.

$\square$

Proposition 14.3 is a special case of this theorem.

### 14.2.1 Procedure for determining singular-set envelopes

We adopt the following procedure, which involves parametrising the singular set.

1. Compute the singular set of $Z(\lambda, t)$.

2. Choose a suitable parameter $u$ which parametrises the singular set[†]. Often, as for evolutes, we can choose $u = \lambda$. Sometimes we can choose $u = t$. Sometimes we must choose a parameter other than $\lambda$ or $t$. The singular set may consist of two or more curves which must be parametrised separately. The singular set may also contain isolated points.

The resulting curve $\epsilon : M \to \mathbf{C}$ or $\epsilon : M \to \mathbf{R}^2$ is an envelope by Theorem 14.7.

---

[†] The general proof that such a parametrisation exists and the conditions under which it exists are quite technical and depend on considering certain higher singularities. We do not include the proof here.

In practice many of the families of which we determine the envelopes are families of lines. We now give some examples of envelopes of families of lines. The second one is for a general family. A family of lines is often called a *pencil* of lines.

**Worked example 14.8.** Find the singular-set envelope of the family of lines $L_\lambda$, where $L_\lambda$ passes through the points $(0, \sin \lambda)$ and $(\cos \lambda, 0)$ (see Figure 14.2).

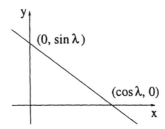

Figure 14.2 *An envelope of a family of lines.*

*Solution.* The line $L_\lambda$ is parametrised by $z_\lambda(t) = (t \cos \lambda, (1-t) \sin \lambda)$. The singular set is given by

$$\begin{vmatrix} \dfrac{\partial X}{\partial \lambda} & \dfrac{\partial X}{\partial t} \\[2mm] \dfrac{\partial Y}{\partial \lambda} & \dfrac{\partial Y}{\partial t} \end{vmatrix} = (-t \sin \lambda)(-\sin \lambda) - (1-t) \cos \lambda \cos \lambda = t - \cos^2 \lambda = 0,$$

that is by $t = \cos^2 \lambda$. We choose $u = \lambda$ and obtain the geometrical envelope $\epsilon(\lambda) = (\cos^3 \lambda, \sin^3 \lambda)$. This curve is the astroid $x^{\frac{2}{3}} + y^{\frac{2}{3}} = 1$ (see Figure 13.10).  $\square$

**Worked example 14.9.** Find the singular-set envelope of the pencil $\{z_\lambda\}$ of lines, where $z_\lambda(t) = a(\lambda) + t b(\lambda)$ $\quad (b(\lambda) \neq 0)$.

*Solution.* The singular set is defined by the relation that

$$\frac{\partial Z}{\partial \lambda} = a' + t b'$$

and $\dfrac{\partial Z}{\partial t} = b$ are parallel, that is $a' + t b' = \theta b$, where $\theta$ is a function of $t$. Eliminating $\theta$ from the $x$–component and the $y$–component, $a_1' + t b_1' = \theta b_1$ and $a_2' + t b_2' = \theta b_2$, gives

$$\frac{a_1' + t b_1'}{b_1} = \frac{a_2' + t b_2'}{b_2},$$

and we obtain a relation of the form $f(\lambda)t = g(\lambda)$. In general we can solve the equation obtaining $t = \dfrac{g(\lambda)}{f(\lambda)}$; thus we choose $u = \lambda$. The envelope is

$$\epsilon(u) = a(u) + \frac{g(u)}{f(u)} b(u).$$

The case where $f(\lambda)$ is identically zero corresponds to

$$\frac{d}{dt}\left(\frac{b_1}{b_2}\right) = 0,$$

that is to all the lines being parallel. In this case the singular set is given by $a'$ is parallel to $b$, and is either empty or consists of a number of the lines $z_\lambda$. $\qquad\square$

**Worked example 14.10.** Find the singular-set envelope of the family $\{z_\lambda\}$ of lines where $z_\lambda(t) = (t, \lambda^3 - \lambda t)$. Find an equation for the envelope as an algebraic curve.

*Solution.* The singular set is

$$0 = \begin{vmatrix} X_t & X_\lambda \\ Y_t & Y_\lambda \end{vmatrix} = \begin{vmatrix} 1 & 0 \\ -\lambda & 3\lambda^2 - t \end{vmatrix} = 3\lambda^2 - t.$$

We parametrise this by $u = \lambda$ and obtain the geometrical envelope

$$\epsilon(\lambda) = (3\lambda^2, -2\lambda^3).$$

This is the semi-cubical parabola

$$\left(\frac{x}{3}\right)^3 = \left(\frac{y}{2}\right)^2.$$

$\qquad\square$

We next give two examples of envelopes for families of conics.

**Worked example 14.11.** Find the singular-set envelope of the family of parabolae given by $z_\lambda(t) = \lambda + t + i\lambda t^2$: thus $z_\lambda$ is the parabola

$$t \mapsto (\lambda + t, \lambda t^2), \quad \text{or} \quad \lambda(x - \lambda)^2 = y.$$

*Solution.* The singular set is given by

$$\begin{vmatrix} 1 & 1 \\ t^2 & 2\lambda t \end{vmatrix} = 0,$$

that is by

$$t(t - 2\lambda) = 0.$$

There are two geometrical envelopes. For $t = 0$, $u = \lambda$ gives

$$\epsilon(u) = z_u(0) = u,$$

which gives the real axis $y = 0$. For $t - 2\lambda = 0$, $u = \lambda$ gives

$$\epsilon(u) = z_u(2u) = 3u + 4iu^3,$$

which gives the cubic curve

$$\left(\frac{x}{3}\right)^3 = \frac{y}{4}.$$

The real axis is the curve $z_\lambda$ where $\lambda = 0$. See Figure 14.3.                                  □

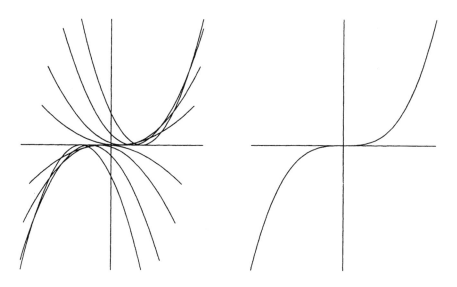

Figure 14.3 *A family of parabolae and their envelope.*

**Worked example 14.12.** Find the singular-set envelope of the family of ellipses given by $z_\lambda(t) = (\lambda \cos t, (1 - \lambda) \sin t)$, whose principal axes are the coordinate axes.

*Solution.* The singular set is given by

$$\begin{vmatrix} \cos t & -\lambda \sin t \\ -\sin t & (1 - \lambda) \cos t \end{vmatrix} = 0,$$

that is by $\lambda = \cos^2 t$. We choose $u = t$ and $\lambda = \cos^2 u$. The envelope is the curve $u \mapsto (\cos^3 u, \sin^3 u)$, which is the astroid $x^{\frac{2}{3}} + y^{\frac{2}{3}} = 1$ as in Example 14.8. See Figure 14.4.                                  □

The above are simply examples for the classes of curves indicated. We can have envelopes of families of many other types of curves. In §14.2.2 we prove that a curve is the envelope of its circles of curvature. In §14.2.3 we show how parallels of a curve arise as envelopes of families of circles.

### 14.2.2 Singular sets using complex multiplication

The formula for the singular set can also be written

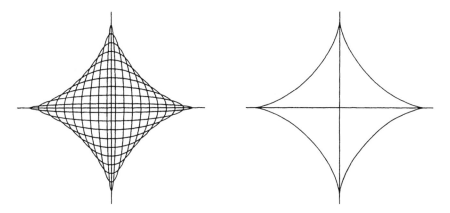

Figure 14.4 *A family of ellipses and their envelope.*

$$\boxed{\operatorname{im}\left(Z_\lambda\,\overline{Z}_t\right) \;=\; -\,(X_\lambda Y_t - X_t Y_\lambda) \;=\; 0\,.}$$

The use of this version of the formula can make calculations easier in some situations.

As an example of the use of this we prove the following important theorem.

**Theorem 14.13.** *Any regular smooth parametrised curve on which $\kappa \neq 0$ is the envelope of its circles of curvature.*

*Proof.* The circle of curvature at $z(t)$ is

$$\lambda \mapsto Z(\lambda, t) = z(t) + \frac{1}{\kappa}\frac{iz'}{|z'|} + \frac{1}{\kappa}e^{i\lambda}.$$

We have

$$\frac{\partial Z}{\partial \lambda} = \frac{1}{\kappa}ie^{i\lambda} \quad \text{and}$$

$$\frac{\partial Z}{\partial t} = z' - \frac{\kappa'}{\kappa^2}\frac{iz'}{|z'|} + \frac{1}{\kappa}i\left(i\kappa z'\right) - \frac{\kappa'}{\kappa^2}e^{i\lambda}$$

$$= -\frac{\kappa'}{\kappa^2}\left(\frac{iz'}{|z'|} + e^{i\lambda}\right).$$

The singular set is given by

$$\operatorname{im}\left(\frac{\partial Z}{\partial t}\,\frac{\overline{\partial Z}}{\partial \lambda}\right) = 0$$

and thus by

$$\operatorname{re}\left(\frac{iz'}{|z'|}e^{-i\lambda} + 1\right) = 0$$

or re $\left(ie^{i\psi}e^{-i\lambda} + 1\right) = 0$, where $z' = |z'|e^{i\psi(t)}$. Thus the singular set is given by $\sin(\psi - \lambda) = 1$, that is by

$$\psi(t) - \lambda = \frac{\pi}{2} + 2n\pi.$$

Taking $u = t$ to be the parameter and substituting for $\lambda$ we have

$$\epsilon(t) = Z(\lambda(t), t)$$

$$= z(t) + \frac{1}{\kappa}\frac{iz'}{|z'|} + \frac{1}{\kappa}e^{i\lambda(t)}$$

$$= z(t) + \frac{1}{\kappa}\frac{iz'}{|z'|} + \frac{1}{\kappa}e^{i\psi}e^{-i\frac{\pi}{2}}$$

$$= z(t),$$

which proves the theorem. □

### 14.2.3 Parallel curves

An important example of envelopes of familes of circles is the construction as envelopes of the parallels to a curve.

**Theorem 14.14.** *Let* $t \mapsto z(t)$ *be a regular curve. Then the envelope of the family of circles of fixed radius* $d > 0$ *whose centres lie on* $t \mapsto z(t)$ *consists of the two parallels of* $t \mapsto z(t)$ *at distance* $d$.

*Proof.* We consider the circles $z_\lambda(t) = z(\lambda) + de^{it}$ which have centres on the curve $\lambda \mapsto z(\lambda)$ and radius $d > 0$. The singular set is given by

$$\frac{\partial Z}{\partial \lambda} = z'(\lambda)$$

is parallel to

$$\frac{\partial Z}{\partial t} = die^{it}.$$

Dividing each by its length, we therefore have that the singular set is given by

$$e^{it} = \pm i\,\frac{z'(\lambda)}{|z'(\lambda)|}.$$

Taking $u = \lambda$ we have the envelope

$$\epsilon(\lambda) = z_\lambda(t(\lambda)) = z(\lambda) \pm d\,\frac{iz'(\lambda)}{|z'(\lambda)|}.$$

This consists of the two parallels distant $d$ from the curve $\lambda \mapsto z(\lambda)$. (See Figure 14.5.) □

A second important example of envelopes of families of circles is where $d(\lambda)$, the radius of the circle, is the distance from a fixed point $P$ to $z(\lambda)$,

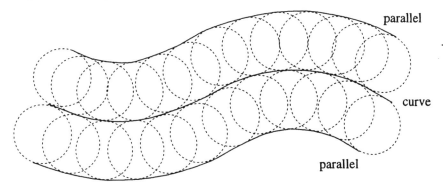

Figure 14.5 *Parallels as envelopes.*

that is, where all the circles $z_\lambda(t) = z(\lambda) + d(\lambda)\,e^{it}$ pass through $P$ and their centres lie on the given curve $\lambda \mapsto z(\lambda)$. The resulting envelope is the orthotomic of $\lambda \mapsto z(\lambda)$. We discuss this in §14.6.

### 14.2.4 Singular envelopes in the singular set

Not all curves arising from the singular set are geometrical envelopes. There are two types of singular[†] envelopes as follows.

i) A locus as $\lambda$ varies of non-regular points of the curves $t \mapsto z_\lambda(t)$. This could for example be a locus of cusps. At such a non-regular point of the curve $t \mapsto z_\lambda(t)$ we have

$$\frac{\partial Z}{\partial t} = z'_\lambda(t) = 0$$

and therefore $(\lambda, t)$ lies in the singular set. The singular set is given by $\dfrac{\partial Z}{\partial t}$ is parallel to $\dfrac{\partial Z}{\partial \lambda}$. The singular-set envelope is defined as having $\geq$ 2–point contact with each curve $z_\lambda$. This certainly happens at cusps and other non-regular points.

ii) One of the curves $z_\lambda$ lies in the singular set, and is not a geometrical envelope. The singular-set envelopes in Example 14.11 contains the line $y = 0$ which is also the member of the family $z_\lambda$ for $\lambda = 0$. However in this case it is a geometrical envelope, since the curve $z_\lambda$ touches $y = 0$ at $(\lambda, 0)$ for each $\lambda$. In general a member of the family of curves $\{z_\lambda\}$ which lies in the singular set may not be a geometrical envelope.

---

[†] The uses of 'singular' in singular set and singular envelope have different meanings. The singular envelope consists of the points of the singular-set envelope which do not lie on the geometrical envelope.

**Mnemonic**

> $\lambda$ is constant in the singular set gives the curve $z_\lambda$.

We now give examples to illustrate these two types of singular envelopes.

**Worked example 14.15.** Find the singular-set envelope of the family of cuspidal cubics $z_\lambda = (\lambda t^2, \lambda t^3 - \lambda)$. Their algebraic equations are

$$\lambda(y + \lambda)^2 = x^3.$$

*Solution.* The singular set is

$$\begin{vmatrix} t^2 & 2\lambda t \\ t^3 - 1 & 3t^2\lambda \end{vmatrix} = \lambda t(t^3 + 2) = 0,$$

that is $\lambda = 0$, $t = 0$ or $t^3 = -2$. The isolated point $\epsilon(t) = (0,0)$ is given by $\lambda = 0$; this is the degenerate curve $t \mapsto z_0(t) = 0$. The axis $x = 0$, $\epsilon(\lambda) = (0, -\lambda)$, is given by $t = 0$; this is a cusp locus. Also $t = -2^{\frac{1}{3}}$ gives the line $\epsilon(\lambda) = (\lambda 2^{\frac{2}{3}}, -3\lambda)$, that is

$$2^{-\frac{2}{3}}x + \frac{y}{3} = 0.$$

This is the geometrical envelope of the family. See Figure 14.6.     □

**Worked example 14.16.** Find the singular-set envelope of the pencil of lines

$$\lambda^2(x - 1) + (2\lambda + 1)(x + y - 1) = 0$$

through $(1, 0)$.

*Solution.* The equation of the line $z_\lambda$ is

$$(z_\lambda - (1,0)) \cdot (\lambda^2 + 2\lambda + 1, 2\lambda + 1) = 0.$$

The parametric equation is

$$z(\lambda) = (1 - t(2\lambda + 1), t(\lambda^2 + 2\lambda + 1)).$$

Notice that

$$\frac{\partial Z}{\partial \lambda} = (-2t, 2t(\lambda + 1)) = 0$$

if, and only if, $t = 0$ and for all $\lambda$. The singular set is

$$\begin{vmatrix} -2t & -(2\lambda + 1) \\ t(2\lambda + 2) & \lambda^2 + 2\lambda + 1 \end{vmatrix} = 0,$$

that is $t = 0$, $\lambda = 0$ or $\lambda = -1$. The isolated point $\epsilon(\lambda) = (1, 0)$ is given by $t = 0$. This point lies on all the lines in the pencil. The curves given by $\lambda = 0$ and $\lambda = -1$ are the curves $z_0$ and $z_{-1}$ of the pencil, and are not geometrical envelopes. This pencil of lines does not have a geometrical envelope.     □

A similar pencil of lines to that of the previous example is given by

$$\lambda(x - 1) + (1 - \lambda)(x + y - 1).$$

In this case the singular set consists only of the isolated point $\epsilon(\lambda) = (1, 0)$. The additional part of the singular set in the above example arises from the parametrisation of the family by $\lambda$ not being sufficiently 'lean'.

A locus of non-regular points can exceptionally be part of the geometrical envelope: this happens for example where a cusp locus is tangent to the cuspidal tangent line at each cusp.

## 14.3 Discriminant envelopes

We next consider the envelopes of parametrised families of unparametrised curves, such as parametrised families of algebraic curves. The envelope we describe is called the discriminant envelope.

In the case of a family of parametrised lines, we can determine the singular-set envelope. In case the lines are given in non-parametric form, for example by $f(\lambda)x + g(\lambda)y = h(\lambda)$, we could find parametric equations. This must be done, of course, in such a way that $Z(\lambda, t)$ is a smooth function. We now consider an alternative method where a family $\{C_\lambda\}$ of curves is given by an equation $F(x, y, \lambda) = 0$, where $F$ is a real valued function; thus $C_\lambda$ is the curve given for fixed $\lambda$ by $F(x, y, \lambda) = 0$. In many of the cases we consider, for each fixed $\lambda$, $F(x, y, \lambda)$ will be a polynomial in $x$ and $y$, and therefore will determine an algebraic curve.

**Definition 14.17.** The *discriminant* or *discriminant envelope* is the result of eliminating $\lambda$ between the two equations[†]

$$\boxed{F(x, y, \lambda) = 0 \quad \text{and} \quad \frac{\partial F}{\partial \lambda}(x, y, \lambda) = 0.}$$

Sometimes the discriminant envelope will consist of several curves and several degenerate envelopes consisting of an isolated point.

In the next example both the singular-set envelope and the discriminant envelope are calculated.

**Worked example 14.18.** Find the envelope of the family of lines given by

$$F(x, y, \lambda) = \lambda x + y - \lambda^3 = 0.$$

*Solution – the discriminant envelope.* We have

$$\frac{\partial F}{\partial \lambda} = x - 3\lambda^2$$

---

[†] The proof that this elimination can be done under quite general conditions and that it results in a curve is quite technical and depends on considering certain higher singularities. We do not include it here.

and therefore

$$\lambda = \pm\sqrt{\frac{x}{3}}.$$

Substituting into $F(x, y, \lambda) = 0$ gives the discriminant

$$\pm\sqrt{\frac{x}{3}}\left(x - \frac{x}{3}\right) + y = 0.$$

These two curves are the 'upper part' $(y \geq 0)$ and 'lower part' $(y \leq 0)$ of the semi-cubical parabola $27y^2 = 4x^3$. This algebraic equation is given by squaring each side of

$$y = \pm\frac{2x}{3}\sqrt{\frac{x}{3}}.$$

Alternatively, since $\dfrac{\partial F}{\partial \lambda} = 0$, we have $x = 3\lambda^2$ and, since $F = 0$, on substituting we have

$$y = \lambda^3 - \lambda x = -2\lambda^3,$$

from which we have the parametrisation $\lambda \mapsto (3\lambda^2, -2\lambda^3)$ of the semi-cubical parabola.                                                                          □

*Solution – the singular-set envelope.* The line $y = -tx + t^3$ passes through $(0, t^3)$ and is parallel to $(1, -t)$. Therefore the family of lines is given by

$$Z(t, s) = z_t(s) = (0, t^3) + s(1, -t).$$

The singular set is given by

$$\begin{vmatrix} 0 & 1 \\ 3t^2 - s & -t \end{vmatrix} = 0,$$

that is by $s = 3t^2$. Taking $u = t$, we obtain the singular-set envelope $\epsilon(u) = (3u^2, -2u^3)$, which has the algebraic equation $27y^2 = 4x^3$.   □

Notice that, in this example, the discriminant envelope and the singular-set envelope are identical.

### 14.3.1 Singular envelopes in the discriminant

Not all curves arising from the discriminant are geometrical envelopes. There are two types of singular envelopes as follows.

i) A locus of singular points. The singular points of $\{C_\lambda\}$ are given by

$$F = \frac{\partial F}{\partial x} = \frac{\partial F}{\partial y} = 0.$$

The points of a singular envelope must also satisfy $\dfrac{\partial F}{\partial \lambda} = 0$. Singular points of algebraic curves include nodes, cusps, and isolated points. In many cases only isolated members of the family $\{C_\lambda\}$ have singular points; such isolated singular points may or may not belong to the discriminant.

   ii) A member of the family $\{C_\lambda\}$ is in the discriminant; that is, for fixed $\lambda$, $\dfrac{\partial F}{\partial \lambda} = 0$ for all points of $C_\lambda$. In general such a curve $C_\lambda$ is not a geometrical envelope, though in exceptional cases it may be.

**Mnemonic**

> $\lambda$ is constant in the discriminant gives the curve $C_\lambda$.

   The discriminant often contains the geometrical envelope as a factor which is not iterated, and, where appropriate, the 2–node locus as a squared factor, and the cusp locus as a cubed factor.

   We give two examples of discriminants which contain the locus of singular points.

**Worked example 14.19.** Find the discriminant envelope of the family of cuspidal cubics

$$F(\lambda, x, y) = \lambda(y + \lambda)^2 - x^3 = 0.$$

*Solution.* The curve $C_\lambda$ has a cusp at $(0, -\lambda)$ for $\lambda \neq 0$. For $\lambda = 0$, $C_\lambda$ is the iterated line $x^3 = 0$. We have

$$\frac{\partial F}{\partial \lambda} = (y + \lambda)^2 + 2\lambda(y + \lambda) = (y + \lambda)(y + 3\lambda).$$

The discriminant, given by substituting $y + \lambda = 0$ and $y + 3\lambda = 0$ into $F(\lambda, x, y)$, is $x^3 = 0$ and $4y^3 = -27\,x^3$. The iterated line $x^3 = 0$, which occurs as a cube, is a cusp locus and the line $4^{\frac{1}{3}}y = -3x$ is the geometrical envelope (see Figure 14.6). □

**Worked example 14.20.** Find the discriminant envelope of the family of cubics

$$F(\lambda, x, y) = \lambda(y + \lambda)^2 - x^2(x + 1) = 0.$$

*Solution.* The curve $C_\lambda$ has a singular point at $(0, -\lambda)$. This is a 2–node in case $\lambda > 0$ and is an isolated singular point if $\lambda < 0$. In case $\lambda = 0$, $C_\lambda$ consists of the line $x + 1 = 0$ and the iterated line $x^2 = 0$. We have

$$\frac{\partial F}{\partial \lambda} = (y + \lambda)^2 + 2\lambda(y + \lambda) = (y + \lambda)(y + 3\lambda).$$

The discriminant, given by substituting $y + \lambda = 0$ and $y + 3\lambda = 0$ into $F(\lambda, x, y)$, is $x^2(x + 1) = 0$ and $4y^3 = -27\,x^2(x + 1)$. The iterated line $x^2 = 0$, which occurs as a square, is a 2–node locus for $\{x = 0, y < 0\}$ and is the locus of the isolated singular points for $\{x = 0, y > 0\}$. The geometrical envelope consists of the line $x + 1 = 0$ and the cubic curve $4y^3 = -27\,x^2(x + 1)$ (see Figure 14.7). □

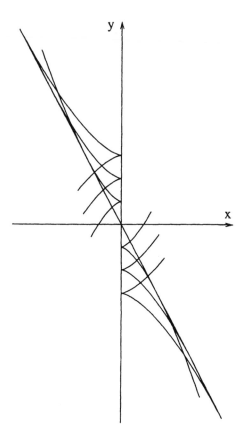

Figure 14.6 *Family of cuspidal cubics with a singular envelope.*

The curves in Example 14.20 can be parametrised for $\lambda > 0$ by

$$z_\lambda(t) = (t^2 - 1, -\lambda + \lambda^{-\frac{1}{2}}t(t^2 - 1)).$$

The singular-set envelope of this family is precisely the geometrical envelope. Singular set envelopes do not contain loci of 2–nodes and do not contain loci of isolated singular points.

A locus of singular points can exceptionally be part of the geometrical envelope. This happens for example where a cusp locus is tangent to the cuspidal tangent line at each cusp.

We consider next an example where all the curves in the discriminant are members of the family $\{C_\lambda\}$.

**Worked example 14.21.** Find the discriminant envelope of the family

$$F(\lambda, x, y) = \lambda^2 f + (2\lambda + 1)\, g = 0,$$

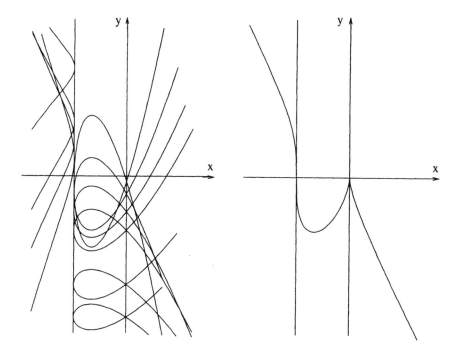

Figure 14.7 *Family of crunodal and acnodal cubics with a singular envelope.*

where $f = f(x, y) = 0$ and $g = g(x, y) = 0$ are two non-singular curves.

*Solution.* We have

$$\frac{\partial F}{\partial \lambda} = 2(\lambda f + g).$$

Substituting $\lambda f = -g$, we obtain the discriminant $g(g - f) = 0$. However $g = 0$ is $C_0$ and $g - f = 0$ is $C_{-1}$. Thus each curve in the discriminant is also a member of the family. In general the geometrical envelope is empty.

□

Compare this last example with Example 14.16 where the singular set consists of members of the family $\{z_\lambda\}$.

## 14.4 Different definitions and singularities of envelopes

There are many different definitions of envelopes: some give rise to different sets (see the 'limiting-positions envelope' below). An example is given above where the singular-set envelope and the discriminant differ because the singular-set envelope does not contain loci of 2–nodes for example. Some authors define 'the envelope' as being obtained by removing the singularities and singular curves from the singular-set envelope or the discriminant envelope. Singularities are points at which certain forms, involving second order partial derivatives, vanish: they are sometimes associated with cusps. The theory of envelopes is fraught with problems caused by non-equivalence of definitions, singularities etc. Our approach here is a simplistic one: in particular we do not define singularities of envelopes.

## 14.5 Limiting-position envelopes

We next consider limiting-position envelopes. These are defined as a limiting position of geometrical intersections. The curves $z_\lambda$ and $z_\mu$, where $\lambda$ and $\mu$ are close together, will in general meet in several points; the number of intersections will depend not only on the type of curve but also on the pencil (or family). Assuming that $z_\lambda$ and $z_\mu$ are distinct, for families of lines we may have one or no points of intersection, for familes of circles we may have two, one or no distinct points of intersection, and for families of ellipses we may have four or fewer distinct points of intersection. We now consider the limiting positions of these intersection points as $\mu \to \lambda$ for fixed $\lambda$.

**Definition 14.22.** The *limiting-positions envelope* of a family $\{z_\lambda\}$ of parametric curves is the set of limiting positions of the intersection points of $z_\mu$ and $z_\lambda$ as $\mu$ tends to $\lambda$ for each fixed $\lambda$.

This definition will serve equally well for either a family of parametric curves or a family of non-parametric curves.

**Proposition 14.23.** *The limiting-positions envelope of a smooth family $\{z_\lambda\}$ of regular parametric curves is a set of points at which members of the family are tangent to the singular-set envelope.*

*Sketch of proof.* The condition for the singular set is that $z_\lambda'(t)$, the tangent to the curve $t \mapsto z_\lambda(t)$ for fixed $\lambda$, is parallel to $\dfrac{\partial}{\partial \lambda} z_\lambda(t)$, which is tangent to the curve $\lambda \mapsto Z(\lambda, t)$ for fixed $t$. Let $P_\mu$ be a point of intersection of $z_\lambda$ and $z_\mu$ and let $P_\mu$ tend to $K$ along the curve $z_\lambda$ as $\mu$ tens to $\lambda$, then the limiting direction of the chord $\overrightarrow{P_\mu K}$ is parallel to the tangent $z_\lambda'(t_0)$ at $K$, where $K$ corresponds to $z_\lambda(t_0)$. On the other hand $P_\mu$ lies on $z_\mu$ at a point with parameter $T(\mu)$, and $T(\mu)$ tends to $T(\lambda) = t_0$ as $\mu$ tends to $\lambda$. The chord $\overrightarrow{P_\mu K}$ is parallel to $z_\mu(T(\mu)) - z_\lambda(T(\lambda))$. The limiting direction of this as $\mu$ tends to $\lambda$ is

$$\frac{\partial}{\partial \lambda} Z(\lambda, T(\lambda)) = \frac{\partial Z}{\partial \lambda} + \frac{\partial Z}{\partial t}\frac{dT}{d\lambda}.$$

The result follows.                                                                                     □

In general, neglecting singularities, the limiting-positions envelope is a subset of the singular-set envelope (and also a subset of the discriminant envelope). Examples exist where a geometrical envelope exists but where the limiting-positions envelope is empty; for example, let

$$t \mapsto z(t) \ (a < t < b)$$

be a curve on which $\rho'(t) < 0$, that is the radius of curvature is decreasing, then no two circles of curvature intersect so the limiting-positions envelope is empty, but the singular-set envelope is the whole curve by Theorem 14.13. One such curve is the portion of the ellipse

$$t \mapsto (a\cos t, b\sin t), \ 0 < t < \frac{\pi}{2} \ (0 < a < b)$$

in the first quadrant.

## Exercises

**14.1.** Find the envelope of the family $\{z_\lambda\}$ of lines in each of the following cases. In each case find an equation for the envelope as an algebraic curve.
  i) The lines $z_\lambda(t) = (-\lambda^2 + \lambda t, t)$. $\qquad\qquad [y^2 = 4x]$
  ii) The lines $z_\lambda(t) = (t, \lambda^2 - \lambda^2 t)$.

**14.2.** Find the envelope of the family of ellipses

$$x = a\cos(t - \lambda), y = b\sin t.$$

[the rectangle $\pm\{(a\sin u, b)\}, \{\pm a, b\sin u)\}]$

**14.3.** The tangent lines $z_\lambda$ to the smooth regular curve $t \mapsto z(t)$ are parametrised by $z_\lambda(t) = z(\lambda) + tz'(\lambda)$. Given that $z''(\lambda)$ is never parallel to $z'(\lambda)$, show that the envelope of the family $\{z_\lambda\}$ is the curve $t \mapsto z(t)$ itself.

**14.4.** The line $L_\lambda$ passes through the point $(\lambda, 0)$. The area enclosed by $L_\lambda$ and the positive $x$–axis and the positive $y$–axis is a constant $A$. Write down the equation of $L_\lambda$ and show that its envelope is part of the hyperbola $2xy = A$.

**14.5.** Find the envelopes, as algebraic curves, of the following families of curves:
  i) the lines $\lambda^2 x - \lambda y - a = 0$, $\qquad\qquad [y^2 = 4ax]$
  ii) the lines $\dfrac{x}{\lambda} + \dfrac{y}{(1 - \lambda)} = 0$,
  iii) the lines $x(\cos\lambda + 1) + y\sin\lambda = 1$, $\qquad\qquad [y^2 = 1 = 2x]$
  iv) the parabolae $y^2 = 4\lambda(x - \lambda)$,
  v) the parabolae $y^2 = \lambda^2(x - \lambda)$, $\qquad\qquad [27y^2 = 4x^3]$
  vi) the cuspidal cubics $(y - \lambda)^2 = x^3$,

**14.6.** Find the envelope of the family of curves

$$\frac{a^2 \cos\lambda}{x} - \frac{b^2 \sin\lambda}{y} = c^2.$$

$\left(\textit{Hint: } \text{Square and add } F \text{ and } \dfrac{\partial F}{\partial \lambda}.\right)$

**14.7.** Let the circle $C$ of radius 2 units with centre at the origin be parametrised by $\lambda \mapsto (2\cos\lambda, 2\sin\lambda) = \mathbf{z}(\lambda)$. Prove that the line through $\mathbf{z}(\lambda)$ which is perpendicular to the vector from $P = (1,0)$ to $\mathbf{z}(\lambda)$ has equation

$$t \mapsto \mathbf{w}_\lambda(t) = \mathbf{z}(\lambda) + t(-2\sin\lambda, 2\cos\lambda - 1).$$

Show that the envelope of this family $\{\mathbf{w}_\lambda\}$ of lines can be parametrised by

$$\lambda \mapsto \mathbf{e}(\lambda) = \frac{1}{2} - \cos\lambda(4\cos\lambda - 2, \ 3\sin\lambda).$$

Prove that all the points of the envelope lie on the ellipse

$$\frac{x^2}{4} + \frac{y^2}{3} = 1.$$

Draw a circle of radius 7cm. Use compasses to mark out the lines $t \mapsto \mathbf{w}_\lambda(t)$, and draw about 18 of these lines. Draw the envelope.
(It is suggested that you use polar graph paper in the landscape position with $P$ at the centre and 1 unit = 3.5 cm.)

**14.8.** Let the circle $C$ of radius 3 units with centre at the origin be parametrised by $\lambda \mapsto (3\cos\lambda, 3\sin\lambda) = \mathbf{z}(\lambda)$. Prove that the line through $\mathbf{z}(\lambda)$ which is perpendicular to the vector from $P = (2,0)$ to $\mathbf{z}(\lambda)$ has equation

$$t \mapsto \mathbf{w}_\lambda(t) = \mathbf{z}(\lambda) + t(-3\sin\lambda, 3\cos\lambda - 2).$$

Show that the envelope of this family $\{\mathbf{w}_\lambda\}$ of lines can be parametrised by

$$\lambda \mapsto \mathbf{e}(\lambda) = \frac{1}{3 - 2\cos\lambda}(9\cos\lambda - 6, \ 5\sin\lambda).$$

Prove that all the points of the envelope lie on the ellipse

$$\frac{x^2}{9} + \frac{y^2}{5} = 1.$$

Using polar graph paper in the landscape position with $P$ at the centre and 1 unit = 2 cm, draw a circle of radius 6 cm. Use compasses to mark out the lines $t \mapsto \mathbf{w}_\lambda(t)$, and draw about 18 of these lines. Draw the envelope.

**14.9.** A family of lines $(z_\lambda)$ is defined by

$$z_\lambda(t) = Z(\lambda, t) = (\lambda^2 - 2\lambda t, 2\lambda + (\lambda^2 - 1)t).$$

Find an envelope $\varepsilon$ of the family and show that every point of the envelope lies on the cubic curve

$$27y^2 = x(x - 9)^2.$$

Prove that this cubic has precisely one singular point.
Verify that $z_\lambda$ passes through the point $(\lambda^2, 2\lambda)$ of the parabola $y^2 = 4x$ and is perpendicular to the line joining this point to the focus $(1,0)$.
Using a drawn parabola make an accurate drawing of at least 20 of the lines $z_\lambda$, positioned so as to make the form of the envelope clear. Mark on

your drawing the position of the singular point, and carefully outline the envelope.

(You may assume that the singular point is a double point.)

$$[(3t^2, t^3 - 3t)]$$

**14.10.** The parabola $y^2 = 4x$ is parametrised by $\lambda \mapsto (\lambda^2, 2\lambda)$. Determine the parametric equation $t \mapsto R(\lambda, t) = r_\lambda(t)$ of the line through $(\lambda^2, 2\lambda)$ which is orthogonal to the line joining the origin to $(\lambda^2, 2\lambda)$. Determine the singular set of the family $\{r_\lambda\}$ and hence find a parametric equation for the envelope. Show that the algebraic equation of the envelope is

$$\left(\frac{x-4}{3}\right)^3 = y^2.$$

Write down the coordinates of the cusp of this semi-cubical parabola.

Using a drawn ellipse, which is drawn at a scale of 1 unit = 7 mm, draw accurately at least 18 of the lines. Calculate and plot accurately, in colour, the point of the cusp. Sketch in the semi-cubical parabola.

**14.11.** Write down the standard parametrisation for the line $t \mapsto Z(\lambda; t)$ which joins the points $e^{i\lambda}$ and $e^{2i\lambda}$ ($\lambda \neq 2n\pi$) on the unit circle, and for which $Z(\lambda; 0) = e^{i\lambda}$. Prove that the singular set of the family $t \mapsto Z(\lambda; t)$ of lines is given by $(1 - 3t)(1 - \cos \lambda) = 0$. Show that the envelope of the family of lines is the cardioid

$$w(t) = \frac{2}{3} e^{it} + \frac{1}{3} e^{2it}.$$

Using polar graph paper, with a scale of 1 unit = 1 cm, draw at least 18 of the lines corresponding to values of $t$ lying in the range $0 < t \leq \pi$. Calculate and plot accurately, in colour, the point of the cusp. Sketch in that part of the cardioid lying in the upper half-plane, using the envelope method.

**14.12.** Write down the standard parametrisation for the line $t \mapsto Z(\lambda; t)$ which joins the points $e^{i\lambda}$ and $e^{3i\lambda}$ ($\lambda \neq 2n\pi$) on the unit circle, and for which $Z(\lambda; 0) = e^{i\lambda}$. Prove that the singular set of the family of lines $t \mapsto Z(\lambda; t)$ is given by $(4t - 1)(1 - \cos 2\lambda) = 0$. Show that the envelope of the family of lines is the nephroid

$$w(t) = \frac{3}{4} e^{it} + \frac{1}{4} e^{3it}.$$

Using polar graph paper, with a scale of 1 unit = 9 cm, draw at least 18 of the lines corresponding to values of $t$ lying in the range $0 < t < \pi$. Calculate and plot accurately, in colour, the point of the cusp. Sketch in that part of the nephroid lying in the upper half-plane, using the envelope method.

**14.13.** Two points $P$ and $Q$ lie on a line $L$ and are separated by distance 1. The line is constrained to move so that $P = (2\cos t, \sin t)$ lies on the ellipse

$$\frac{x^2}{4} + y^2 = 1,$$

and $Q = (X, 0)$ lies on the $x$–axis where $|X| < 2|\cos t|$. Show that the envelope of the family of lines obtained in this way is the astroid

$$x^{\frac{2}{3}} + y^{\frac{2}{3}} = 1.$$

Using a drawn ellipse, carefully draw enough of the lines for the envelope to become clear.

**14.14.** Show that the circle $z_\lambda$ which has its centres at $e^{i\lambda}$ on the unit circle and which touches the $x$–axis can be parametrised as

$$z_\lambda(t) = e^{i\lambda} + \sin\lambda\, e^{it}.$$

Prove that the singular set of the family $\{z_\lambda\}$ of circles is given by $\sin\lambda = 0$ or

$$\sin(\lambda - t) = \cos\lambda = \sin\left(\frac{\pi}{2} \pm \lambda\right).$$

Deduce that the envelope is the nephroid

$$\epsilon(\lambda) = \tfrac{3}{2}\, e^{i\lambda} - \tfrac{1}{2}\, e^{3i\lambda}$$

and part of the $x$–axis. Find the non-regular points of this nephroid and show that they are cusps.

Using a drawn circle (1 unit = 5 cm) in the portrait mode, mark the cusps accurately in colour and draw the envelope using the envelope technique.

## 14.6 Orthotomics and caustics

An important special case of envelopes is given by reflexion caustics. Regarding a curve $\Gamma$ as a mirror, light rays from a point source are reflected in this mirror to give reflected rays. The envelope of the reflected rays is called a (reflexion) caustic of $\Gamma$. Caustic curves can be seen on the surface of the tea in a teacup as light rays fall in from the side. Each caustic is the evolute of an orthotomic of $\Gamma$; the latter is also defined as an envelope.

### 14.6.1 Orthotomics

Let $\Gamma$ be a curve and $P$ a point in the plane. Consider the family of circles having centre on $\Gamma$ and passing through $P$ (see Figure 14.8).

**Definition 14.24.** The *orthotomic* of $\Gamma$ with respect to $P$ is the envelope of the family of circles having centre on $\Gamma$ and passing through P.

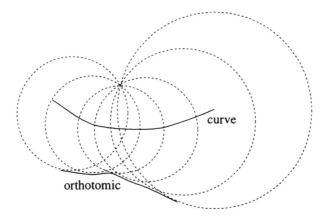

Figure 14.8 *Orthotomics.*

## 14.6.2 Caustics

Consider the pencil of lines through P. Reflect each line in the normals, or, equivalently, the tangents, at the points where it meets the curve. This gives a pencil $\Lambda$ of reflected lines. The point $P$ is called the *radiant* point. Note that the reflected line is regarded as the whole line, not just the half-line lying on one side of the normal or tangent. (See Figure 14.9.)

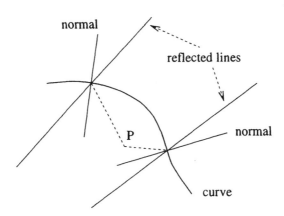

Figure 14.9 *Reflexion caustics.*

**Definition 14.25.** The (reflexion) *caustic* of $\Gamma$ with respect to $P$ is the envelope of the pencil $\Lambda$ of reflected lines.

## 14.7 The relation between orthotomics and caustics

**Theorem 14.26.** *The caustic of* $\Gamma$ *with respect to* $P$ *is the evolute of the orthotomic of* $\Gamma$ *with respect to* $P$.

*Proof.* Let $\Gamma$ be given by $\lambda \mapsto z(\lambda)$. The circle centre $z(\lambda)$ having radius $|p - z(\lambda)|$ is given by

$$t \mapsto r_\lambda(t) = z(\lambda) + |p - z(\lambda)|e^{it}.$$

Let $u \mapsto \epsilon(u)$ be the envelope. The normal lines to $r_{\lambda(u)}$ and the envelope are identical at the point determined by the parameter value $u$, that is at their point of intersection which is given by $\epsilon(u) = r_{\lambda(u)}(t(u))$ (see the definition of singular-set envelope), and this normal line passes through $z(\lambda(u))$. By considering the intersection point of two 'consecutive' circles centred $z(\lambda)$ and $z(\lambda_1)$ and taking the limit as $\lambda_1 \to \lambda$ (see Figure 14.10), we see that $\epsilon(u)$ is the reflection of $P$ in the tangent line to $\lambda \mapsto z(\lambda)$ at $\lambda = \lambda(u)$, and hence the line joining $\epsilon(u)$ to $z(\lambda(u))$ is the reflection of the line joining $P$ to $z(\lambda(u))$ and is the normal to $\epsilon$ at u. Since the evolute is the envelope of the normals, the result follows.                   $\square$

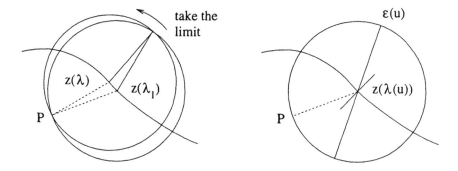

Figure 14.10 *Caustic as evolute of the orthotomic.*

The above proof includes a proof of the following corollary.

**Corollary 14.27.** *The orthotomic of* $\Gamma$ *is the locus of points of reflection of* $P$ *across the tangents of* $\Gamma$.

## 14.8 Orthotomics of a circle

An important special case of orthotomics and caustics is where the original curve is a circle. We consider this case in some detail. First we consider the orthotomics. We consider a circle radius $a$ and a point $P$ distant $b$ from its centre: we choose $P$ as origin. The equation of the circle is $z(t) = b + ae^{it}$ where $a$ and $b$ are real, and $a, b > 0$.

**Theorem 14.28.** *The orthotomic of the circle* $t \mapsto b + ae^{it}$ *with respect to the point* $z = 0$ *(the origin) is the limaçon* $t \mapsto 2(a + b\cos t)e^{it}$.

*Proof.* The circle with centre $z(\lambda) = b + ae^{i\lambda}$ passing through 0 is

$$t \mapsto z_\lambda(t) = b + ae^{i\lambda} + (b + ae^{i\lambda})e^{it} = (b + ae^{i\lambda})(1 + e^{it}) = Z(\lambda, t).$$

Now

$$
\begin{aligned}
X_\lambda Y_t - X_t Y_\lambda &= -\operatorname{im}\left(\frac{\partial Z}{\partial \lambda}\frac{\overline{\partial Z}}{\partial t}\right) \\
&= -\operatorname{im}\left[aie^{i\lambda}(1 + e^{it})(b + ae^{-i\lambda})(-i)e^{it}\right] \\
&= -\operatorname{im}\left[ae^{\frac{it}{2}}2\cos(\frac{t}{2})(be^{i\lambda} + a)e^{-it}\right].
\end{aligned}
$$

Thus $X_\lambda Y_t - X_t Y_\lambda = 0$ if, and only if, either $\operatorname{im}\left[(be^{i\lambda} + a)e^{-\frac{t}{2}}\right] = 0$ or $\cos\frac{t}{2} = 0$; that is, if, and only if,

$$be^{i\lambda} + a = |be^{i\lambda} + a|\, e^{\frac{it}{2}} \tag{14.1}$$

or $\cos\frac{t}{2} = 0$. This then is the singular set. On substituting Equation 14.1 into the equation of the circle in order to eliminate $t$, we have

$$\epsilon(\lambda) = (b + ae^{i\lambda})\left(1 + \frac{(be^{i\lambda} + a)^2}{|(be^{i\lambda} + a)|^2}\right) = (b + ae^{i\lambda}) + (be^{i\lambda} + a)e^{i\lambda},$$

using

$$(b + ae^{i\lambda})(be^{i\lambda} + a) = (b + ae^{i\lambda})(b + ae^{-i\lambda})e^{i\lambda} = |b + ae^{i\lambda}|^2 e^{i\lambda}.$$

Thus we have

$$\epsilon(\lambda) = 2be^{i\lambda}\cos\lambda + 2ae^{i\lambda} = 2(a + b\cos\lambda)e^{i\lambda}.$$

$\square$

The proof of this theorem is considerably shortened by the use of complex numbers. An alternative method of proof would of course be to show that $Z_\lambda$ and $Z_t$ are parallel by some other method.

In the proof, $\cos\frac{t}{2} = 0$, that is $t = \pi$, gives $\lambda \mapsto z_\lambda(\pi) = 0$. In many cases, families of circles have two envelopes, since circles close to one another meet, in many cases, in two (real) points. In this case one envelope is the orthotomic itself and the other is a degenerate envelope consisting of just one point, namely the point P.

For the equation of a limaçon in the form given in Theorem 14.8, see Chapter 10.

### 14.9 Caustics of a circle*

We consider a circle centre $b$ on the real axis where $b > 0$ and radius $a > 0$ as above and the caustic with the origin as radial point (light source). We proved above that the caustic is the evolute of the orthotomic, that is, the evolute of the curve $t \mapsto 2(a + b\cos t)e^{it}$. Notice that $a$ and $b$ in the general form of the equation of the limaçon above are replaced here by $2a$ and $2b$. The equation of the evolute of this curve is given by

$$
\begin{aligned}
z_*(t) &= 2(a + b\cos t)e^{it} + \frac{a^2 + b^2 + 2ab\cos t}{a^2 + 2b^2 + 3ab\cos t}(2i^2)(ae^{it} + be^{2it}) \\
&= 2(a + b\cos t)e^{it} - 2\left(1 - \frac{b^2 + ab\cos t}{a^2 + 2b^2 + 3ab\cos t}\right)(ae^{it} + be^{2it}) \\
&= b(e^{it} + e^{-it})e^{it} - 2be^{2it} + \frac{2b^2 + ab(e^{it} + e^{-it})}{a^2 + 2b^2 + 3ab\cos t}(ae^{it} + be^{2it}) \\
&= \frac{b}{a^2 + 2b^2 + 3ab\cos t}\left((1 - e^{2it})(a^2 + 2b^2 + \frac{3}{2}ab(e^{it} + e^{-it})) + \right. \\
&\qquad\qquad\qquad\qquad \left. + 2b^2 + abe^{it} + e^{-it})(ae^{it} + be^{2it})\right) \\
&= \frac{b}{a^2 + 2b^2 + 3ab\cos t}\left(-\frac{1}{2}abe^{3it} + 3abe^{it} + \right. \\
&\qquad\qquad\qquad\qquad \left. + 2(a^2 + b^2) + \frac{3}{2}abe^{-it}\right).
\end{aligned}
$$

The shape of the caustics is indicated in Figure 14.11.

---

\* For the meaning of \* see the preface.

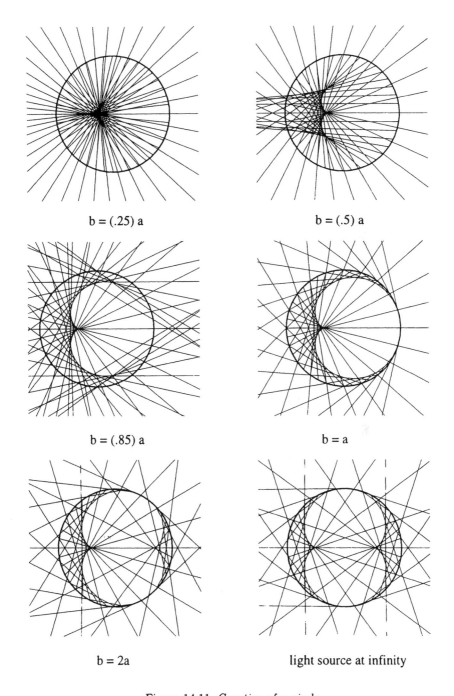

b = (.25) a          b = (.5) a

b = (.85) a          b = a

b = 2a          light source at infinity

Figure 14.11 *Caustics of a circle.*

# 15

# Singular points of algebraic curves

In this chapter we give a further application of contact order. We determine the tangent lines, if any, at singular points of algebraic curves. These tangent lines are tangents to 'branches' of the curve at the singular point. We show that in certain cases these branches can be given smooth local parametrisations. We give conditions in terms of the defining polynomial of the algebraic curve at the singular point that the singular point is a node, acnode, or ordinary cusp. We also give further examples of curves having other types of behaviour at singular points. Our results enable us to classify cubic curves in terms of their singular point behaviour. We show that non-degenerate cubic curves can have at most one singular point and that such a point must be a node, an acnode, or an ordinary cusp. We also show that a non-degenerate cubic curve having a singular point has a rational parametrisation. Finally, in this chapter, we give a formula for determining the curvature of certain branches at singular points. Some of the proofs given in this chapter are technical. It is suggested that the reader, for whom this a first university course in geometry, simply concentrates on solving specific examples, and understands the examples illustrating the behaviour at more general singular points.

## 15.1 Intersection multiplicity with a given line

We first consider the multiplicity of intersection at points of intersection of a line given in parametric form and a curve given in algebraic form. The reader should compare this procedure with that used for parametric curves in Section 5.1 where we used the algebraic equation of the line and the parametric equation of the curve. In Theorem 5.13 we showed that the order of point contact at a common point of two curves does not depend on which of the curves is given in parametric form provided that the point is regular for the parametric curve and non-singular for the algebraic curve. Next we extend our study of the order of point contact of an algebraic curve and a regularly parametrised line to cover non-singular points of the algebraic curve, but for such points we use the term 'intersection multiplicity' instead of 'order of point contact'.

Let the algebraic curve $\Gamma$ be given by $f(x,y) = 0$, where $f(x,y)$ is a polynomial of degree $d$. We parametrise the given line $\ell$ by

$$\mathbf{r}(t) = (x(t), y(t)) = (a, b) + t(p, q),$$

where $(p, q) \neq 0$: since $\mathbf{r}' = (p, q) \neq 0$, the parametrisation is regular. As in Section 5.1, we substitute $(x, y) = \mathbf{r}(t)$ into $f(x, y)$ to obtain

$$\gamma(t) = f(x(t), y(t)) = f(\mathbf{r}(t)).$$

where $\gamma$ in this case is a polynomial of degree less than or equal to $d$. The polynomials $f$ and $\gamma$ are, of course, both smooth functions; indeed they are analytic and their complete (infinite) Taylor series, at each point, have only finitely many non-zero terms. The following lemma plays a key role.

**Lemma 15.1.**
$$\gamma' = \frac{d\gamma}{dt} = \mathbf{grad} f \cdot \mathbf{r}'.$$

*Furthermore, at any singular point, that is any point for which* $\mathbf{grad} f = 0$,

$$\gamma'' = \frac{d^2\gamma}{dt^2} = f_{xx} (x')^2 + 2f_{xy} x'y' + f_{yy} (y')^2.$$

*Proof.* By the chain rule $\dfrac{d\gamma}{dt} = \dfrac{\partial f}{\partial x}\dfrac{dx}{dt} + \dfrac{\partial f}{\partial y}\dfrac{dy}{dt} = \mathbf{grad} f \cdot \mathbf{r}'$, and

$$\gamma'' = f_{xx} (x')^2 + 2f_{xy} x'y' + f_{yy} (y')^2 + f_x x'' + f_y y''.$$

The result follows.                                                         $\square$

We now consider the points of intersection of $f(x, y) = 0$ and the fixed line. These are the points $\mathbf{r}(t)$ for which $\gamma(t) = 0$. Either $\gamma(t)$ is identically 0, in which case $\ell$ is part of $\Gamma$, or, by the fundamental theorem of algebra, $\gamma$ factorizes (essentially uniquely) to give

$$\gamma(t) = (t - t_1)^{m_1} (t - t_2)^{m_2} \ldots (t - t_k)^{m_k} G(t),$$

where $(k \geq 0, t_1, \ldots, t_k$ are real and distinct, $m_i \geq 1$ for each $i$, and $G(t)$ is either constant or a product of irreducible real quadratic factors: these irreducible real quadratic factors can each be factorised into products of complex, but not real, linear factors. The line $\ell$ meets $\Gamma$ in the $k$ distinct points $\mathbf{r}(t_1), \ldots, \mathbf{r}(t_k)$, where $0 \leq k \leq d$, and in $m_1 + m_2 + \cdots + m_k$ points counting multiplicity, where $m_1 + m_2 + \cdots + m_k \leq d$. See Figure 15.1.

**Definition 15.2.** The *intersection multiplicity* of $\Gamma$ and $\ell$ at $\mathbf{r}(t_i)$, for $i = 1, \ldots k$, is $m_i \geq 1$.

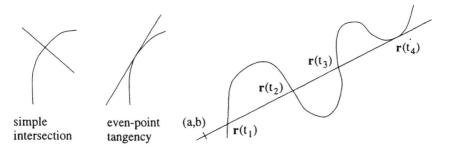

Figure 15.1 *Intersection multiplicity.*

Notice that the intersection multiplicity is always $\leq d$ the degree of $f$. At a point $\mathbf{r}(t_i)$ which is a non-singular point of the algebraic curve, the intersection multiplicity coincides with the order of point contact. We note the following two cases for small $m_i$ (compare §5.4).

**Case 1** : $m_i \geq 2$, that is $\gamma' = \mathbf{grad} f \cdot \mathbf{r}' = 0$ at $t = t_i$. In this case either $\mathbf{r}(t_i)$ is a singular point of $\Gamma$ (i.e., $\mathbf{grad}\, f = \mathbf{0}$), or $\mathbf{r}(t_i)$ is not a singular point and $\ell$ is tangent to $\Gamma$ at $\mathbf{r}(t_i)$. Where $\mathbf{r}(t_i)$ is a singular point, $\ell$ can still be a tangent at $\mathbf{r}(t_i)$.

**Case 2** : $m_i = 3$. In this case either $\mathbf{r}(t_i)$ is a singular point of $\Gamma$ and

$$\gamma'' = f_{xx} \left(x'\right)^2 + 2f_{xy}\, x'y' + f_{yy} \left(y'\right)^2 = 0,$$

or $\mathbf{r}(t_i)$ is not a singular point and both $\ell$ is tangent to $\Gamma$ at $\mathbf{r}(t_i)$ and $\Gamma$ has a point of simple inflexion at $\mathbf{r}(t_i)$. See Figure 5.1. Again $\ell$ can still be a tangent at $\mathbf{r}(t_i)$ where $\mathbf{r}(t_i)$ is a singular point.

In the factorisation above, $t_1, \ldots, t_k$ and $k$ depend on the line $\ell$ and the parametrisation. For a different $\ell$, or different $(a, b)$ or $(p, q)$, $\gamma$ will in general be different.

### Exercise

**15.1.** Show that the intersection multiplicity is unchanged if the line is given a different smooth parametrisation which is regular at the point.

## 15.2 Homogeneous polynomials

We digress to discuss homogeneous polynomials.

**Definition 15.3.** A *homogeneous polynomial of degree $d \geq 0$ in the variables $x$ and $y$* is a polynomial $\theta(x, y) = \sum a_{ij} x^i y^j$ where $a_{ij} \neq 0$ implies that $i + j = d$. A *zero* of $\theta(x, y)$ is an ordered pair $(\alpha, \beta) \neq \mathbf{0}$ for which $\theta(\alpha, \beta) = 0$.

Of course, in a polynomial, only finitely many of the $a_{ij}$ are not zero.

**Example 15.4.** The polynomial $x^3 + 4x^2y - 6xy^2 - 9y^3$ is homogeneous of degree 3, and the polynomial $x^4 - x^3y + 4x^2y^2$ is homogeneous of degree 4.

For a homogeneous polynomial $\theta(x, y)$ of degree $d$ we have

$$\theta(\lambda x, \lambda y) = \lambda^d \theta(x, y)$$

for each real number $\lambda$. Therefore, if $(\alpha, \beta) \neq \mathbf{0}$ is a zero of a homogeneous polynomial $\theta$, then so is $(\lambda\alpha, \lambda\beta)$: thus for all points $(x, y)$ on the line through $(\alpha, \beta)$ and the origin, we have $f(x, y) = 0$.

By the fundamental theorem of algebra, each real polynomial in one variable is a product of real linear, and real irreducible quadratic factors. Similarly each real homogeneous polynomial in two variables is a product of real homogeneous linear, and real homogeneous irreducible quadratic factors. The following lemma is equivalent to the fundamental theorem of Algebra.

**Lemma 15.5.** *Let $\theta$ be a real homogeneous polynomial of degree $d \geq 1$. Then*

$$\theta(x, y) = x^s (y - c_1 x)^{m_1} \dots (y - c_k x)^{m_k} K(x, y)$$

*where $c_1, \dots, c_k$ are real and distinct, $0 \leq k \leq d$, $s \geq 0$, $m_i \geq 1$ for each $i$, and $K(x, y)$ is constant or is a product of real homogeneous irreducible quadratic factors.*

*Proof.* Dividing $\theta(x, y)$ by $x^d$, we obtain the polynomial $\theta(1, z) = \sum b_j z^j$ where $z = \dfrac{y}{x}$ and $b_j = a_{ij}$ for $i = d - j$. By the fundamental theorem this polynomial in one variable factorises as

$$\theta(1, z) = (z - c_1)^{m_1} \dots (z - c_k)^{m_k} P(z)$$

where each $c_i$ is real and where $P$ is constant or is a product of real irreducible quadratic factors. The lemma follows on multiplying the equation through by $x^d$.                                                                    □

Notice that, in this lemma, we have $s + m_1 + \dots + m_k \leq d$.

**Example 15.6.** There are the factorisations

$$x^5 y - xy^5 = -yx(y - x)(y + x)(x^2 + y^2), \text{ and}$$

$$x^6 + y^6 = (x^2 + y^2)(x^2 - \sqrt{3}xy + y^2)(x^2 + \sqrt{3}xy + y^2)$$

into the products of real linear and real irreducible quadratic factors.

## 15.3 Multiplicity of a point

The multiplicity of a point on an algebraic curve is the lowest intersection multiplicity which the curve has at the point with regularly parametrised lines through the point. The tangent lines are among the lines having intersection multiplicity strictly greater than the multiplicity of the point.

As above we consider a line through $(a, b)$, but now we choose the point $(a, b)$ to be on the curve. Taking $t_1 = 0$, we then have, in the case where $\ell$ is not part of $\Gamma$,

$$\gamma(t) = t^{m_1}(t - t_2)^{m_2} \ldots (t - t_k)^{m_k} G(t) = t^{m_1} H(t) \quad (m_1 \geq 1)$$

where $H$ is a polynomial and $H(0) \neq 0$.

**Definition 15.7.** The *multiplicity of a point* $(a, b)$ *on* $\Gamma$ is $m$ if every line $\ell$ through $(a, b)$ has intersection multiplicity $\geq m$ with $\Gamma$ at $(a, b)$, and at least one such line has intersection multiplicity equal to $m$ at $(a, b)$.

We now consider the pencil of lines through the fixed point $(a, b)$ on $\Gamma$. This pencil will consist of the lines of the form

$$\mathbf{r}(t) = (x(t), y(t)) = (a, b) + t(p, q)$$

for different $(p, q) \neq 0$. The polynomial $\gamma$ can be written

$$\boxed{\gamma(t) = f(\mathbf{r}(t)) = \theta(p, q)t^m + \text{higher powers of } t,}$$

where $m \geq 1$ is the multiplicity of $(a, b)$, $m \leq d$, and $\theta(p, q)$ is a homogeneous polynomial in $p$ and $q$ of degree $m$ where $\theta(p, q)$ is not identically zero.

**Definition 15.8.** A *singular line* at a point $(a, b)$ of multiplicity $m \geq 1$ is a (real) line whose intersection multiplicity with the curve at $(a, b)$ is $> m$.

**Theorem 15.9.** *A point on* $\Gamma$ *has multiplicity* 1 *if, and only if, it is a non-singular point. At each point of* $\Gamma$ *the multiplicity is* $\leq d$. *Through a point of multiplicity* $m \geq 1$ *there are* $\leq m$ *singular lines; they are parallel to the non-zero real solutions of* $\theta(p, q) = 0$. *If* $m$ *is odd, there is at least one singular line, and, if* $m = 1$, *there is precisely one singular line.*

*Proof.* The number of 'directions' (real ratios $p : q$ where $(p, q) \neq 0$) for which $\theta(p, q)$ vanishes is $\leq m$. For the finite number of corresponding lines we have $\theta(p, q) = 0$, where $(p, q) \neq 0$, and therefore their intersection multiplicity at $(a, b)$ is $> m$. For all other lines we have $\theta(p, q) \neq 0$, and the intersection multiplicity is precisely $m$. The theorem follows easily. $\square$

It is among the singular lines that we shall find the tangent lines, though not all singular lines are (real) tangent lines (see Example 15.33).

**Worked example 15.10.** Find the singular lines of

$$y^2 + 2y - x^3 + 3x - 1 = 0$$

at $(1, -1)$ and their intersection multiplicities with the curve.

*Proof.* We substitute $(x, y) = (1, -1) + t(p, q)$ into $f(x, y)$ to obtain

$$\gamma(t) = t^2(q^2 - 3p^2) - t^3 p^3.$$

Thus the singular lines have directions $(p, q) = (1, \pm\sqrt{3})$ their equations are

$$\mathbf{r}(t) = (1, -1) + t(1, \pm\sqrt{3}), \text{ or}$$
$$\frac{x-1}{1} = \frac{y+1}{\sqrt{3}} \text{ and } \frac{x-1}{1} = \frac{y+1}{-\sqrt{3}}.$$

They each have intersection multiplicity 3 with the curve, and the intersection multiplicity with the curve of any other line is 2. □

**Exercises**

**15.2.** Find the singular lines of the following curves at the points indicated and their intersection mutiplicities with the curve.

i) $x^6 + x^4(x - y) + (x - y)^2(x + y) = 0$, $(0, 0)$
$$[y = x, 6; \ y = -x, 5]$$
ii) $(x^2 + 2x + 1)(y - x) - (y + 1)^4 = 0$, $(-1, -1)$

**15.4 Singular lines at the origin**

The singular lines and their multiplicities at the origin can be found using the method of §15.3, that is using the polynomial $\gamma(t)$ given by substituting $\mathbf{r}(t) = (x(t), y(t)) = t(p, q)$ into $f(x, y)$. However we now give an alternative method which can be shorter.

Let the algebraic curve $f(x, y) = 0$ pass through the origin. First we consider the case where the origin is a non-singular point of the curve. In this case we have

$$f(x, y) = \alpha x + \beta y + H(x, y)$$

where $(\alpha, \beta) \neq 0$ and $H$ consists of terms of degree $\geq 2$.

**Theorem 15.11.** *Let the origin be a non-singular point of the algebraic curve $f(x, y) = 0$. Then the tangent at the origin is $\alpha x + \beta y = 0$.*

*Proof.* Since $\mathbf{grad} \ f|_{(0,0)} = (\alpha, \beta)$, the result is immediate. □

**Mnemonic**

> Equating the linear terms to zero gives the tangent at the origin.

**Worked example 15.12.** Find the tangent to the circle

$$(x - 1)^2 + (y + 2)^2 - 5 = 0$$

at the origin.

*Solution.* We have $f(x, y) = -2x + 4y + x^2 + y^2$, and therefore the tangent at the origin is $2y - x = 0$. □

## Exercises

**15.3.** Find the tangent to the circle $(x + 3)^2 + (y - 1)^2 - 10 = 0$ at the origin.
$$[y - 3x = 0]$$

**15.4.** Find the tangent to the circle $x^2 + y^2 - 25 = 0$ at $(3, 4)$.

**15.5.** Find the tangent to the parabola $y^2 - 4x = 0$ at $(1, 2)$.
$$[x - y + 1 = 0]$$

We now consider the case where the algebraic curve $f(x, y) = 0$ has a singular point at the origin: thus the order of each monomial of $f$ is $\geq 2$. Let

$$\boxed{f(x, y) = \sigma(x, y) + \rho(x, y)}$$

where $\sigma$ is a homogeneous polynomial of degree $m \geq 2$ consisting of the non-zero terms of lowest degree $m \geq 2$ and the degree of each term of $\rho$ is $\geq m + 1$. Substituting $(x, y) = t(p, q)$, we have

$$\gamma(t) = t^m \sigma(p, q) + \text{ terms of higher order in } t.$$

From this we have the following lemma.

**Lemma 15.13.** *Let the origin be a singular point of $f(x, y) = 0$, then, at the origin, $\sigma = 0$.*

**Theorem 15.14.** *Let the origin be a singular point of the algebraic curve $f(x, y) = 0$, then the singular lines at the origin are the real linear factors of $\sigma$.*

*Proof.* We have $f = \sigma + \rho$ where degree $\sigma = m \geq 2$. By the lemma $\sigma = 0$, and the result follows from Theorem 15.9. $\square$

**Worked example 15.15.** Find the singular lines at the origin of

$$x^5 y - xy^5 + 3x^7 - 8y^8 = 0.$$

*Solution.* $\sigma = x^5 y - xy^5 = -yx(y - x)(y + x)(x^2 + y^2)$. The singular lines are $x = 0$, $y = 0$, $y - x = 0$ and $y + x = 0$. $\square$

The above method can be modified, for singular points at $(a, b)$, by translating the singular point to the origin using the substitution

$$(x, y) = (a, b) + (X, Y),$$

finding the singular lines of the resulting algebraic curve, and using the reverse translation to obtain the singular lines of the original curve.

**Worked example 15.16.** Find the singular lines at $(0, -1)$ of the curve

$$x^4 + x^5 - y^5 - 4y^4 - 6y^3 - 4y^2 - y = 0.$$

*Solution.* The substitution $(x, y) = (0, -1) + (X, Y)$ gives

$$X^4 - Y^4 + X^5 + Y^5.$$

Thus
$$\sigma = X^4 - Y^4 = (X - Y)(X + Y)(X^2 + Y^2).$$
The singular lines are $X - Y = 0$ and $X + Y = 0$, that is $x - (y + 1) = 0$
and $x + (y + 1) = 0$.                                                      □

## Exercises

**15.6.** Find the singular points, and the singular lines and their intersection
multiplicities at the singular points, for the following curves.
  a)  $y^3 - y^2 + x^3 - x^2 + 3xy^2 + 3x^2y + 2xy = 0$      $[(0,0), \ y - x = 0]$
  b)  $x^4 + y^4 - x^2y^2 = 0$
  c)  $x^3 + y^3 - 3x^2 - 3y^2 + 3xy + 1 = 0$          $[(1,1), \ x = 1, y = 1]$

**15.7.** Show that the algebraic curve with polar equation $r^2 = \sin^2 2\theta$ is of
degree 6. Find the singular lines at the origin.                    $[x = 0, y = 0]$

**15.8.** Let $\Gamma$ be an algebraic curve of degree $n$ and let $P$ be a point on $\Gamma$
of multiplicity $n$. Show that $\Gamma$ consists of $\leq n$ lines through $P$.

## 15.5 Isolated singular points

Before discussing tangent lines to algebraic curves at singular points, we
consider singular points which are 'isolated'. At an isolated singular point
the algebraic curve has no tangent lines.

**Definition 15.17.** An *isolated* singular point or *acnode* of the algebraic
curve $f(x, y) = 0$ is a point $(a, b)$ for which $f(a, b) = 0$ but $f(x, y) \neq 0$ for
$0 < |(x, y) - (a, b)| < \epsilon$ for some $\epsilon > 0$.

Thus a point is isolated if there is a small circle with centre $(a, b)$ such
that the only point of the algebraic curve within the circle is $(a, b)$ itself.
The following is a useful criterion for determining isolated singular points.

**Theorem 15.18.** *A singular point for which* $\theta(p, q) = 0$ *has no real solution[†] (for the ratio $p : q$) is isolated.*

*Proof\*.* In the following we repeatedly use the triangle inequalities
$$|P + Q| \leq |P| + |Q| \quad \text{and} \quad |P - Q| \geq ||P| - |Q||.$$
Each irreducible quadratic homogeneous polynomial
$$p(x, y) = ax^2 + 2hxy + by^2$$
can be diagonalized by an orthogonal transformation to $AX^2 + BY^2$, where
$A$ and $B$ are non-zero and have the same sign. Thus we have
$$|p(x, y)| = |A|X^2 + |B|Y^2$$
$$\geq \min\{|A|, |B|\}(X^2 + Y^2) = \min\{|A|, |B|\}(x^2 + y^2).$$

---
[†] This cannot happen if $m$ is odd.
\* For the meaning of \* see the preface.

It follows that, if $\sigma(x, y)$, of degree $m = 2u$, is a product of irreducible quadratic homogeneous polynomials, then there is a number $K > 0$ such that

$$| \sigma(x, y) | \geq K(x^2 + y^2)^u$$

for all $x, y$. On the other hand if $\phi(x, y)$ is a homogeneous polynomial of degree $r$, there is a number $L > 0$ such that

$$| \phi(x, y) | \leq L(x^2 + y^2)^{r/2}$$

for all $x, y$ (we use the inequalities $|x| \leq (x^2 + y^2)^{1/2}$ and $|y| \leq (x^2 + y^2)^{1/2}$). Thus if $\rho(x, y)$ is a sum of finitely many homogeneous polynomials each of degree $\geq 2u + 1$, there is a number $M > 0$ such that

$$| \rho(x, y) | \leq M(x^2 + y^2)^{(2u+1)/2}$$

for all $x, y$ satisfying $(x^2 + y^2) \leq 1$. Thus we have

$$| \sigma(x, y) + \rho(x, y) | \geq (x^2 + y^2)^u (K - M(x^2 + y^2)^{1/2}) > 0$$

for all $x, y$ with $(x^2 + y^2) < \epsilon$ where $\epsilon = \min (K^2/M^2, 1)$. In the case where the singular point is at the origin, the result now follows since $f = \sigma + \rho$: more generally the result follows after translating the singular point to the origin. ☐

Notice that $\theta(p, q) = 0$ has no non-zero real solution if, and only if, $\theta$ is either constant or is a product of real irreducible quadratic factors. This is the case if, and only if, either $\theta(p, q) > 0$ for all $(p, q) \neq 0$, or $\theta(p, q) < 0$ for all $(p, q) \neq 0$: also, in this case, $m$ must be even.

**Corollary 15.19.** *A singular point is isolated if, and only if, there are no singular lines at the singular point.*

**Worked example 15.20.** Show that the acnodal cubic $y^2 = x^3 - x^2$ has an isolated point at the origin.

*Solution.* Let $\mathbf{r}(t) = t(p, q)$ be a line through $(0, 0)$. We have

$$f(x, y) = y^2 + x^2 - x^3$$

and $\gamma(t) = t^2(p^2 + q^2) - t^3 p^3$. The multiplicity of $(0, 0)$ is 2, and in this case $\theta(p, q) = p^2 + q^2$ has no real homogeneous linear factor. Therefore $(0, 0)$ is an isolated point, and there is no tangent line at $(0, 0)$. The intersection multiplicity with the curve of any line through $(0, 0)$ is 2. ☐

**Worked example 15.21.** Show that the curve

$$f(x, y) = x^4 + y^4 + x^5 y^5$$

has an isolated singular point at the origin.

*Solution.* In this case

$$\sigma(x, y) = x^4 + y^4 = (x^2 + \sqrt{2}xy + y^2)(x^2 - \sqrt{2}xy + y^2)$$

is positive for all real $(x, y) \neq 0$. There are no real linear factors. The multiplicity of $(0, 0)$ is 4, but the origin is an isolated point. There are no tangent lines. All lines through the origin have intersection multiplicity 4 with the curve.                                                                    □

## Exercises

**15.9.** Show that for $a > b > 0$ the limaçon

$$(x^2 + y^2 - bx)^2 - a^2(x^2 + y^2) = 0$$

has an isolated singular point at the origin.

**15.10.** Show that for $b > a > 0$ the conchoid of Nicomedes

$$(x^2 + y^2)(x - b)^2 - a^2 x^2 = 0$$

has an isolated singular point at the origin.

**15.11.** Show that for $a > b > 0$ the hippopede of Proclus

$$(x^2 + y^2)^2 + 4b(b - a)(x^2 + y^2) - 4b^2 x^2 = 0$$

has an isolated singular point at the origin.

The converse of Theorem 15.18 is not true. We give in Example 15.33 a curve for which the singular point is isolated even though $\theta$ has a repeated real linear factor.

## 15.6 Tangents and branches at non-isolated singular points

In the case where the origin is a singular point, we have

$$f(x, y) = \sigma(x, y) + \rho(x, y)$$

where $\sigma(x, y)$ is homogeneous of degree $m \geq 2$ and all terms of $\rho$ are of degree $> m$. If there are real homogeneous linear factors of $\sigma(x, y)$, we investigate which of them correspond to tangent lines at the origin.

In case the singular point is at $(a, b)$, we can use the substitution $(x, y) = (a, b) + (X, Y)$ to move the singular point to the origin; then, $f$ is replaced by $g$ where $g(x, y) = f(x + a, y + b))$. Again we have $g(x, y) = \sigma(x, y) + \rho(x, y)$. Alternatively we consider

$$\gamma(t) = \theta(p, q)t^m + \text{ higher powers of } t,$$

given by making the substitution $(x, y) = (a, b) + t(p, q)$ into $f(x, y)$ as above. Generally we have $\theta = \sigma$.

**Definition 15.22.** A *branch* at a singular point $(a, b)$ of the algebraic curve $f(x, y) = 0$ is a part of the curve including the singular point given by a continuous local solution of $f(x, y) = 0$, either as $y = \phi(x)$ where $\phi(x)$ tends to $b$ as $x$ tends to $a$ and $\phi'(a)$ exists, or as $x = \psi(y)$ where $\psi(y)$ tends to $a$ as $y$ tends to $b$ and $\psi'(b)$ exists. A branch is *ordinary* if $\phi$ is defined for $a - \epsilon < x < a + \epsilon$, respectively $\psi$ is defined for $b - \epsilon < y < b + \epsilon$, where $\epsilon > 0$.

In some cases $\phi(x)$ will be defined, for example, for $a \leq x < a + \epsilon$ where $\epsilon > 0$, and there will be no local solution defined for $a - \epsilon < x < a + \epsilon$. The derivative $\phi'(a)$ in this case is the right-derivative of $\phi$ at the end point $a$. A branch may be parametrised by parameters other than $x$ or $y$.

**Lemma 15.23.** *The tangent line to a branch $y = \phi(x)$ at the singular point $(a, b)$ is $t \mapsto (a, b) + t(1, \phi'(a))$. The tangent line at a branch $x = \psi(y)$ at the singular point $(a, b)$ is $t \mapsto (a, b) + t(\psi'(b), 1)$.*

*Proof.* The branch is parametrised by $\mathbf{r}(x) = (x, \phi(x))$. Thus

$$\mathbf{r}' = (1, \phi'(x)),$$

and the result is immediate. □

In general $\phi$ and $\psi$ are not smooth functions, though they are smooth in case the branch has a non-iterated real linear factor of $\theta$ as its tangent (see Theorem 15.28 below). Even where $\mathbf{r}(x) = (x, \phi(x))$ is not smooth, it is often theoretically possible to give a smooth parametrisation to the branch using a different parameter (see for example the proof of Theorem 15.43).

**Example 15.24.** The cuspidal cubic $y^2 = x^3$ has the two branches

$$y = x^{\frac{3}{2}} \quad (x \geq 0), \quad \text{and}$$

$$y = -x^{\frac{3}{2}} \quad (x \geq 0).$$

At each branch the derivative $\phi'$ is continuous for $x \geq 0$, but the second derivative $\phi''(0)$ does not exist, and therefore the parametrisations by $x$ are not smooth at the singular point. However the parametrisation $t \mapsto (t^2, t^3)$ covers both branches and is smooth.

**Example 15.25.** The crunodal cubic $y^2 = x^3 + x^2$ has the two branches

$$y = x\sqrt{1 + x}, \quad \text{and}$$

$$y = -x\sqrt{1 + x},$$

each defined for $x \geq -1$, and the parametrisation of each by $x$ is regular and smooth for $x > -1$. The singular point of the algebraic curve on each branch is given by $x = 0$. The whole curve can be parametrised by

$$t \mapsto (t^2 - 1, t(t^2 - 1)),$$

which covers both branches and is smooth and regular.

**Theorem 15.26.** *Let $f(x, y) = 0$ have a singular point at $(a, b)$. Then the tangent line at $(a, b)$, to any branch at $(a, b)$, is parallel to a real solution $(p, q)$ of $\theta(p, q) = 0$.*

*Proof\**. We first prove the theorem in case the singular point is at the origin. We wish to show that, if $y = \phi(x)$ is a branch, then $y - \phi'(0)x$ is one of the linear factors of $\sigma$. Suppose that this is not the case. Then we have

$$\frac{\sigma(x, \phi(x))}{x^m} = \sigma\left(1, \frac{\phi(x)}{x}\right) \quad \text{tends to} \quad \sigma(1, \phi'(0)) = T \neq 0$$

as $x$ tends to 0. On the other hand we have

$$\frac{\rho(x, \phi(x))}{x^m} \quad \text{tends to} \quad 0$$

as $x$ tends to 0 since

$$\frac{\phi(x)}{x} \quad \text{tends to} \quad \phi'(0)$$

as $x$ tends to 0 and since the degree of each monomial in $\rho$ is $\geq m + 1$. Since

$$0 = f(x, \phi(x)) = \sigma(x, \phi(x)) + \rho(x, \phi(x)),$$

on dividing by $x^m$ and letting $x$ tend to 0 we obtain $0 = T + 0$ which gives a contradiction. Thus $y - \phi'(0)x$ is one of the linear factors of $\sigma$. The proof in the case of a branch $x = \psi(y)$ is similar. In case the singular point is at $(a, b)$, we translate the origin to $(a, b)$ by $(x, y) = (a, b) + (X, Y)$ and apply the above argument to $g(x, y) = f(x + a, y + b)$, which has a singular point at the origin. Since the tangent lines to $f(x, y) = 0$ at $(a, b)$ are parallel to the tangent lines to $g(x, y) = 0$ at $(0, 0)$, the result follows.     □

**Corollary 15.27.** *The number of tangent lines at a point of an algebraic curve $f(x, y) = 0$ of degree $d \geq 1$ is less than or equal to $d$.*

In Theorem 15.28 we obtain a partial converse of Theorem 15.26, by showing that each non-repeated linear factor of $\sigma$ is tangent to a unique branch of $f(x, y) = 0$. In the case of repeated linear factors the situation is more complicated. For the case degree $\sigma = 2$, we give conditions in Theorem 15.43 involving the third derivative of $f(x, y)$ at the singular point, which ensure that the two branches joined together give a simple cusp. However other types of cusps can occur, and not all repeated linear factors give rise to a (real) branch of the curve.

## 15.7 Branches for non-repeated linear factors

In §4.4, we showed that algebraic curves can be parametrised locally at non-singular points. We now show that, for each non-repeated linear factor of $\theta$, there is a corresponding ordinary branch passing through the singular point, which has the linear factor as its tangent there. We also show that the branch has a smooth local parametrisation.

* For the meaning of * see the preface.

**Theorem 15.28.** *Let the algebraic curve $f(x, y) = 0$ have a singular point at $(a, b)$. For each non-iterated real linear factor $q - cp$ or $p$ of $\theta(p, q)$ at $(a, b)$, there is a unique branch of the curve whose tangent is parallel to $(1, c)$ or $(0, 1)$ respectively, and therefore to $y - cx = 0$ or $x = 0$ respectively. Furthermore such a branch with tangent parallel to $y - cx$ has a regular smooth parametrisation by the $x$-coordinate near (on both sides of) $x = a$, and such a branch with tangent parallel to $x - dy$ has a regular smooth parametrisation by the $y$-coordinate near (on both sides of) $y = b$.*

*Proof*\*. We first prove the theorem in case the singular point is at the origin. We have

$$f(x, y) = (y - cx)\tau(x, y) + \sum_{i+j \geq m+1} a_{ij} x^i y^j,$$

where $\tau$ is homogeneous of degree $m - 1$. Substituting $(x, y) = (x, cx + xz)$, we obtain

$$f(x, cx + xz) = xz\tau(x, cx + xz) + \sum_{i+j \geq m+1} a_{ij} x^{i+j} (c + z)^j$$

$$= x^m z\tau(1, c + z) + \sum_{i+j \geq m+1} a_{ij} x^{i+j} (c + z)^j$$

$$= x^m g(x, z)$$

where

$$g(x, z) = z\tau(1, c + z) + \sum_{i+j \geq m+1} a_{ij} x^{i+j-m} (c + z)^j.$$

We have $g(0, 0) = 0$, and

$$\left. \frac{\partial g}{\partial z} \right|_{(0,0)} = \tau(1, c) \neq 0,$$

since the factor $y - cx$ is not repeated in $\theta(x, y)$. By Theorem 4.22, there is a smooth function $\phi : J = (-\epsilon, +\epsilon) \to \mathbf{R}$ such that $\phi(0) = 0$, and $g(x, \phi(x)) = 0$ for $x \in J$. Now

$$x \mapsto \mathbf{r}(x) = (x, cx + x\phi(x))$$

is a regular smooth local parametrisation of the branch of $f(x, y)$ at $(0, 0)$ having $y = cx$ as its tangent, since, at $(0, 0)$,

$$\mathbf{r}' = (1, c + \phi + x\phi') = (1, c) \neq 0.$$

The proof that the branch having $x = dy$ as tangent has a regular local parametrisation at $(0, 0)$ is similar. In case the singular point is at $(a, b)$, we translate the singular point to the origin, use the result proved above, then

---

\* For the meaning of \* see the preface.

use the reverse translation. A parametrisation of the branch corresponding to the factor $q - cp$, for example, is then

$$x \mapsto x - a \mapsto (a,b) + (x - a, c(x - a) + (x - a)\phi(x - a))$$
$$= (x, b + c(x - a) + (x - a)\phi(x - a)).$$

$\square$

**Example 15.29.** Let $f(x,y) = x^3 - y^3 + x^5y^5$, then

$$\sigma(x,y) = -(y - x)(x^2 + xy + y^2).$$

Thus the origin is a singular point with multiplicity 3, and the *unique* tangent line to the curve at the origin is the line $y - x = 0$ corresponding to the non-iterated factor $y - x$ of $\sigma$. The tangent line has intersection multiplicity 10 with the curve. All lines other than the tangent line have intersection multiplicity 3 with the curve. The curve has precisely one branch at the origin, and this ordinary branch can be smoothly parametrised by either one of $x$ and $y$.

### 15.7.1 Nodal singular points

We give a complete determination of the branches and tangent lines at nodal singular points.

**Definition 15.30.** A *nodal* singular point or *node* is a singular point at which $\theta$ has two or more real linear factors but has no repeated real linear factors. An *m–node* is a node of multiplicity $m \geq 2$.

By Theorem 15.28, passing through a nodal singular point there is a unique branch of the curve corresponding to each real linear factor of $\theta$.

At a singular point of an algebraic curve, the intersection multiplicity of a line can be $\geq 2$ without the line in question being a tangent line.

**Example 15.31.** The crunodal cubic $y^2 = x^3 + x^2$. Let $r(t) = t(p,q)$ be a line through $(0,0)$. We have $f(x,y) = y^2 - x^2 - x^3$ and

$$\gamma(t) = t^2(q^2 - p^2) - t^3 p^3.$$

Thus the only linear factors of $\theta$ are the two non-iterated linear factors $q - p$ and $q + p$. Therefore $(0,0)$ is a nodal singular point of multiplicity 2. There are precisely two ordinary branches at the origin with tangent lines $y = \pm x$, parallel to the two vectors $(p,q) = (1, \pm 1)$. These tangent lines have intersection multiplicity 3 with the curve at $(0,0)$, and all other lines have intersection multiplicity 2. Notice that, since

$$\mathbf{grad} f = (f_x, f_y) = (-3x^2 - 2x, 2y),$$

all points on the curve, other than $(0,0)$, are non-singular, and therefore have multiplicity 1. The point $(-\frac{2}{3}, 0)$ is not on the curve.

**Worked example 15.32.** Classify the singular point at the origin of the algebraic curve $x(x^2 - y^2) - x^6 + 3y^5$.

*Solution.* $\sigma = -x(y-x)(y+x)$. Thus there is a node of multiplicity 3 at the origin. There are precisely three ordinary branches with tangent lines $x = 0$, $y - x = 0$ and $y + x = 0$. The tangent lines have intersection multiplicity 5 with the curve. All other lines have intersection multiplicity 3. ☐

### Exercises

**15.12.** Show that the following curves each have a 2–node at the origin, and find the tangent lines there.

a) $x^3 + y^3 - 3axy = 0$ (folium of Descartes) $\qquad$ $[x = 0, y = 0]$
b) $y^2(a - x) - x^2(x + 3a) = 0$ (trisectrix of Maclaurin)
c) $y^2(a - x) - x^2(x + a) = 0$ (right strophoid) $\qquad$ $[y = \pm x]$

**15.13.** Show that for $b > a > 0$ the limaçon

$$(x^2 + y^2 - bx)^2 - a^2(x^2 + y^2) = 0$$

has a 2–node at the origin, and find the tangent lines there.
$$[ay = \pm\sqrt{b^2 - a^2}\, x]$$

**15.14.** Show that for $a > b > 0$ the conchoid of Nicomedes

$$(x^2 + y^2)(x - b)^2 - a^2x^2 = 0$$

has a 2–node at the origin, and find the tangent lines there.

**15.15.** Show that the only singular points of the following curves are 2–nodes and find them. In each case write down the tangent lines.

a) $(x^2 - 1)^2 - y^2(2y + 3) = 0$
$[(1,0), (0,-1), (-1,0); 4x - 4 = \pm\sqrt{3}y,$ isolated, $4x + 4 = \pm\sqrt{3}y]$
b) $x^2(x - 2)^2 + y^2(3 - 2y) - 1 = 0$
c) $x^2y^2 - 36x - 24y + 108 = 0$
$[(2,3); (6 \pm \sqrt{3})(x - 2) + 2(y - 3) = 0]$
d) $(x^2 - 1)^2y - (y^2 - 1)^2x = 0$
e) $(y - x - 1)^3 - 27xy = 0$ $\qquad$ $[(1,-1); 2x - 2 = (3 \pm \sqrt{5})(y + 1)]$
f) $x^2y^2 - 2(x + y) + 3 = 0$

**15.16.** Show that the folium of Kepler

$$(x^2 + y^2)(x(x + b) + y^2) - 4axy^2 = 0$$

has a 3–node at the origin if $4a > b > 0$ and only one ordinary branch if $b > 4a > 0$. Find the tangent lines.

**15.17.** Find an algebraic curve of degree 4 for which $x = 0$, $x + y = 0$ and $x + 2y = 0$ are tangent lines at the origin, but which contains none of these lines as part of the curve. $\qquad$ $[x(x + y)(x + 2y) + x^4 + y^4 = 0]$

**15.18.** Show that the algebraic curve with polar equation $r = \sin 3\theta$ is of degree 4 and has a 3–node at the origin.
(Multiply $r = \sin 3\theta = \text{im } e^{3i\theta} = \text{im } (\cos \theta + i \sin \theta)^3$ by $r^3$ to get the equation.)

**15.19.** Let $n \geq 3$ be odd. Show that the algebraic curve with polar equation $r = \sin n\theta$ is of degree $n + 1$ and has an $n$–node at the origin.
(Multiply $r = \sin n\theta = \text{im } \left(e^{i\theta}\right)^n$ by $r^n$ to get the equation. Then show that $y = 0$ is a tangent line, and note that the curve is unchanged under a rotation through $\dfrac{\pi}{n}$ about the origin.)

## 15.8 Branches for repeated linear factors

In the case where $\theta$ has a repeated real linear factor at a singular point of multiplicity $m \geq 2$, it is necessary to consider the terms of order $> m$ in order to determine the branches, if any, corresponding to this real linear factor. The determination of the form of the branches is less simple than when there are no repeated real linear factors. If the linear factor is squared, either there is no branch corresponding to it, or the branches corresponding to it may give a cusp or a tachnode (similar to a node but where the two branches have the same tangent at the singular point). If the linear factor is cubed, there may only be one branch. Later in this section we give conditions under which a linear factor which occurs twice in $\sigma$ corresponds to an ordinary cusp.

We first give some simple examples to illustrate the types of branches which may occur. The first example is cautionary.

**Example 15.33.** The curve $x^2 + y^4 = 0$ has an isolated point at the origin, and therefore no (real) tangent line there, even though $\sigma(x, y) = x^2$ is a product of real linear factors.

**Example 15.34.** The semi-cubical parabola $x^2 - y^3 = 0$ has a cusp at the origin. There is an 'iterated' tangent line $x = 0$ at the origin.

**Example 15.35.** The curve $x^2 - y^4 = 0$ consists of the two parabolae $x - y^2 = 0$ and $x + y^2 = 0$ which touch at their vertices. There is an 'iterated' tangent line $x = 0$ at the origin. The origin is a tachnode.

**Example 15.36.** The curve $x^3 - xy^4 = 0$ consists of the two parabolae $y^2 = \pm x$ which touch at the origin together with the tangent $x = 0$ line of the resulting tachnode. At the singular point there are three branches, each having $x = 0$ as tangent. However on the branch $x = 0$, $\rho = -xy^4$ is identically zero; this corresponds to a sense of 'infinite' point contact, since the tangent line is part of the algebraic curve.

**Example 15.37.** The curve $x^2 y - y^5 = 0$ consists of the two parabolae $y^2 = \pm x$ which touch at the origin, giving a tachnode, together with the line $y = 0$. At the singular point there are three branches, two having $x = 0$ as tangent, and the third having $y = 0$ as tangent.

**Example 15.38.** The curve

$$x^3y^3 - x^5 - y^5 + x^2y^2 = (x^2 - y^3)(y^2 - x^3) = 0$$

consists of the two semi-cubical parabolae $x^2 - y^3 = 0$ and $y^2 - x^3 = 0$. At the origin there are four branches comprising two cusps, which have the iterated tangent lines $x = 0$ and $y = 0$ respectively.

**Example 15.39.** The curve $x^3 - y^4 = 0$ has only one branch at the origin, since

$$x^3 - y^4 = (x - y^{\frac{4}{3}})(x - \omega y^{\frac{4}{3}})(x - \omega^2 y^{\frac{4}{3}})$$

where $\omega$ is a complex cube root of unity. The branch coincides with the curve $x - y^{\frac{4}{3}} = 0$, and can be parametrised by $y \mapsto \mathbf{r}(y) = (y^{\frac{4}{3}}, y)$. The curve has a tangent line at the origin. The tangent line is parallel to $\mathbf{r}'(0) = (0, 1)$. A smooth parametrisation of the curve is given by $t \mapsto (t^4, t^3)$.

**Example 15.40.** At the origin the curve $x^4 - y^5 = 0$ has precisely two branches $x = y^{\frac{5}{4}}$ and $x = -y^{\frac{5}{4}}$, each defined for $y \geq 0$, and each having $x = 0$ as tangent. A smooth parametrisation of the whole curve is given by $t \mapsto (t^5, t^4)$. The shape of the curve is similar to that of the cuspidal cubic $x^2 - y^3 = 0$.

At a tachnode there is an iterated tangent line, and thus $\theta$ will have repeated linear factors. In the next example we give an *ad hoc* method of determining whether a point with repeated linear factors is isolated.

**Example 15.41.** (Harder) Consider the algebraic curve

$$(x - y)^2(x^2 + 1) - 9x^2y^2 = 0.$$

For the line $t \mapsto t(p, q)$, we have

$$\gamma(t) = t^2(p - q)^2 + t^4(p^4 - 2qp^3 - 8p^2q^2)$$

for which $\theta$ has repeated linear factors. We consider the line $L_\epsilon$ given by $t \mapsto t(1, 1 + \epsilon)$; for small $\epsilon$ this is 'close' to the line $t \mapsto t(1, 1)$. For $L_\epsilon$ we have

$$\gamma_\epsilon = t^2\epsilon^2 + t^4(-9 - 18\epsilon - 8\epsilon^2).$$

Thus, for small $\epsilon$, $L_\epsilon$ intersects the algebraic curve where $t = 0$ or

$$t^2 = \frac{\epsilon^2}{9 + 18\epsilon + 8\epsilon^2}.$$

For small $\epsilon \neq 0$, these latter two values $t_\epsilon$ and $t_{-\epsilon}$ are not zero; indeed their approximate values are $\pm\frac{\epsilon}{3}$. Also $t_\epsilon, t_{-\epsilon} \to 0$ as $\epsilon \to 0$. It follows that there are points on the curve arbitrarily close to the origin other than the origin itself. Therefore the origin is not an isolated point and the tangent line $x = y$ has intersection mutiplicity 4 with the curve at the origin. The above analysis also shows that the curve behaves like $t \to t(1, 1 \pm 3t)$ close to the origin. The algebraic equation of the latter curve is that of the two parabolas $y = x \pm 3x^2$, and therefore the origin is a tachnode.

**Example 15.42.** The degenerate cubic $x^2 - x^3y = 0$ consists of the iterated line $x = 0$ and the rectangular hyperbola $xy = 1$. The factor $x^2$ occurs in the defining polynomial. For all points on the line $x = 0$ we have $\mathbf{grad} f = (2x - 3x^2y, -x^3) = 0$. This phenomena, of all points on a continuous part of a degenerate curve satisfying $\mathbf{grad} f = \mathbf{0}$, occurs more generally where the defining polynomial $f(x, y)$ has as a factor a squared term.

### 15.8.1 Factors of order two: ordinary cusps

We now show, where $m \geq 2$ and $\sigma$ has a linear factor which occurs precisely of degee 2, that the two branches corresponding to the linear factor form an ordinary cusp, unless a certain coefficient, which involves terms of degree $\geq m + 1$ in the defining polynomial, is zero. Also the cusp has a smooth local parametrisation. We first translate the singular point to the origin. The linear factor of $\sigma(x, y) = \theta(x, y)$ is either $y - cx$ or $x$ according to our convention in §15.2.

**Theorem 15.43.** *Let the algebraic curve $f(x, y) = 0$ have a singular point at the origin for which $m \geq 2$, let $y - cx$, respectively $x$, occur in $\sigma(x, y)$ as a factor precisely of degree 2, and let $\rho_{m+1}(1, c) \neq 0$, respectively*

$$\rho_{m+1}(0, 1) \neq 0.$$

*Then at the origin the curve has precisely two branches having $y - cx$, respectively $x$, as tangent. These two branches form an ordinary cusp having $y - cx = 0$, respectively $x = 0$, as cuspidal tangent. Furthermore the two branches together can be given a smooth local parametrisation near the cusp.*

*Proof\*.* We rotate the curve about the origin so that the repeated linear factor becomes $Y$ up to a constant multiple, where $(X, Y)$ are the coordinates after rotation. We then have

$$f(x, y) = F(X, Y) = Y^2 G(X, Y) + \rho_{m+1}(X, Y) + L(X, Y) \qquad (15.1)$$

where $G(X, Y)$ is a homogeneous polynomial of degree $m - 2$ which satisfies $G(1, 0) \neq 0$, $\rho_{m+1}(X, Y)$ is a homogeneous polynomial of degree $m+1$ which satisfies $\rho_{m+1}(1, 0) \neq 0$, and each term of $L(X, Y)$ has degree $\geq m + 2$. If a branch associated to the factor $Y$ exists then $Y = 0$ is tangent to it, and consequently $\dfrac{Y}{X}$ tends to 0 along the branch as $X$ tends to 0. Therefore we substitute $Y = uX$ in Equation 15.1, and divide by $X^m$ to obtain

$$g(X, u) = u^2 G(1, u) + X\rho_{m+1}(1, u) + X^2 \zeta(X, u) \qquad (15.2)$$

where $\zeta(X, u)$ is a polynomial in $X$ and $u$. Now

$$\frac{\partial g}{\partial X}(0, 0) = \rho_{m+1}(1, 0) \neq 0$$

---

\* For the meaning of \* see the preface.

and therefore, by Theorem 4.22, there is a smooth function

$$\phi : J = (-\epsilon, +\epsilon) \to \mathbf{R}$$

such that $\phi(0) = 0$, and $g(\phi(u), u) = 0$ for each $u$ in $J$. From the equation

$$0 = u^2 G(1, u) + X(\rho_{m+1}(1, u) + \phi(u)\zeta(\phi(u), u)),$$

we deduce that $X = \phi(u) = u^2 H(u)$ where $H$ is smooth near $u = 0$ and

$$H(0) = -\frac{G(1,0)}{\rho_{m+1}(1,0)} \neq 0.$$

Thus $X$ has constant sign for $u$ near 0 and we can substitute $X = H(0) t^2$, to obtain

$$t^2 = u^2 \frac{H(u)}{H(0)}.$$

This equation has the two solutions

$$t = \pm u \sqrt{\frac{H(u)}{H(0)}},$$

each defined near $u = 0$. However we want to obtain $u$ as a function of $t$. This prompts us to consider the algebraic curve

$$h(t, u) = g\left(H(0)t^2, u\right) = 0.$$

The terms of degree 2 in $h(t, u)$ are $G(1, 0)(u^2 - t^2)$ and therefore the curve $h(t, u) = 0$ has a nodal double point at the origin, where the tangents are $u = \pm t$. By Theorem 15.28 the branch, at this nodal double point, having $u = t$ as tangent can be parametrised locally near $t = 0$ by $u = tv(t)$ where $v(0) = 1$ and where $v$ is a smooth function. Replacing $t$ by $-t$ in $g\left(H(0) t^2, tv(t)\right) = 0$ we have $g\left(H(0) t^2, -tv(-t)\right) = 0$, and therefore the branch having $u = -t$ as tangent is $u = -tv(-t)$. Since $v(0) = 1$, we have $v(t) > 0$ for $t$ near 0. Therefore $F(X, Y) = 0$ has precisely two branches having $Y = 0$ as tangent, and they have the differentiable parametrisations $X \mapsto (X, Y)$, where

$$Y = X \left(\frac{X}{H(0)}\right)^{\frac{1}{2}} v\left(\left(\frac{X}{H(0)}\right)^{\frac{1}{2}}\right)$$

for one branch, and

$$Y = -X \left(\frac{X}{H(0)}\right)^{\frac{1}{2}} v\left(-\left(\frac{X}{H(0)}\right)^{\frac{1}{2}}\right)$$

for the other. In these parametrisations, $X$ takes only values $\geq 0$ or only values $\leq 0$ depending on the sign of $H(0)$. Alternatively, using a parameter $s$, the union of the two branches has the smooth parametrisation given near the origin of the $(X, Y)$–plane by

$$\mathbf{r}(s) = \begin{cases} H(0)\,(s^2, -s^3 v(s)) & (s \geq 0), \text{ and} \\ \\ H(0)\,(s^2, -s^3 v(s)) & (s < 0). \end{cases}$$

We have easily

$$\mathbf{r}'(0) = \mathbf{0},$$
$$\mathbf{r}''(0) = (H(0), 0), \quad \text{and}$$
$$\mathbf{r}'''(0) = (0, -6H(0)).$$

Since $\mathbf{r}''(0) \neq \mathbf{0}$ and $\mathbf{r}'''(0)$ is not parallel to $\mathbf{r}''(0)$, the singular point is an ordinary cusp. The parametrisation of the curve in the $(x, y)$–plane can now be obtained by applying the appropriate transformation. ☐

**Remark 15.44.** The condition of the hypotheses that $\rho_{m+1}(1, c) \neq 0$, respectively $\rho_{m+1}(0, 1) \neq 0$, is that $Y - cX$, respectively $X$, does not occur as a factor of $\rho_{m+1}(X, Y)$. A solution of the equation $\alpha x - \beta y = 0$ is $(\beta, \alpha)$.

**Mnemonic**

> $\alpha x - \beta y$ is a factor of $\rho_{m+1}$ if, and only if, $\rho_{m+1}(\beta, \alpha) = 0$.

**Worked example 15.45.** Show that

$$(3xy + 2x^3 + 1)^2 - 4(y + x^2)^3 = 0$$

has an ordinary cusp at $(1, -1)$.

*Solution.* Substituting $x = X + 1$ and $y = Y - 1$ gives a new defining polynomial for which $\sigma = 9(X - Y)^2$ and

$$\rho_3 = -4(2X + Y)^3 - 18(X - Y)X(2X + Y).$$

Since $\rho_3(1, 1) = -108 \neq 0$, there is, by Theorem 15.43, an ordinary cusp at $(X, Y) = (0, 0)$, and therefore at $(x, y) = (1, -1)$. ☐

**Example 15.46.** At the origin, the curve $x^2 y^2 + x^5 + y^5 = 0$ has four branches comprising two ordinary cusps with tangents $x = 0$ and $y = 0$. This follows immediately from Theorem 15.14, Theorem 15.26, and Theorem 15.43.

**Exercises**

**15.20.** Show that the only singular point of the semi-cubical parabola $y^2 - x^3 = 0$ is an ordinary cusp at the origin.

**15.21.** Show that the cissoid of Diocles $y^2(2a - x) - x^3 = 0 \quad (a > 0)$ has an ordinary cusp at the origin.

**15.22.** Show that the cardioid $(x^2 + y^2 - ax)^2 - a^2(x^2 + y^2) = 0$ $(a > 0)$ has an ordinary cusp at the origin.

**15.23.** Show that for $a = b > 0$ the conchoid of Nicomedes

$$(x^2 + y^2)(x - b)^2 - a^2 x^2 = 0$$

has an ordinary cusp at the origin.

**15.24.** Show that piriform curve (pear-shaped quartic)

$$a^4 y^2 - b^2 x^3 (2a - x) = 0 \qquad (a, b \neq 0)$$

has an ordinary cusp at the origin.

**15.25.** Show that the only singular point of

$$y(x + 3)^2 - 4(4x - 3y)(2x - 3y - 6) = 0$$

is an ordinary cusp at $(-3, -4)$.

**15.26.** Show that $4(x - 1)^3 + (y - 3x + 2)^2 = 0$ has an ordinary cusp at $(1, 1)$.

**15.27.** Show that $y^3(4-y)^3 - 4x^4(x+3)^2 = 0$ has ordinary cusps at $(-3, 0)$ and $(-3, 4)$.

**15.28.** Show that $x^3(x - 1)^2(x + 1)^2(x - 2) + y^2 = 0$ has singular points at $(-1, 0)$, $(0, 0)$, and $(1, 0)$. Determine their nature and find the tangent lines. Sketch the curve using $x$ as parameter.

[acnode, ordinary cusp, 2–node: $y = 0$, $y = \pm 2(x - 1)$]

**15.29.** Show that $(x^2 + y^2)^2 - 4axy^2 = 0$ (a folium of Kepler) has three branches at the origin; two of which form an ordinary cusp.

**Further reading.** In case $y - cx$, respectively $x$, occurs in $\sigma$ as a factor precisely of degree 2 and $\rho_{m+1}(1, c) = 0$, further analysis is needed. A fuller analysis of branches corresponding to squared linear factors is given in §6.46 of *The Elementary Differential Geometry of Plane Curves*, R. H. Fowler, Cambridge University Press 2nd edition 1929, reprinted by Stechert Hafner 1964. A technique which can be used more generally for determining all the branches at higher singular points is the method of Newton's polygon, though care needs to be taken with this method since it gives the complex branches as well as the real ones. We do not consider Newton's polygon here. Some branches at a singular point corresponding to iterated factors of $\theta(x, y)$ can be *parametrised* in the form $x = t^q, y = a_1 t^p + \ldots$ (Puiseux fractional power series for $y$). See, for example, the book by Semple and Kneebone: *Algebraic Curves*.

## 15.9 Cubic curves

We are now in a position to determine the types of branches which can occur at singular points of cubic curves. The only singular points of cubic curves which can occur are 2–nodes, acnodes, and ordinary cusps, except that certain degenerate cubics can have a tachnode or a 3–node. Furthermore a non-degenerate cubic curve can have at most one singular point, and a non-degenerate cubic which has a singular point can be parametrised by rational functions.

We say that a cubic curve is *non-degenerate* if the defining polynomial $f(x, y)$ does not factorise and the plane curve has points on it which are not isolated, otherwise the curve is *degenerate*.

**Example 15.47.** The cubic $(x - y - 1)(x^2 + 2y^2 - 2) = 0$ is degenerate. It is the union of the line $x - y - 1 = 0$ and the ellipse $x^2 + 2y^2 = 2$.

We assume that the singular point has been moved to the origin. Thus

$$f(x, y) = \sigma(x, y) + \rho(x, y), \quad \text{where}$$
$$\sigma(x, y) = a_2 x^2 + b_2 xy + c_2 y^2, \quad \text{and}$$
$$\rho(x, y) = a_3 x^3 + b_3 x^2 y + c_3 xy^2 + d_3 y^3.$$

First suppose that the singular point has multiplicity 2.

**Case 1: 2–node.** Let $\sigma$ be the product of two distinct linear factors, that is $\sigma(x, y) = c_2(y - \alpha x)(y - \beta x)$ or $\sigma(x, y) = b_2 x(y - \alpha x)$. By Theorem 15.28 the singular point is a 2–node with tangent lines $y - \alpha x = 0$ and $y - \beta x = 0$, or, respectively, $x = 0$ and $y - \alpha x = 0$. A degenerate cubic, whose defining polynomial factorises as $f(x, y) = g(x, y)h(x, y)$, where $g(x, y) = 0$ is the equation of a line which meets the conic $h(x, y) = 0$ in two distinct points has two such 2–nodes. The product of three distinct lines, each pair of which intersect giving three different points of intersection, has three 2–nodes.

**Case 2: acnode.** Let $\sigma$ be an irreducible real quadratic. By Theorem 15.18 the singular point is isolated.

**Case 3: cusps.** Let $\sigma$ have a repeated (real) linear factor. By first performing a rotation, if necessary, we can assume that $\sigma = c_2 y^2$. By Theorem 15.43 there is an ordinary cusp at the origin provided that $a_3 \neq 0$. In case $a_3 = 0$ and $b_3 \neq 0$, the curve is

$$y(c_2 y + b_3 x^2 + c_3 xy + d_3 y^2) = 0,$$

which is the union of the line $y = 0$ and the curve

$$c_2 y + b_3 x^2 + c_3 xy + d_3 y^2 = 0.$$

The second curve is a conic with tangent $y = 0$ at the origin, or is degenerate; in the former case the (degenerate) cubic has a tachnode at the singular point. In case $a_3 = b_3 = 0$ and either $c_3 \neq 0$ or $d_3 \neq 0$, the curve is

$$y^2(c_2 + c_3 x + d_3 y) = 0,$$

which consists of the iterated line $y^2 = 0$, together with a line not passing through the origin.

It remains to consider the case where the singular point has multiplicity 3. In this case

$$f(x,y) = \sigma(x,y) = a_3 x^3 + b_3 x^2 y + c_3 xy^2 + d_3 y^3.$$

The curve consists three lines through the origin in the case where $\sigma$ is the product of three real linear factors; two or all three of these lines may be coincident. In the case where $\sigma$ is the product of a real linear and a real irreducible quadratic factor, the (real) curve consists of one line through the origin.

Typical examples of nodal, acnodal and cuspidal cubics are given in Chapter 3.

**Theorem 15.48.** *A non-degenerate cubic has at most one singular point, and such a singular point is a 2–node, an acnode, or a cusp. A non-degenerate cubic with a singular point can be parametrised by a rational parametrisation.*

*Sketch of proof\*.* If the curve has a singular point, first translate the singular point to the origin. Since the cubic is non-degenerate, it does not contain a line. Consider the pencil of lines $L_t : y = tx$ through the singular point. A line can meet a (non-degerate) cubic curve in at most three points (see §15.1). The line $L_t$ has intersection multiplicity $\geq 2$ with the cubic at the origin, and therefore has intersection multiplicity at most 1 at a second point of intersection, which cannot therefore be a singular point. The same conclusion applies for the line $L_\infty : x = 0$. It follows that the cubic has at most one singular point. To obtain the parametrisation, we substitute $y = tx$. Solving the equation $f(x, tx) = 0$ after cancelling $x^2$ gives $x = R(t)$ where $R$ is a rational function of $t$. The cubic therefore has the rational parametrisation $t \mapsto (R(t), tR(t))$. One point of the curve, namely the point on the line $x = 0$ and not at the origin, will not correspond to a value of $t$, unless we rotate the curve before we substitute $y = tx$. We need to rotate the curve so that a real linear factor of $\rho(x, y)$ is replaced by a factor $\alpha x$. The plane curve will then only meet the line $x = 0$ at the origin. In case $\rho$ is the product of a real linear and a real irreducible quadratic factor, the parametrisation will then give a smooth function from the whole of $\mathbf{R}$ to the plane curve. This assigns one value of the parameter to each point of the curve except the double point, to which it assigns two values. In general the lines $\alpha x + \beta y = 0$, where $\alpha x + \beta y$ is a real linear factor of $\rho(x, y)$, will only meet the plane curve at the origin, but will meet the associated projective curve at a point at infinity. (See Chapter 16.) The values of $t$ corresponding to these lines will not therefore determine points of the plane curve. The values of $t$ for the two points at a 2–node correspond to the cases where $y = tx$ is a tangent line. The tangent lines are given by equating to zero the real linear factors of $\theta(1, t)$. The parameter value at a

---

\* For the meaning of \* see the preface.

cusp corresponds to the case where $y = tx$ is the cuspidal tangent line; this is given by equating to zero the linear factor of $\theta(1, t)$. An acnode does not lie on the parametrised acnodal curve. ☐

**Worked example 15.49.** Parametrise the cubic

$$f(x, y) = x^2 + y^2 + 6xy + (x^2 + y^2)(x + y) = 0.$$

*Solution.* Since

$$x^2 + y^2 + 6xy = (x + 3y)^2 - 8y^2 = (x + (3 - 2\sqrt{2})y)(x + (3 + 2\sqrt{2})y),$$

there is a 2-node at the origin. Putting $y = tx$, we obtain

$$f(x, tx) = x^2(1 + 6t + t^2) + x^3(1 + t^2)(1 + t) = 0,$$

and we obtain the parametrisation

$$(x, y) = \left( -\frac{(1 + 6t + t^2)}{(1 + t^2)(1 + t)}, -\frac{t(1 + 6t + t^2)}{(1 + t^2)(1 + t)} \right).$$

The line $x = 0$ meets the conic at the point $(0, -1)$. This point is not covered by the parametrisation, whereas $t = -1$ does not give a point on the plane curve, but gives the point at infinity of the associated projective curve (see Chapter 16). To obtain a parametrisation covering all the points of the cubic in the plane, we first rotate the curve. Let

$$x = \frac{1}{\sqrt{2}}(u - v), \quad \text{and}$$

$$y = \frac{1}{\sqrt{2}}(u + v),$$

or, equivalently,

$$u = \frac{1}{\sqrt{2}}(x + y), \quad \text{and}$$

$$v = \frac{1}{\sqrt{2}}(-x + y).$$

The rotation gives

$$4u^2 - 2v^2 + \sqrt{2}u(u^2 + v^2) = 0;$$

thus, the real linear factor of $\rho$ is now $u$. The line $u = 0$ meets the plane curve only at the origin. Substituting $v = su$ gives the parametrisation

$$(u, v) = \left( \frac{\sqrt{2}(s^2 - 2)}{s^2 + 1}, \frac{s\sqrt{2}(s^2 - 2)}{s^2 + 1} \right), \quad \text{or}$$

$$(x, y) = \left( \frac{\sqrt{2}(1 - s)(s^2 - 2)}{s^2 + 1}, \frac{s\sqrt{2}(1 + s)(s^2 - 2)}{s^2 + 1} \right).$$

Under this parametrisation each real number $s$ determines a point of the curve. Conversely each point of the curve corresponds to a unique real number $s$, except that a 2–node corresponds to two real numbers. □

## 15.10 Curvature at singular points*

We obtained a formula for the curvature of algebraic curves at non-singular points in §8.3. We now obtain a formula for curvature at a singular point of a branch corresponding to a non-iterated factor of $\theta(x, y)$. Again we assume that the singular point has been moved to the origin by a translation. Let

$$f(x,y) = \sigma(x,y) + \rho_{m+1}(x,y) + \sum_{i+j \geq m+2} a_{ij} x^i y^j$$

where $\sigma(x,y) = (y - cx)\tau(x,y)$ is homogeneous of degree $m \geq 2$, and $\rho_{m+1}(x,y)$ is homogeneous of degree $m + 1$.

**Theorem 15.50.** *Let the algebraic curve $f(x,y) = 0$ have a singular point at the origin, and let $y - cx$ be a non-iterated factor of the homogeneous polynomial $\sigma(x,y)$ consisting of terms of lowest degree $m \geq 2$ of $f$. Then the curvature at the origin of the branch of the curve which has $y - cx = 0$ as tangent is*

$$\kappa(0,0) = -\frac{2\rho_{m+1}(1,c)}{\tau(1,c)(1 + c^2)^{\frac{3}{2}}}.$$

*The sign of the curvature is that given using $x$ as parameter.*

*Proof.* From the proof of Theorem 15.28, the branch is parametrised by

$$\mathbf{r}(x) = (x, cx + x\phi(x)).$$

Therefore

$$\mathbf{r}' = (1, c + \phi + x\phi'),$$
$$\mathbf{r}'' = (0, 2\phi' + x\phi''),$$

and

$$g(x,z) = z\tau(1,c+z) + x\rho_{m+1}(1,c+z) + \sum_{i+j \geq m+2} a_{ij} x^{i+j-m}(c+z)^j.$$

Thus $\mathbf{r}'(0) = (1, c)$, $\mathbf{r}''(0) = (0, 2\phi'(0))$, and

$$\phi'(0) = -\frac{g_x(0,0)}{g_z(0,0)} = -\frac{\rho_{m+1}(1,c)}{\tau(1,c)}.$$

The result follows. □

A similar result holds where the tangent line is $x - dy = 0$.

* For the meaning of * see the preface.

**Worked example 15.51.** Find the curvature at the origin of the two branches of the crunodal cubic $y^2 = x^3 + x^2$.

*Solution.* We have $f = x^2 - y^2 + x^3$, $\sigma = x^2 - y^2$, and $\rho_{m+1} = x^3$. For the branch with tangent line $y - x = 0$, we have $c = 1$, $\tau = -(x + y)$, and

$$\kappa(0,0) = -\frac{2}{(-2)2^{\frac{3}{2}}} = \frac{1}{2\sqrt{2}}.$$

For the branch with tangent line $y + x = 0$, we have $c = -1$, $\tau = x - y$, and

$$\kappa(0,0) = -\frac{2}{2(2^{\frac{3}{2}})} = -\frac{1}{2\sqrt{2}}.$$

□

For the crunodal cubic, the curvature can be calculated by the parametric formula using the explicit parametrisation given in Chapter 3. For general cubic curves having a 2–node or a cusp, the curvature at the singular point can be calculated using a parametrisation as given in Theorem 15.48.

# 16

# Projective curves

Algebraic curves in the plane are often 'incomplete', in that certain 'points at infinity' are missing from them. We show here how plane algebraic curves can be extended to projective curves in the projective plane; any additional points obtained in the extension are the points at infinity for the original plane curve. The plane curve becomes an affine view of the projective curve. The asymptotes to the plane curve are tangents to the projective curve at the points at infinity. Different affine views of the projective curve can give further information about the original plane curve. A projective curve has many different affine views; for example, ellipses, hyperbolae, and parabolae are all affine views of the same projective curve. We give a classical definition of elements of a projective space as lines through the origin of the appropriate coordinate space. We show that points of the projective plane have a 'coordinate' representation, and use coordinates to obtain subsequent results and for solving examples. The asymptotes of the plane curve can be determined by studying the projective curve. Another application is to determine the boundedness of the plane curve.

## 16.1 The projective line

The projective curves we consider lie in the projective plane. The projective plane is given by adding certain points at infinity to the plane. We first consider the projective line, which is given by adding a point at infinity to the line.

**Definition 16.1.** The *projective line* $\mathbf{RP}^1$ is the set of lines in $\mathbf{R}^2$ which pass through the origin $(0,0)$. A *point* in $\mathbf{RP}^1$ is a line through the origin of $\mathbf{R}^2$.

We now show that the projective line identifies with (is bijective with) any line in $\mathbf{R}^2$, not through the origin, together with the point at infinity for that line. Consider the line $\Lambda : ax + by = d$ in $\mathbf{R}^2$ which does not pass through the origin, that is $d \neq 0$; for any line we have $(a, b) \neq \mathbf{0}$. Then any point $P$ on $\Lambda$ determines a unique element $L$ of $\mathbf{RP}^1$, where $L$ is the line through $P$ and $(0,0)$. Conversely any element $L$ of $\mathbf{RP}^1$, other than the line $\Lambda_0 : ax + by = 0$, determines a point $P$ of $\Lambda$. The line $\Lambda_0 : ax + by = 0$ is parallel to $\Lambda$. We thus have a bijection

$$\mathbf{RP}^1 \leftrightarrow \Lambda + \{\text{one extra point}\},$$

between $\mathbf{RP}^1$ and the union of the line $\Lambda$, which does not pass through the origin, and an extra point. The extra point corresponds to the line $\Lambda_0$ under the bijection. The extra point is called the *point at infinity* of $\Lambda$ (see Figure 16.1). The line $\Lambda$ is an *affine line*; it lies in the plane $\mathbf{R}^2$. Two lines $\Lambda : ax + by = d$ and $\Omega : px + qy = r$, which do not pass through the origin, will have different points at infinity (that is $ax + by = 0$ and $px + qy = 0$ are different lines) if they are not parallel, and will have the same point at infinity (that is $ax + by = 0$ and $px + qy = 0$ are the same line) if they are parallel.

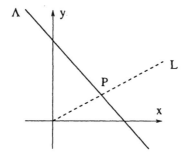

Figure 16.1 *The point at infinity of a line.*

**Remark 16.2.** Each element of $\mathbf{RP}^1$ determines two (opposite) points on the unit circle $x^2 + y^2 = 1$. Indeed $\mathbf{RP}^1$ is bijective with the set[†] obtained from the upper semi-circle $\{x^2 + y^2 = 1, y \geq 0\}$ after 'gluing together' the points $(1,0)$ and $(-1,0)$. See Figure 16.2.

## 16.2 The projective plane

The definition given above for the 'one-dimensional' projective line can be used as a model for defining projective spaces of 'higher dimensions'. We consider here the projective plane, which is two-dimensional projective space.

**Definition 16.3.** The *projective plane* $\mathbf{RP}^2$ is the set of lines in $\mathbf{R}^3$ which

---

[†] The projective line is topologically a circle, that is its shape can be deformed or stretched out to form a circle.

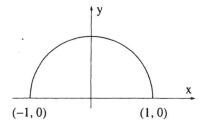

Figure 16.2 *The projective line as a circle.*

pass through the origin $(0,0,0)$. A point of $\mathbf{RP}^2$ is a line passing through the origin of $\mathbf{R}^3$.

We denote by $(x:y:z)$ the element of $\mathbf{RP}^2$ determined by the unique line through $(x,y,z)$ and $(0,0,0)$ where $(x,y,z) \neq 0$. Note that

$$(\lambda x : \lambda y : \lambda z) = (x:y:z)$$

for $(x,y,z) \neq 0$ and $\lambda \neq 0$. The algebraic equation of the line in $\mathbf{R}^3$ corresponding to the point $(a:b:c)$ of $\mathbf{RP}^2$ is

$$\frac{x}{a} = \frac{y}{b} = \frac{z}{c}.$$

This form of the equation allows up to two of $a,b,c$ to be zero; for example if $a = 0$ the line is given by $x = 0$ and $cy = bz$. The parametric equation of the line is $(x,y,z) = \lambda(a,b,c)$. There is no point of $\mathbf{RP}^2$ which corresponds to the symbol $(0:0:0)$.

**Mnemonic**

> The projective plane is the set of points $(x:y:z)$
> with $(\lambda x : \lambda y : \lambda z) = (x:y:z)$ and $(x,y,z) \neq (0,0,0)$

The coordinates $x, y, z$ are called *homogeneous coordinates*. The numbers $x, y, z$ are not themselves fixed for a point of the projective plane. The ratios between them are fixed for a given point.

### 16.2.1 Geometrical interpretation in terms of the unit sphere

Each element of $\mathbf{RP}^2$, that is each line through the origin, meets the unit sphere $x^2 + y^2 + z^2 = 1$ in $\mathbf{R}^3$ in two (opposite) points. Indeed $\mathbf{RP}^2$ can be considered as the space obtained from the unit sphere by identifying such pairs $(x,y,z)$ and $-(x,y,z)$ of opposite points to give the point

$$(x:y:z).$$

(Formally $\mathbf{RP}^2$ is the set of equivalence classes under the equivalence relation $(x,y,z) \sim -(x,y,z)$.) Similarly $\mathbf{RP}^2$ can be considered as the space obtained from the upper hemisphere $\{x^2 + y^2 + z^2 = 1, z \geq 0\}$ by identifying $(x, y, 0)$ with $-(x, y, 0)$ for all points on the circle $\{x^2 + y^2 = 1, z = 0\}$. The space obtained from this identification cannot be deformed to give a sphere[†], unlike the corresponding space given by identifying opposite points of the circle, which can be deformed to give a circle. Rather than choosing the upper hemisphere as above, we could instead choose the hemisphere on one side of any plane through the origin; identifying opposite points on the boundary circle would then give the projective plane. A different choice of axes for $\mathbf{R}^3$ will result in different homogeneous coordinates for the projective plane.

### 16.2.2 Points at infinity for an affine plane in $\mathbf{R}^3$

Consider now a plane $\Pi : ax + by + cz = d$ which does not pass through the origin of $\mathbf{R}^3$ (that is $d \neq 0$). For any plane we have $(a, b, c) \neq \mathbf{0}$. Each element $P$ of $\Pi$ determines a unique element $L$ of $\mathbf{RP}^2$ ($L$ is the line through $(0, 0, 0)$ and $P$). Conversely each element $L$ of $\mathbf{RP}^2$, other than those elements representing lines in the plane $\Pi_0 : ax + by + cz = 0$, determines a unique element $P$ of $\Pi$. Under this bijection, the affine plane $\Pi$ can be regarded as a subset of $\mathbf{RP}^2$. The points of $\mathbf{RP}^2$ not lying in this subset are the *points at infinity* for the affine plane $\Pi$. This set of points at infinity for the affine plane $\Pi : ax + by + cz = d$ is the projective line obtained from the great circle[‡]

$$\{x^2 + y^2 + z^2 = 1, ax + by + cz = 0\}$$

by identifying opposite points. We thus have a bijection

$$\boxed{\mathbf{RP}^2 \leftrightarrow \Pi + \text{ the projective line of points at infinity}}$$

between $\mathbf{RP}^2$ and the union of $\Pi$, which is a plane not through the origin, and the projective line of points of $\Pi$ at infinity.

**Mnemonic**

$$\boxed{\text{The set of points at infinity for an affine plane is a projective line.}}$$

---

[†] Topologically, the projective plane is a non-orientable closed surface and therefore cannot be represented in three-dimensional Euclidean space without self-intersections; models can be constructed with self-intersections. In four-dimensional Euclidean space, models of the projective plane can be defined without such self-intersections.

[‡] A great circle on a sphere is the circle of intersection of the sphere and any plane through its centre.

*16.2.3 Duality of lines and points*

In the plane $\mathbf{R}^2$, any two distinct points determine a unique line which passes through them. However it is not generally true that any two distinct lines determine a unique common point, since parallel lines in general have no point in common. In the projective plane any two distinct lines do have a unique common point, and therefore a duality exists in the 'geometry' of the projective plane between lines and points.

**Theorem 16.4.** *In the projective plane*
*i) two distinct lines meet in precisely one point, and*
*ii) there is a unique line through any two distinct points.*

*Proof.* We give a geometric proof. Projective lines are obtained from great circles on the sphere by identifying opposite points. Two distinct great circles intersect at two opposite points, which become identified to one point of the projective plane. Two distinct points of the sphere, which are not opposite, determine a unique great circle which passes through both, and this circle becomes identified as a projective line. The result follows. □

*16.2.4 Diagrammatic representation of the projective plane*

It will sometimes be convenient to represent the projective plane diagrammatically by a portion of the plane. We represent the axes $x = 0$, $y = 0$ and $z = 0$ by 'lines' in the plane (see Figure 16.3).

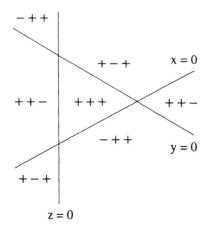

Figure 16.3 *The axes of the projective plane.*

The notation '$- + +$' for example means that the points in this region can be written $(x : y : z)$ where $x < 0$ and $y, z > 0$. The axes divide $\mathbf{RP}^2$ into four such regions. Notice that the regions determined by '$- + +$' and by '+ − −' are identical for example. The axes come from great circles, on the sphere $x^2 + y^2 + z^2 = 1$ in $\mathbf{R}^3$, given by intersection with the planes $x = 0$, $y = 0$ and $z = 0$: these great circles bound the 'first octant'

$$\{x \geq 0, y \geq 0, z \geq 0\}$$

of the sphere. The diagram should be thought of as lying within a closed curve, where 'roughly opposite' points of the closed curve are identified to give $\mathbf{RP}^2$. Care should be taken in drawing conclusions from this diagrammatic representation, particularly those relating to lines. For example there is no sense of 'parallel lines' in the projective plane; indeed, any pair of lines meet. In the affine plane given by $z = 1$, parallel lines meet on the line at infinity $z = 0$ in the projective plane. Notice in the diagram that crossing over the $x$–axis for example changes the sign of the $x$–coordinate.

## 16.3 Projective curves

Recall that a *homogeneous function of degree n* is a function $F : \mathbf{R}^3 \to \mathbf{R}$ which satisfies

$$F(\lambda x, \lambda y, \lambda z) = \lambda^n F(x, y, z)$$

for each real number $\lambda$ and for each $(x, y, z)$. We are especially interested in homogeneous polynomials of degree $n$. Such a homogeneous polynomial will have the form

$$F(x, y, z) = \sum a_{ij} x^i y^j z^{n-(i+j)}.$$

The *degree* of each monomial

$$a_{ij} x^i y^j z^{n-(i+j)} \text{ is } i + j + (n - (i + j)) = n.$$

**Definition 16.5.** A *projective curve* in $\mathbf{RP}^2$ of degree $n \geq 1$ is the set of points $(x : y : z)$ in $\mathbf{RP}^2$ for which $F(x, y, z) = 0$, where $F(x, y, z)$ is a homogeneous polynomial of degree $n$.

**Definition 16.6.** A *projective line* in $\mathbf{RP}^2$ is a projective curve of degree one.

This definition of 'projective line' coincides with the one defined geometrically above. The projective line $ax + by + cz = 0$, where $(x : y : z)$ lies in $\mathbf{RP}^2$, corresponds to the (geometrical) projective line given by identifying opposite points on the circle of intersection of the plane $ax + by + cz = 0$ and the unit sphere $x^2 + y^2 + z^2 = 1$ in $\mathbf{R}^3$. More generally the set of points in $\mathbf{R}^3$ satisfying $F(x, y, z) = 0$, for a homogeneous polynomial $F$, form a surface in $\mathbf{R}^3$. This surface is the union of the lines in a pencil of lines through the origin. Each of the lines in this pencil determines a point of $\mathbf{RP}^2$, and the set of such points in $\mathbf{RP}^2$ is the projective curve $F(x, y, z) = 0$ where $(x : y : z)$ lies in $\mathbf{RP}^2$.

Note the special case above of the projective line $ax + by + cz = 0$ of points at infinity for the affine plane $ax + by + cz = d$, considered as a subset of $\mathbf{RP}^2$.

In general a line meets a curve of degree $n \geq 2$ in the plane in $\leq n$ points. We now prove that a similar result holds in the projective plane.

**Theorem 16.7.** *A projective line meets a projective curve of degree $n$, which does not contain it, in at most $n$ distinct points.*

*Proof.* We give a proof in the case where $z = 0$ is the line; the proof in the general case then follows by choosing a rotation of $\mathbf{R}^3$ which maps the plane determined by the given line to the plane $z = 0$. Let the projective curve be $F(x, y, z) = 0$. Now $F(x, y, 0)$ is the sum of the terms of $f(x, y) = 0$ of highest degree. Therefore $F(x, y, 0)$ is not identically zero. Thus $F(x, y, 0)$ is a homogeneous polynomial of degree $n$ which factorises as

$$Ky^m \prod_{i=1}^{n-m} (x - c_i y)$$

where some $c_i$ may not be real and some real $c_i$ may be repeated (we are interested here only in real $c_i$). The only possible points of intersection are therefore the points $(c_i\!:\!1\!:\!0)$ and $(1\!:\!0\!:\!0)$. The result follows. $\square$

## 16.4 The projective curve determined by a plane curve

Given a plane algebraic curve of degree $n$ with equation $f(x, y) = 0$, we define a homogeneous polynomial $F(x, y, z)$ of degree $n$ which determines a projective curve $F(x, y, z) = 0$.

The homogeneous polynomial $F$ is formally determined by

$$F(x, y, z) = z^n f\left(\frac{x}{z}, \frac{y}{z}\right).$$

The easiest way to find $F(x, y, z)$ is to determine the degree $n$ of $f$, then to multiply each monomial in $f$ by a suitable power of $z$ in order to obtain, in each case, a monomial of degree $n$.

**Definition 16.8.** The projective curve *determined by* or *associated to* the plane curve $f(x, y) = 0$ in $\mathbf{R}^2$ is $F(x, y, z) = 0$ in $\mathbf{RP}^2$.

**Mnemonic**

> A projective curve has elements $(x\!:\!y\!:\!z)$ in $\mathbf{RP}^2$.
> Affine curves in $\mathbf{R}^3$ have elements $(x, y, z)$ in $\mathbf{R}^3$.
> A plane curve has elements $(x, y)$ in $\mathbf{R}^2$.

**Worked example 16.9.** Find the equation of the projective curve determined by the plane curve

$$y^2 = x^3 + x.$$

*Solution.* The plane curve has degree 3, therefore we must multiply each monomial by a suitable power of $z$ to give a homogeneous polynomial of degree 3. The projective curve therefore is $y^2 z = x^3 + xz^2$.                □

## 16.5 Affine views of a projective curve

Let the projective curve $\Gamma$ in $\mathbf{RP}^2$ be given by $F(x, y, z) = 0$; thus, the points of $\Gamma$ are

$$\{(x{:}y{:}z) : F(x, y, z) = 0\}.$$

We consider the affine plane in $\mathbf{R}^3$ given by $\{(x, y, z) : ax + by + cz = d\}$ where $d \neq 0$. In this plane in $\mathbf{R}^3$, the projective curve $\Gamma$ determines the affine view $\Pi$ given by

$$\{(x, y, z) : F(x, y, z) = 0,\ ax + by + cz = d\} :$$

the affine curve $\Pi$ is the intersection in $\mathbf{R}^3$ of the affine plane $ax + by + cz = d$ and the surface $\{(x, y, z) : F(x, y, z) = 0\}$. The projective curve can be regarded as the union of the affine curve and the set of points at infinity for the affine curve. The set of points at infinity for the affine curve in the plane $ax + by + cz = d$ is

$$\{(x{:}y{:}z) : ax + by + cz = 0,\ F(x, y, z) = 0\}.$$

**Definition 16.10.** The *affine view* (or *affine section*) of the projective curve

$$\{(x{:}y{:}z) : F(x, y, z) = 0\}$$

for the plane $ax + by + cz = d$ $(d \neq 0)$ is the affine curve in $\mathbf{R}^3$,

$$\{(x, y, z) : F(x, y, z) = 0,\ ax + by + cz = d\}.$$

Notice that the projective curve lies in $\mathbf{RP}^2$ and the affine view lies in the plane $ax + by + cz = d$ in $\mathbf{R}^3$. This plane does not pass through the origin of $\mathbf{R}^3$.

**Mnemonic**

> Projective curve ↔ affine view + points at infinity for affine view.

**Worked example 16.11.** Find the affine view for $y + z = 1$ of the projective curve

$$x^3 - y^3 - y^2 z = 0.$$

*Solution.* Substituting $z = 1 - y$ gives $x^3 = y^2$. The equation $x^3 = y^2$ in $\mathbf{R}^3$ gives the cylinder passing through the curve $x^3 = y^2$ in the $xy$–plane and with generators parallel to the $z$–axis. The affine view is the intersection of this cylinder with the plane $y + z = 1$. It has a cusp at $(0, 0, 1)$.                □

## 16.6 Plane curves as views of a projective curve

An important case of affine views is where the affine plane is taken to be $z = 1$. The affine view in $\mathbf{R}^3$ is then

$$\{(x, y, z) : z = 1, F(x, y, 1) = 0\}$$

in the plane $z = 1$. We may identify this with the plane curve

$$F(x, y, 1) = 0$$

in $\mathbf{R}^2$. Notice that this plane curve is not given by putting $z = 0$ in $F(x, y, z) = 0$. The points at infinity for the plane curve are then

$$\{(x : y : z) : z = 0, F(x, y, 0) = 0\}$$

in $\mathbf{RP}^2$.

**Theorem 16.12.** *Let $F(x, y, z) = 0$ be the projective curve determined by the plane curve $f(x, y) = 0$, then the affine view for $z = 1$ of the projective curve $F(x, y, z) = 0$ is $f(x, y) = 0$. Conversely let $f(x, y) = 0$ be the affine view for $z = 1$ of a projective curve $F(x, y, z) = 0$, for which $z$ is not a factor of $F(x, y, z)$, then the projective curve determined by $f(x, y) = 0$ is $F(x, y, z) = 0$.*

*Proof.* We assume that the curves are of degree $n \geq 1$.

i) We have formally

$$F(x, y, z) = z^n f\left(\frac{x}{z}, \frac{y}{z}\right).$$

The affine view for $z = 1$ is

$$F(x, y, 1) = f(x, y).$$

ii) We have $f(x, y) = F(x, y, 1)$. This determines the projective curve

$$z^n f\left(\frac{x}{z}, \frac{y}{z}\right) = z^n F(\frac{x}{z}, \frac{y}{z}, 1) = F(x, y, z).$$

$\square$

The affine views for the planes $x = 1$ and $y = 1$ can similarly be identified with plane curves in $\mathbf{R}^2 = \{(y, z)\}$ and $\mathbf{R}^2 = \{(x, z)\}$ respectively, and a result similar to Theorem 16.12 holds.

**Worked example 16.13.** Determine the affine views and their points at infinity of the projective curve $y^2 z = x^3$ for the planes $x = 1$, $y = 1$, and $z = 1$.

*Solution.* To find the affine view for the plane $y = 1$, we put $y = 1$ in the equation $y^2 z = x^3$. The affine view is $z = x^3$. To find the point at infinity we put $y = 0$ in the equation $y^2 z = x^3$; the plane curve therefore has just one point $(0 : 0 : 1)$ at infinity. The affine view for $z = 1$ is the semi-cubical parabola $y^2 = x^3$, which also has just one point $(0 : 1 : 0)$ at infinity. The affine view for $x = 1$ is $y^2 z = 1$, which has two points $(0:1:0)$ and $(0:0:1)$ at infinity. Recall that these points at infinity are in $\mathbf{RP}^2$.                        □

We now give an example involving conic sections.

**Worked example 16.14.** Show that ellipses, hyperbolae, and parabolae can occur as affine views of the projective curve given by $F = x^2 + y^2 - z^2$.

*Solution.* The homogeneous polynomial $F = x^2 + y^2 - z^2$ determines the right circular cone

$$\{(x, y, z) \in \mathbf{R}^3 : x^2 + y^2 - z^2 = 0\},$$

which is a surface in $\mathbf{R}^3$ (see Figure 16.4), and determines the projective curve

$$\Gamma = \{(x:y:z) : x^2 + y^2 - z^2 = 0\}$$

in $\mathbf{RP}^2$. The affine view of this projective curve depends on the choice of the affine plane $ax + by + cz = d$. We note three of the views which can occur.

i)   The affine view for $z = 1$. In this case the line at infinity is $z = 0$, and the affine view is $x^2 + y^2 = 1$, which is an ellipse. There are no points at infinity.

ii)   The affine view for $x = 1$. In this case the line at infinity is $x = 0$, and the affine view is $z^2 - y^2 = 1$, which is a hyperbola. There are two points $(0:1:\pm 1)$ at infinity.

iii)   The affine view for $y + z = 1$. In this case the line at infinity is $y + z = 0$, and the affine view is given by

$$x^2 + y^2 - (1 - y)^2 = 0 \quad \text{and} \quad y + z = 1,$$

that is

$$x^2 = 1 - 2y = -2 \left( y - \tfrac{1}{2} \right) \quad \text{and} \quad y + z = 1.$$

This is a parabola in the plane $y + z = 1$. It is the intersection of the parabolic cylinder $x^2 = -2 \left( y - \tfrac{1}{2} \right)$ in $\mathbf{R}^3$ with the plane $y + z = 1$. To find the points at infinity, we put $y + z = 0$. This gives $x = 0$, $y + z = 0$; therefore there is precisely one point $(0:1:-1)$ at infinity. (Compare Theorem 16.34 below.)                        □

**Mnemonic**

> Ellipses, hyperbolae, and parabolae are
> affine views of the same projective curve.

The set of points at infinity for a projective curve depends on the view chosen. Different views will give different sets of points at infinity, as illustrated in the above examples.

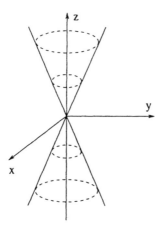

Figure 16.4 *A right circular cone in 3-space.*

## 16.7 Tangent lines to projective curves

The tangent lines to a plane algebraic curve $f(x, y) = 0$ are most readily defined at non-singular points, that is at points at which **grad** $f \neq 0$. We consider the analogous situation for projective curves.

**Definition 16.15.** A *non-singular point* of the projective curve

$$F(x, y, z) = 0$$

is a point on the curve at which

$$\mathbf{grad}\, F = \left( \frac{\partial F}{\partial x}, \frac{\partial F}{\partial x}, \frac{\partial F}{\partial x} \right) \neq 0.$$

A *singular point* is a point at which

$$\mathbf{grad}\, F = \left( \frac{\partial F}{\partial x}, \frac{\partial F}{\partial x}, \frac{\partial F}{\partial x} \right) = 0.$$

The singular points of a plane curve correspond to singular points of the associated projective curve as in the following theorem.

**Theorem 16.16.** $(x, y)$ *is a singular point of the plane curve* $f(x, y) = 0$ *if, and only if,* $(x : y : 1)$ *is a singular point of the associated projective curve* $F(x, y, z) = 0$.

*Proof.* Differentiating $F(x, y, z) = z^n f\left(\dfrac{x}{z}, \dfrac{y}{z}\right)$ we have

$$F_x(x, y, z) = z^n \frac{1}{z} f_x\left(\frac{x}{z}, \frac{y}{z}\right),$$

$$F_y(x, y, z) = z^n \frac{1}{z} f_y\left(\frac{x}{z}, \frac{y}{z}\right), \quad \text{and}$$

$$F_z(x, y, z) = nz^{n-1} f\left(\frac{x}{z}, \frac{y}{z}\right) +$$

$$z^n\left(-\frac{x}{z^2}\right) f_x\left(\frac{x}{z}, \frac{y}{z}\right) + z^n\left(-\frac{y}{z^2}\right) f_y\left(\frac{x}{z}, \frac{y}{z}\right).$$

Thus, for a point $(x, y)$ on the curve $f(x, y) = 0$, we have

$$F_x(x, y, 1) = F_y(x, y, 1) = F_z(x, y, 1) = 0$$

if, and only if, $f_x(x, y) = f_y(x, y) = 0$. $\qquad\square$

Similar results apply for the affine views given by $x = 1$ and $y = 1$, and more generally for the affine views given by $ax + by + cz = 1$.

We now determine the tangent lines at non-singular points of a projective curve. The tangent line of a projective curve is a projective line.

**Theorem 16.17.** *Let* $(x_0, y_0, z_0)$ *be a non-singular point of the projective curve* $F(x, y, z) = 0$, *then the tangent line (a projective line in* $\mathbf{RP}^2$ *) at* $(x_0, y_0, z_0)$ *is*

$$\boxed{\mathbf{r} \cdot \mathbf{grad}\, F = x F_x(x_0, y_0, z_0) + y F_y(x_0, y_0, z_0) + z F_z(x_0, y_0, z_0) = 0.}$$

*Proof\*.* We give a geometrical proof. The tangent plane to the surface $F(x, y, z) = 0$ in $\mathbf{R}^3$ at $(x_0, y_0, z_0)$ is

$$(x - x_0)F_x + (y - y_0)F_y + (z - z_0)F_z = 0$$

where

$$F_x = \frac{\partial F}{\partial x}, \quad F_y = \frac{\partial F}{\partial y} \quad \text{and} \quad F_z = \frac{\partial F}{\partial z}$$

at $(x_0, y_0, z_0)$. This equation is of the form

$$x F_x + y F_y + z F_z = \text{constant}.$$

Since this plane must pass through the origin (the surface is given by a pencil of lines through the origin), it has the form

$$x F_x + y F_y + z F_z = 0.$$

$\qquad\square$

> The partial derivatives $F_x$, $F_y$, and $F_z$ must be evaluated at the fixed point $(x_0, y_0, z_0)$.

\* For the meaning of \* see the preface.

**Theorem 16.18.** *The associated projective line of the tangent line to the plane curve* $f(x,y) = 0$ *at a non-singular point* $(x_0, y_0)$ *is the tangent line to the associated projective curve at* $(x_0 : y_0 : 1)$.

*Proof.* The tangent line at $(x_0, y_0)$ is

$$(x - x_0) f_x(x_0, y_0) + (y - y_0) f_y(x_0, y_0) = 0.$$

The associated projective line is

$$(x - x_0 z) f_x(x_0, y_0) + (y - y_0 z) f_y(x_0, y_0) = 0.$$

The tangent line to the projective curve at $(x_0 : y_0 : 1)$ is

$$x F_x(x_0 : y_0 : 1) + y F_y(x_0 : y_0 : 1) + z F_z(x_0 : y_0 : 1) = 0,$$

that is, as in the proof of Theorem 16.16,

$$x f_x(x_0, y_0) + y f_y(x_0, y_0) + z(n f(x_0, y_0) - x_0 f_x(x_0, y_0) - y_0 f_y(x_0, y_0)) = 0.$$

The result follows since $(x_0, y_0)$ lies on the curve $f(x, y) = 0$. □

The tangent line $x F_x + y F_y + z F_z = 0$ to the projective curve

$$F(x, y, z) = 0$$

at $(x_0 : y_0 : 1)$ determines the tangent line $x F_x + y F_y + F_z = 0$ at $(x_0, y_0, 1)$ to the affine view given by $z = 1$; that is the tangent line $x F_x + y F_y + F_z = 0$ at $(x_0, y_0)$ in the case of the corresponding plane curve. We consider later the tangent lines to the projective curve at the points at infinity for the plane curve.

**Worked example 16.19.** Find the singular points of the projective curve

$$x^2 z = y^3$$

and determine the tangent lines at the non-singular points $(x_0 : y_0 : z_0)$.

*Solution.* $F_x = 2xz, F_y = -3y^2, F_z = x^2$, thus $\mathbf{grad}\, F = 0$ if, and only if, $x = y = 0$. There is exactly one singular point at $(0:0:1)$. The tangent at a non-singular point $(x_0 : y_0 : z_0)$ is

$$2 x_0 z_0 x - 3 y_0^2 y + x_0^2 z = 0.$$

□

## 16.8 Boundedness of the associated affine curve

One of the important applications of projective geometry to the study of plane curves is to determine boundedness.

**Definition 16.20.** A curve in $\mathbf{R}^2$ or $\mathbf{R}^3$ is *bounded* if there is a number $K$ such that $|\mathbf{r}| \leq K$ for each point $\mathbf{r}$ of the curve.

**Theorem 16.21.** *Let a given affine or plane curve have no points at infinity, then the plane curve is bounded (as a curve in* $\mathbf{R}^3$ *or* $\mathbf{R}^2$).

*Sketch of proof\**. Our proof uses similar techniques to the proof of Theorem 15.18. Given a polynomial $f(x,y)$ of degree $N$, we may write it in the form $f(x,y) = g(x,y) + h(x,y)$ where $g(x,y)$ is the homogeneous polynomial consisting of all monomials of $f(x,y)$ of degree $N$, and $h(x,y)$ has degree $\leq N-1$. We construct the homogeneous polynomial $F(x,y,z)$ associated with $f(x,y)$ as above. Clearly $F(x,y,0) = g(x,y)$, and, since there are no points at infinity, the equation $F(x,y,0) = 0$ has no non-trivial solution, that is $g(x,y)$ has no real linear factors. Therefore $g(x,y)$ is a product of irreducible quadratic homogeneous polynomials and consequently has even degree $N = 2n$. As in the proof of Theorem 15.18, there is a $K > 0$ such that

$$| \, g(x,y) \, | \geq K(x^2 + y^2)^n$$

for all real $x$ and $y$. On the other hand if $\phi(x,y)$ is a homogeneous polynomial of degree $r$, there is an $L > 0$ such that

$$| \, \phi(x,y) \, | \leq L(x^2 + y^2)^{r/2}$$

for all real $x$ and $y$ (we use the inequalities

$$|x| \leq (x^2 + y^2)^{1/2} \quad \text{and} \quad |y| \leq (x^2 + y^2)^{1/2}).$$

Now $h(x,y)$ is a sum of finitely many homogeneous polynomials each of degree $\leq 2n - 1$, therefore there is an $M > 0$ such that

$$| \, h(x,y) \, | \leq M(x^2 + y^2)^{(2n-1)/2}$$

for all real $x$ and $y$ satisfying $(x^2 + y^2) \geq 1$. Thus we have

$$|f(x,y)| = |g(x,y) + h(x,y)| \geq (x^2 + y^2)^{(2n-1)/2}(K(x^2 + y^2)^{1/2} - M)) > 0$$

for all real $x$ and $y$ with

$$x^2 + y^2 > \frac{M^2}{K^2}.$$

Therefore $f(x,y) = 0$ has no solutions for

$$x^2 + y^2 > \max\left\{1, \frac{M^2}{K^2}\right\},$$

which proves that the curve $f(x,y) = 0$ is bounded. $\qquad\square$

**Worked example 16.22.** Show that the affine view for $z = 1$ of the projective curve $(x^2 + y^2)^2 - (x^2 - y^2)z^2 = 0$ is bounded.

*Solution.* Substituting $z = 0$ into $(x^2 + y^2)^2 - (x^2 - y^2)z^2 = 0$, we obtain $x = y = z = 0$. Therefore the affine view for $z = 1$ has no points at infinity $((0{:}0{:}0)$ does not represent a point in $\mathbf{RP}^2)$. Thus the affine curve

$$(x^2 + y^2)^2 = x^2 - y^2$$

is bounded. $\qquad\square$

---

\* For the meaning of \* see the preface.

The converse of Theorem 16.21 is not true. The existence of points at infinity does not imply that the affine curve is not bounded. The points at infinity could be isolated points of the projective curve. More generally it can be shown that an affine or plane curve is bounded if, and only if, the only points at infinity are isolated.

**Worked example 16.23.** Find the singular points of the projective curve

$$y^2 z^2 = x^2 z^2 - x^4.$$

Show that the affine view for $z = 1$ has a point at infinity but is bounded.

*Solution.* The affine view for $z = 1$ is the figure–8 curve

$$y^2 = x^2(1 - x^2).$$

It has singular points at $(0:0:1)$ and $(0:1:0)$. Now $(1 - x^2) \geq 0$ and

$$\frac{d}{dx}(x^2(1 - x^2)) = 2x(1 - 2x^2).$$

Thus $y$ is zero at $x = 0$ and $x = \pm 1$ and has critical points at

$$x = \pm \frac{1}{\sqrt{2}}.$$

It follows that the affine view lies in the rectangle

$$\{-1 \leq x \leq 1, -\tfrac{1}{2} \leq y \leq \tfrac{1}{2}\}.$$

However this affine view has an isolated singular point $(0:1:0)$ at infinity. Notice that for the affine view given by $y = 1$ this singular point of the projective curve becomes the origin of the $(x, z)$–plane, and the quadratic terms of the equation have an iterated linear factor which does not correspond to a (real) tangent at the origin. □

## 16.9 Summary of the analytic viewpoint

1. The projective plane $\mathbf{RP}^2$ is the set of triples $(x:y:z)$ where
   i) $(x, y, z) \neq 0$ that is $x, y, z$ are not all zero, and
   ii) $(\lambda x : \lambda y : \lambda z) = (x:y:z)$ for each $\lambda \neq 0$.
   For $(x, y, z) \neq 0$, the set $\{\lambda(x, y, z)\}$ for varying $\lambda$, including $(0, 0, 0)$ given by $\lambda = 0$, is a line in $\mathbf{R}^3$ through $(x, y, z)$. Excluding $(0, 0, 0)$ from the line determines a point of $\mathbf{RP}^2$ in this analytic sense by ii.

2. A curve of degree $n$ in $\mathbf{RP}^2$ is the set of points of $\mathbf{RP}^2$ satisfying $F(x, y, z) = 0$, where $F$ is a homogeneous polynomial of degree $n \geq 1$.

3. A plane $ax + by + cz = d$ in $\mathbf{R}^3$, for which $d \neq 0$, determines (bijectively) an affine plane regarded as a subset of $\mathbf{RP}^2$; and the set of points of $\mathbf{RP}^2$, which are not on this plane, constitute the (projective) line at infinity of the affine plane. The projective line given by the equation $ax + by + cz = 0$

in $\mathbf{RP}^2$ is the line is the line at infinity of the affine plane $ax + by + cz = d$.

4. The affine plane $ax + by + cz = d$ $(d \neq 0)$ intersects a projective curve of degree $n$ in an algebraic curve of degree $\leq n$.

5. Often the affine plane considered is $z = 1$ and the projective line of points at infinity is then precisely the set $\{(x:y:0)\}$, where $x$, $y$ are not both zero. Each point in this latter set can be written $(x:1:0)$ except the point $(1:0:0)$.

## 16.10 Asymptotes

The tangent lines to a projective curve, excluding tangent lines at points at infinity, correspond bijectively with tangent lines to the affine or plane curve. We now consider tangent lines to the projective curve at the points at infinity (see Figure 16.5).

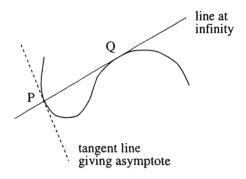

Figure 16.5 *Tangents at points at infinity.*

**Definition 16.24.** An *asymptote* of a plane curve $\Gamma$ is a tangent to the projective curve determined by it at a point at infinity, which tangent is not the line at infinity.

**Theorem 16.25.** *A plane or affine curve of degree $n$ has at most $n$ distinct points at infinity.*

*Proof.* The points at infinity are the points of intersection of the associated projective curve, which has the same degree, and the line at infinity. The result follows from Theorem 16.7.                                                    □

**Example 16.26.** $\prod_{r=1}^{n}(x - ry) + 1 = 0$ has $n$ distinct points $(r:1:0)$ at infinity. For $n = 2$ this is a hyperbola.

For points at infinity on a projective curve there are two cases to consider as follows.

1.   $P$ is a point at infinity and at least one tangent at $P$ is not the line at infinity. In this case each tangent, at $P$, which is not the line at infinity will determine an asymptote.

2.   $P$ is a point at infinity and the unique tangent at $P$ is the line at infinity. In this case the point at infinity does not give rise to an asymptote.

### 16.10.1 Procedure for determining the asymptotes

We demonstrate the method for finding the asymptotes to a plane curve

$$f(x,y) = 0.$$

First find the points at infinity on the related projective curve

$$F(x,y,z) = z^n f\left(\frac{x}{z}, \frac{y}{z}\right) = 0.$$

Find the tangents at these points: these tangents will be projective lines. For those tangents which are not the line at infinity, determine the related affine line in the plane $z = 1$. The asymptotes to the plane curve will have the same equations as these affine lines.

**Worked example 16.27.** Find the points at infinity and the asymptotes of the hyperbola

$$\frac{x^2}{a^2} - \frac{y^2}{b^2} = 1.$$

*Solution.* The plane curve is the affine view for $z = 1$ of the projective curve

$$F(x,y,z) = \frac{x^2}{a^2} - \frac{y^2}{b^2} - z^2 = 0.$$

The points at infinity for the plane curve are given by $z = 0$ and

$$\frac{x^2}{a^2} - \frac{y^2}{b^2} = 0 :$$

the points at infinity are therefore $(a:b:0)$ and $(a:-b:0)$. The tangent line to the projective curve at $(p,q,r)$ is

$$\frac{2p}{a^2}x - \frac{2q}{b^2}y - 2rz = 0.$$

For the two points at infinity the tangents are the projective lines

$$\frac{x}{a} = \pm\frac{y}{b}.$$

Putting $z = 1$, we now obtain the plane lines

$$\frac{x}{a} = \pm\frac{y}{b}$$

in $\mathbf{R}^2$; these are the asymptotes.   □

In the above example the equations for the projective lines and plane lines corresponding to the asymptotes are the same. This will not happen in general since substituting $z = 1$ will usually give a constant term.

**Example 16.28.** In Example 16.14 we considered the affine views of

$$x^2 + y^2 - z^2 = 0.$$

We now investigate the asymptotes of these views.

  i)  The affine view for $z = 1$. In this case the line at infinity is $z = 0$. There are no points at infinity and therefore no asymptotes.

  ii)  The affine view for $x = 1$. In this case the line at infinity is $x = 0$. There are two points $(0:1:\pm1)$ at infinity. The tangents to the projective curve at these two points are $y \mp z = 0$. The asymptotes to the plane curve, given by substituting $x = 1$, are $y \mp z = 0$.

  iii)  The affine view for $y + z = 1$. In this case the line at infinity is $y + z = 0$. There is precisely one point $(0:1:-1)$ at infinity. The tangent to the projective curve at this point is $y + z = 0$, which is the line at infinity. Therefore the affine curve has no asymptote.

Recall that the three homogeneous-coordinate 'axes' $x = 0$, $y = 0$ and $z = 0$ are determined by great-circles on the sphere. The diagrams in Figure 16.6 demonstrate what is happening.

## 16.11 Singular points and inflexions of projective curves

A singular point of a projective curve $F(x, y, z) = 0$ of degree $n$ is a point at which

$$\mathbf{grad}\, F = (F_x, F_y, F_z) = 0.$$

By Euler's theorem (see §4.7.6),

$$nF = xF_x + yF_y + zF_z$$

and therefore any point $(x:y:z)$, which satisfies $F_x = F_y = F_z = 0$, lies on the curve. By Theorem 16.16 above, a point in an affine view is a singular point if, and only if, the corresponding point of the projective curve is a singular point.

**Definition 16.29.** A singular point of a projective curve is *isolated, nodal,* or a *cusp* according as the associated point (not at infinity) in an affine view is isolated, nodal, or a cusp.

It can be shown that the above definition is independent of the view chosen.

**Definition 16.30.** A *singular point at infinity* of a plane algebraic curve $f(x, y) = 0$, is a point at infinity which is also a singular point of the associated projective curve $F(x, y, z) = 0$ defined by

$$F(x, y, z) = z^n f\left(\frac{x}{z}, \frac{y}{z}\right).$$

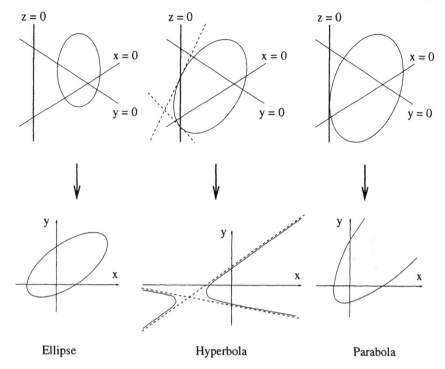

Figure 16.6 *Asymptotes of second order projective curves.*

In order to classify a singular point at infinity for a plane or affine curve, we can take a different affine view of the associated projective curve for which the singular point which we are considering is not at infinity, and classify it (cusp, nodal, isolated) in the usual way. It can be shown that if a singular point is a particular type in one affine view (for example ordinary cusp, 3–node, etc.), then it is of the same type in other affine views.

The tangent lines at singular points of a projective curve can be found as follows. First we take an affine view (usually given by one of $x = 1$, $y = 1$ or $z = 1$), for which the singular point is not the point at infinity, and then find the tangents to the corresponding plane curve at this singular point in one of the usual ways (see for example Chapter 15). On making homogeneous the equations of these tangent lines to the plane curve, then finding the projective lines determined by them, we obtain the equations of the required tangent lines to the projective curve. It can be shown that the tangent lines found in this way do not depend on the affine view used in determining them.

**Worked example 16.31.** Determine the singular points at infinity and their tangent lines in $\mathbf{RP}^2$ for the plane curve $y = x^3$. Does the plane curve have an asymptote?

*Solution.* The associated projective curve is

$$F(x, y, z) = yz^2 - x^3 = 0.$$

This has one point $(0:1:0)$ at infinity. Also

$$\mathbf{grad}\, F = (-3x^2, z^2, 2yz),$$

so the only singular point of the projective curve is $(0:1:0)$. We take the affine view $z^2 - x^3 = 0$ for $y = 1$. The origin for the plane curve corresponds to the singular point $(0:1:0)$ of the associated projective curve. The origin of this plane curve is an ordinary cusp, and the cuspidal tangent for the plane curve is $z = 0$. Therefore the projective curve has cuspidal tangent $z = 0$, given by making the equation homogeneous. This is the line at infinity for the plane curve $y = x^3$, and therefore this plane curve has no asymptote.                                                                 □

We next consider inflexions. Similar results to those above are valid for inflexions.

**Definition 16.32.** A *point of simple inflexion* of a projective curve

$$F(x, y, z) = 0$$

is a point which corresponds to a point of simple inflexion in some affine view.

Again it can be shown that the above definition is independent of the view chosen. Points of higher inflexion and points of simple and higher undulation can be defined in a similar way, and again it can be shown that the definitions are independent of the view chosen.

## 16.12 Equivalence of curves*

Equivalence of curves is dependent on the type of transformations allowed. Two curves in the plane are congruent if there is a rigid map sending one to the other; congruence is one type of equivalence. In Chapter 2 we showed how conics can be moved to canonical position using a rigid map. We now describe another form of equivalence of plane curves, and a form of equivalence for projective curves.

### 16.12.1 Projective equivalence of plane curves

**Definition 16.33.** Two plane curves given by $f(x, y) = 0$ and $g(x, y) = 0$ are *projectively equivalent* if there is a non-singular linear transformation

---

* For the meaning of * see the preface.

of $\mathbf{R}^3$ sending the homogeneous defining polynomial of the associated curve of one into that of the other, that is $G(x, y, z) = F(u, v, w)$ where

$$\begin{pmatrix} u \\ v \\ w \end{pmatrix} = M \begin{pmatrix} x \\ y \\ z \end{pmatrix}$$

for some non-singular matrix $M$.

**Theorem 16.34.** *Ellipses, hyperbolae, and parabolae are projectively equivalent.*

*Proof.* Consider first the following special case: in the array the curves on the left are the plane curves, and the curves on the right are the associated projective curves.

$$\begin{aligned} x^2 + y^2 = 1 &\mapsto x^2 + y^2 - z^2 = 0, \\ x^2 - y^2 = 1 &\mapsto x^2 - y^2 - z^2 = 0, \\ y^2 = x &\mapsto y^2 = xz. \end{aligned}$$

The homogeneous polynomial representing the last of these becomes

$$X^2 - Y^2 - Z^2 = 0$$

under the non-singular linear transformation

$$x = X + Z, y = Y \text{ and } z = X - Z.$$

It follows readily that these three curves are projectively equivalent: the only further transformation required is a change of coordinate axes in each case. By elementary geometrical considerations any ellipse, hyperbola, or parabola is projectively equivalent to one of these three. To prove this, first move a given (plane) conic to canonical position by a rotation and a translation of $\mathbf{R}^2$. Each of these transformations determines a non-singular linear transformation of $\mathbf{R}^3$. The rotation determines the obvious rotation which has the same effect on the $xy$–plane as the given rotation and which leaves the $z$–axis fixed. The translation of $\mathbf{R}^2$, given by

$$x \mapsto x + a \text{ and } y \mapsto y + b,$$

determines the non-singular linear transformation of $\mathbf{R}^3$ given by

$$x \mapsto x + az, y \mapsto y + bz \text{ and } z \mapsto z.$$

A further non-singular linear transformation (consisting of compressions along the axes) of $\mathbf{R}^2$ then reduces the (Euclidean) canonical form (see Chapter 1) of the ellipse, hyperbola, or parabola to the appropriate one of the forms given above. The three affine equations on the left of the above array are, respectively, the affine canonical forms of the ellipse, hyperbola, and parabola. □

**Worked example 16.35.** Prove that the acnodal cubic $x^3 - x^2 - y^2 = 0$ is projectively equivalent to the witch of Agnesi $1 - x(1 + y^2) = 0$.

*Solution.* The two associated projective curves are respectively

$$x^3 - x^2 z - y^2 z = 0$$

and $z^3 - xz^2 - xy^2 = 0$. The non-singular linear transformation

$$x \mapsto z, z \mapsto x, y \mapsto y$$

transforms one into the other.                                        □

### 16.12.2 Equivalence of projective curves

**Definition 16.36.** Two projective curves $\Phi$ and $\Gamma$ given respectively by $F(x, y, z) = 0$ and $G(x, y, z) = 0$ are *equivalent* if $G(x, y, z) = F(u, v, w)$ where

$$\begin{pmatrix} u \\ v \\ w \end{pmatrix} = M \begin{pmatrix} x \\ y \\ z \end{pmatrix}$$

for some non-singular matrix $M$.

'Equivalence of projective curves' is clearly an equivalence relation. Under this equivalence, singular points are mapped to singular points, tangents to tangents etc. Two affine curves are *projectively equivalent* if they are affine views of two equivalent projective curves (see §16.12.1). Two plane curves are *affinely equivalent* if one can be obtained from the other by a translation together with a non-singular (i.e., invertible) linear map (the latter need not be a rotation). Also shearing maps, such as $x \mapsto x$, $y \mapsto x + y$, are non-singular linear maps but do not preserve angles or distance. Such affine maps do preserve 'tangents', 'two lines being parallel', 'cusps', and 'singular points', but do not preserve 'normals'.

## 16.13 Examples of asymptotic behaviour

We give examples of asymptotic behaviour in Figure 16.7 as follows.

i) The projective curve does not cross its tangent at a point at infinity and the tangent is not the line at infinity  ⇒  two distant parts of the affine curve typically approach the asymptote on opposite sides of the asymptote.

ii) The projective curve has, as its tangent at a point at infinity, the line at infinity  ⇒  two distant parts of the affine curve tend to infinity, though not along an asymptote.

iii) The projective curve has a simple inflexion at a point at infinity and its tangent there is not the line at infinity  ⇒  two distant parts of the affine curve approach the asymptote on the same side of the asymptote.

iv) The projective curve has a simple double point at infinity and neither

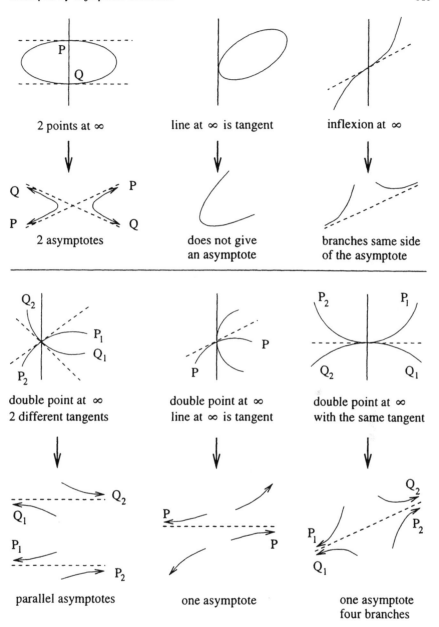

Figure 16.7 *More examples of asymptotic behaviour.*

of its tangents there is the line at infinity $\Rightarrow$ the resulting asymptotes of the affine curve are parallel. Note the sides of these asymptotes on which the curve in this example approaches its asymptotes. This assumes that the double point is a point of inflexion of neither branch.

v) The projective curve has a simple double point at infinity and one of its tangents there is the line at infinity $\Rightarrow$ the point at infinity gives rise to only one asymptote, but two other parts of the curve tend to infinity in a non-asymptotic manner.

vi) The projective curve has a cusp at infinity and the cuspidal tangent there is not the line at infinity $\Rightarrow$ two distant parts of the affine curve tend to infinity in the same direction along the asymptote, either on the same side or on opposite sides depending on the type of cusp.

vii) The projective curve has a tachnode at infinity and the cuspidal tangent there is not the line at infinity $\Rightarrow$ four distant parts of the affine curve tend to infinity in the manner indicated.

Notice the two cases where a tangent to a branch of the curve at a point at infinity is not the line at infinity.

a) The branch crosses its tangent. In this case the two distant portions of the curve having the given asymptote are on the same side of the asymptote.

b) The branch does not cross its tangent. In this case the two distant portions are on different sides of their asymptote.


## 16.14 Worked example

We now give a longer worked example using the results of this chapter.

**Worked example 16.37.** Write down the associated projective curve $\Pi$ for the plane curve $\Gamma$ given by $x^3 + x^2 - y^2 = 0$. Prove that $\Gamma$ has precisely one point at infinity, and find the tangent there. Deduce that $\Gamma$ has no asymptotes. Show that $\Pi$ has precisely one singular point, and that this corresponds to a nodal double point of $\Gamma$. Find the equations of the tangents to $\Gamma$ at this nodal point and deduce the equations of the tangents to $\Pi$ at the corresponding singular point. Let $\Delta$ be the affine view of $\Pi$ for $x = 1$. Show that $\Delta$ has three asymptotes, and give a rough sketch of $\Delta$. Let $\Sigma$ be the affine view of $\Pi$ for $y = 1$. Show that $\Sigma$ has two asymptotes, and give a rough sketch of $\Sigma$. Give a rough sketch of the projective curve in $\mathbf{RP}^2$.

*Solution.* The solution is indicated in Figure 16.8. $\Pi$ is given by

$$F = x^3 + x^2 z - y^2 z = 0.$$

We have

$$\mathbf{grad}\, F = (3x^2 + 2xz, -2yz, x^2 - y^2).$$

Now $\Pi$ meets $z = 0$ where $x^3 = 0$, that is only at $(0:1:0)$. The tangent to $\Pi$ at $(0:1:0)$ is $z = 0$ (given by $(x, y, z) \cdot \mathbf{grad}\, F = 0$). Thus the tangent to

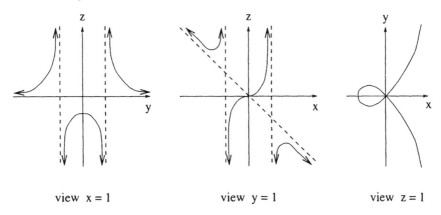

view x = 1         view y = 1         view z = 1

Figure 16.8 *Different affine views of a curve.*

$\Pi$ at the (only) point at infinity for $\Gamma$ is the line at infinity, and therefore $\Gamma$ has no asymptotes.

We have **grad** $F = 0$ if, and only if,

$$3x^2 + 2xz = -2yz = x^2 - y^2 = 0.$$

The only point of $\mathbf{RP}^2$ which satisfies these equations is $(0:0:1)$, and we have easily that this point lies on $\Pi$. Thus the projective curve has precisely one singular point. The point of $\Gamma$ corresponding to this singular point is the origin of $\mathbf{R}^2$. Notice that the point at infinity for $\Gamma$ is not a singular point. The homogeneous polynomial of terms of lowest degree in the defining equation for $\Gamma$ is $x^2 - y^2 = (x + y)(x - y)$. Since there are no repeated real linear factors, the origin is a double point and the tangent lines there are $x + y = 0$ and $x - y = 0$. The tangents to $\Pi$ at the singular point (given by making these equations homogeneous) are $x + y = 0$ and $x - y = 0$.

We now consider the affine view of $\Pi$ for $x = 1$. Now $\Pi$ meets $x = 0$ where $y^2 z = 0$. Thus, for this view, there are two points at infinity, namely $(0:1:0)$ and $(0:0:1)$. The tangent to $\Pi$ at $(0:1:0)$ is $z = 0$, and the tangents at $(0:0:1)$ are $x + y = 0$ and $x - y = 0$, as noted above. The affine view for $x = 1$ is given by $1 + z - y^2 z = 0$ that is by

$$z = \frac{1}{y^2 - 1}$$

and the asymptotes for this view (given by substituting $x = 1$ in the equations of the above tangents) are $z = 0$, $1 + y = 0$ and $1 - y = 0$. Note that this plane curve has no singular point in the plane, but has a singular point (a double point) at infinity.

We now consider the affine view of $\Pi$ for $y = 1$. Now $\Pi$ meets $y = 0$

where $x^2(x + z) = 0$. Thus, for this view, there are two points at infinity, namely $(1:0:-1)$ and $(0:0:1)$. The tangent to $\Pi$ at $(1:0:-1)$ is $x + z = 0$, and, as noted above, the tangents at $(0:0:1)$ are $x + y = 0$ and $x - y = 0$. The affine view for $y = 1$ is given by $x^3 + x^2z - z = 0$ that is by

$$z = \frac{x^3}{1 - x^2}$$

and the asymptotes for this view (given by substituting $y = 1$ in the equations of the above tangents) are $x + z = 0$, $x + 1 = 0$ and $x - 1 = 0$. Note that this plane curve is regular, but has a singular point (a double point) at infinity. Also the plane curve has an inflexion (indeed a stationary point of inflexion) at the origin (check that

$$\frac{dz}{dx} = \frac{d^2z}{dx^2} = 0).$$

Note in the sketch of the projective curve that no affine view of the projective curve is bounded, since any projective line (as a line infinity) meets the projective curve in at least one point. (Compare this with the situation for the projective curve associated with a hyperbola.)  □

In the two views considered above the affine curves are graphs of functions, that is their equations take on the form $z = \varphi(y)$ and $z = \psi(x)$ respectively. Thus the curve is parametrised by the $y$ or $x$ variable respectively. The crunodal cubic is not the graph of a function even after a rotation.

The projective curve given by a quadratic homogeneous polynomial such as $x^2 + y^2 - z^2 = 0$ can be drawn in such a way that it 'resembles' a closed curve in the projective plane. This is not true for all projective curves as the above example shows. If it cannot be drawn to resemble a closed curve in the plane, we can usually note in our sketch in the projective plane that certain points must be identified in order to close the curve – recall that projective space is obtained from the sphere by identifying each point with its opposite. The projective curve $\Pi$ sketched here, regarded as a curve on the sphere, will have as its 'opposite' on the sphere a similar looking curve, and when the two are joined together, they will form one continuous curve. On the other hand the quadratic cone projective curve determines on the sphere two disjoint closed curves, one of which is 'opposite' to the other.

## Exercises

**16.1.** Find the singular points of $(x^2 + y^2)^2 - (x^2 - y^2)z^2 = 0$.
$$[(1:1:\pm 2), (1:-1:\pm 2), (0:0:1)]$$

**16.2.** Show that the cubic

$$(x - y)(x^2 + y^2) + 2xyz = 0$$

intersects the line $z = 0$ in precisely one point and find the tangent line at this point. Show that the projective curve has precisely one singular point and that this is a node.

**16.3.** Find the asymptotes to the hyperbola

$$(y - \alpha x)(y - \beta x) + 2fx + 2gy + d = 0 \quad (\alpha \neq \beta).$$

[one is $-\alpha(\alpha - \beta)x + (\alpha - \beta)y + 2f + 2\alpha g = 0$]

**16.4.** Show that the only singular point of the algebraic curve

$$x^4 - y^4 = 4xy^2$$

is at the origin. Find the asymptotes of the curve.

**16.5.** Show that the quartic curve $y^2(x^2 + z^2) - x^4 = 0$ has precisely two singular points, and that one of these is an isolated point. Find the asymptotes to the affine view given by $z = 1$.

[$(0{:}0{:}1), (0{:}1{:}0); x^2 + z^2 - x^4 = 0; x - y = 0, x + y = 0$]

**16.6.** Let $f(x, y) = g(x, y) + h(x, y) + k(x, y)$ be a polynomial, where $g$ is non-trivial and homogeneous of degree $n \geq 2$, $h$ is homogeneous of degree $n - 1$, and the degree of $k$ is $\leq n - 2$. Show that the points at infinity $(x_0 : y_0 : 0)$ of $f(x, y) = 0$ are given by the real solutions of $g(x, y) = 0$. Show further that, where $g(x, y)$ has no repeated real linear factors, the asymptotes to the plane curve are the lines

$$xg_x(x_0, y_0) + yg_y(x_0, y_0) + h(x_0, y_0) = 0$$

corresponding to these solutions.

**16.7.** The projective curve $\Pi$ given by

$$x^4 + x^2yz - y^3z + y^4 = 0$$

has as affine view for $z = 1$ the curve $\Sigma$, given by

$$x^4 + x^2y - y^3 + y^4 = 0.$$

Prove that the origin is a triple point of the affine curve $\Sigma$, and find the tangent lines to the curve there. Show that $\Pi$ meets the axes $x = 0$, $y = 0$ and $z = 0$ altogether in precisely two points, and find the equations of the four tangents to $\Pi$ at these two points. Deduce also that $\Sigma$ is a bounded curve.

Show that the affine view $\Gamma$, for $x = 1$, of the projective curve $\Pi$ is a graph. Write down the coordinates of the points of infinity for $\Gamma$, and determine the equations of its asymptotes. Make a rough sketch of $\Gamma$ indicating the asymptotes and carefully indicating any intersection of the curve with the axes.

[$x = \pm y, y = 0 : (0{:}0{:}1), (0{:}1{:}1); x = \pm y, y = 0, y = z : (0{:}0{:}1), (0{:}1{:}1); y = \pm 1, y = 0, y = z$]

**16.8.** The algebraic curve $\Sigma$ with equation

$$y^2(1 - x) - x^2(1 + x) = 0$$

has a double point at the origin.

a) Let $\Pi$ be the associated projective curve of $\Sigma$. Show that $\Sigma$ has precisely one point at infinity, namely $(x:y:z) = (0:1:0)$ on $\Pi$. Let $\Gamma$ be the affine view of $\Pi$ for $y = 1$. Determine the tangent line to $\Gamma$ at the origin, and show that it has 3-point contact with $\Gamma$ at the origin. Write down the equation of the asymptote to $\Sigma$.

$$[z = x, x = 1]$$

b) Each line $y - m = mx$ through $V = (-1,0)$ meets $\Sigma$ in points $Q$ and $R$ and meets the $y$-axis in a point $P = (0,m)$. Show that $OP = QP = PR$, where $O$ is the origin. The curve $\Sigma$ is the right strophoid.

Draw this strophoid, in the usual way, using a drawn line and marked points $V$ and $O$. For a line $L$ through $V$ which meets the drawn line in a point $P$, mark on $L$ the points $Q$ and $R$ which satisfy $OP = QP = PR$. Repeat this for a suitable collection of lines through the origin, and draw the curve in the usual way.

**16.9.** The projective curve $\Pi$ given by

$$y^4 + x^2y^2 + xy^2z - x^3z = 0$$

has as affine view for $z = 1$ the curve $\Sigma$, given by

$$y^4 + x^2y^2 + xy^2 - x^3 = 0.$$

Prove that the origin is a triple point of the affine curve $\Sigma$, and find the tangent lines to the curve there. Show that $\Pi$ meets the axes $x = 0$, $y = 0$ and $z = 0$ altogether in precisely two points, and find the equations of the four tangents to $\Pi$ at these two points. Deduce that $\Sigma$ has no asymptotes.

Show that the affine view $\Gamma$, for $x = 1$, of the projective curve $\Pi$ is a graph. Write down the coordinates of the points of infinity for $\Gamma$, and determine the equations of its asymptotes. Make a rough sketch of $\Gamma$ indicating the asymptotes and carefully indicating any intersection of the curve with the axes.

$$[x = \pm y, x = 0 : (0:0:1), (1:0:0); x = \pm y, x = 0, z = 0 : (0:0:1); y = \pm 1]$$

**16.10.** The algebraic curve $\Sigma$ with equation

$$(x^2 + y^2)y^2 - x^2 = 0$$

has a tachnode at the origin.

a) Let $\Pi$ be the associated projective curve of $\Sigma$. Show that $\Sigma$ has precisely one point at infinity, namely $(x:y:z) = (1:0:0)$ on $\Pi$. Let $\Gamma$ be the affine view of $\Pi$ for $x = 1$. Show that $\Gamma$ has a node at the origin and find the tangent lines there. Write down the equations of the asymptotes to $\Sigma$.

$$[y = \pm z, y = \pm 1]$$

b) Each line $y = mx$ through the origin $O = (0,0)$ meets $\Sigma$ in points $P$ and $P'$, and meets the line $y = 1$ in a point $Q$. Prove that $OP = OP' = CQ$, where $C = (0,1)$.

Use the following method to draw the curve $\Sigma$ (Gutschoven's curve) using drawn rectangular axes and line $y = 1$. For a line $L$ through $O$ which meets $y = 1$ in point $Q$, mark on $L$ the points $P$ and $P'$ which satisfy $OP = OP' = CQ$. Repeat this for a suitable collection of lines through the origin, and draw the curve in the usual way.

**16.11.** The algebraic curve given by

$$y^2(1 - x) - x^2(m + x) = 0 \quad (m \neq 0, -1)$$

has a singular point at the origin and no other singular points. Show that the origin is an isolated point in case $m < 0$, and is a nodal double point in case $m > 0$. In the latter case write down the equations of the tangent lines.

The projective curve $\Pi$ given by

$$y^2(z - x) - x^2(4z + x) = 0$$

has as affine view for $z = 1$ the curve $\Sigma$, given by

$$y^2(1 - x) - x^2(4 + x) = 0.$$

Show that $\Pi$ meets the axes $x = 0$, $y = 0$ and $z = 0$ altogether in precisely three points, and find the equations of the four tangents to $\Pi$ at these three points, including the two at the double point. Show that $\Sigma$ has an asymptote and write down its equation.

Show that the affine view $\Gamma$, for $y = 1$, of the projective curve $\Pi$ is a graph. Write down the coordinates of the points at infinity for $\Gamma$, and determine the equations of its asymptotes. Make a rough sketch of $\Gamma$ indicating the asymptotes and carefully indicating any intersection of the curve with the axes.

$[y = \pm\sqrt{m}x; y = \pm 2x, z = x, x + 4z = 0; x = 1; 2x = \pm 1, x + 4z = 0]$

**16.12.** The projective curve $\Pi$ given by

$$(x^2 + y^2)(x - 2y) + z(x^2 - y^2) = 0$$

has, as affine view for $z = 1$, the affine curve $\Sigma$ given by

$$(x^2 + y^2)(x - 2y) + x^2 - y^2 = 0.$$

Show that the origin is a singular point of $\Sigma$.

Find the other points of intersection of the projective curve with the three lines $x = 0, y = 0, z = 0$, and find the equations of the tangents at all four points.

Hence write down the equations of the asymptotes to the related affine curve $\Omega$ with equation

$$(x^2 + 1)(x - 2) + z(x^2 - 1) = 0,$$

given that the tangents to the projective curve at $(0:0:1)$ have equations

$$x = \pm y.$$

Give a rough sketch of $\Omega$, including on your sketch the asymptotes, and the tangent lines at the points where the curve crosses the axes. (A scale of 1 unit = 2 cm is suggested.)

**16.13.** Show that the algebraic curve $\Sigma$ with equation

$$(x - 1)^2 (x^2 + y^2) - 9x^2 = 0$$

has a double point at the origin.

a) Let $\Pi$ be the associated projective curve of $\Sigma$. Show that $\Sigma$ has precisely one point at infinity, namely $(x : y : z) = (0 : 1 : 0)$ on $\Pi$. Let $\Gamma$ be the affine view of $\Pi$ for $y = 1$. The tangent line to $\Gamma$ at the origin is $x = z$. Show that it has 4–point contact with $\Gamma$ at the origin. Write down the equation of the asymptote to $\Sigma$.                                      $[x = 1]$

b) Each line $y = mx$ meets the line $x = 1$ in a point $P$ and the curve $\Sigma$ in precisely 2 points $Q$ and $R$. Show that $Q$ and $R$ lie on opposite sides of the line $x = 1$, and that $QP = PR = 3$. The curve $\Sigma$ is a conchoid of Nicomedes.

Draw this conchoid, in the usual way, using a drawn line and origin. For a line $L$ through the origin which meets the given line at $P$, mark the two points on $L$ distant 6 cm from $P$ (use the scale 1 unit = 2 cm). Repeat this for a suitable collection of lines through the origin, and draw the curve in the usual way.

**16.14.** The projective curve $\Pi$, given by

$$x\left(x^2 + y^2\right) + \left(3x^2 - y^2\right) z = 0,$$

has as affine view for $z = 1$ the trisectrix of MacLaurin $\Sigma$, given by

$$x\left(x^2 + y^2\right) + \left(3x^2 - y^2\right) = 0.$$

Prove that the origin is a double point of the affine curve $\Sigma$ and find the tangent lines to the curve there. Show that $\Pi$ meets the axes $x = 0$, $y = 0$ and $z = 0$ altogether in three points, and find the equations of the four tangents to $\Pi$ at these three points, including the two at the double point. Show that $\Sigma$ has an asymptote and write down its equation.

Let $\Gamma$ be the affine view, for $y = 1$, of the projective curve $\Pi$. Determine the equations of the asymptotes of $\Gamma$. Make a rough sketch of $\Gamma$ indicating the asymptotes and carefully indicating the tangent line where $\Gamma$ crosses the axis $z = 0$. (You may assume that the projective curve $\Pi$ has an inflexion at the point where it meets the axis $z = 0$.)

**16.15.** The algebraic curve $\Sigma$ with equation

$$x^3 + xy^2 - 2y^2 = 0$$

has a cusp at the origin.

a) Let $\Pi$ be the associated projective curve of $\Sigma$. Show that $\Sigma$ has precisely one point at infinity, namely $(x : y : z) = (0 : 1 : 0)$ on $\Pi$. Let $\Gamma$ be the affine view of $\Pi$ for $y = 1$. Determine the tangent line to $\Gamma$ at the origin, and show that it has 3-point contact with $\Gamma$ at the origin. Write down the equation of the asymptote to $\Sigma$. $\qquad [2z = x, x = 2]$

b) Each line $y = mx$ meets $\Sigma$ in a point $P$ and meets the circle

$$(x - 1)^2 + y^2 = 1$$

and the line $x = 2$ in points $Q$ and $R$. By considering the $x$-coordinates of $P$, $Q$ and $R$, or otherwise, show that $OP = QR$, where $O$ is the origin. The curve $\Sigma$ is the cissoid of Diocles.

Draw this cissoid, in the usual way, using a drawn line, drawn circle, and marked origin. For a line $L$ through the origin which meets the given line at $R$ and the given circle at $Q$, mark on $L$ the point $P$ which satisfies $OP = QR$. Repeat this for a suitable collection of lines through the origin, and draw the curve in the usual way.

$$\left[ x \text{ coordinate of } P \text{ is } \frac{2m^2}{1 + m^2} \right]$$

**16.16.** The projective curve $\Pi$, given by

$$y^2(z - x) - x^2(z + x) = 0,$$

has as affine view for $z = 1$, the strophoid of Barrow $\Sigma$, given by

$$y^2(1 - x) - x^2(1 + x) = 0.$$

Prove that the origin is a double point of the affine curve $\Sigma$ and find the tangent lines to the curve there. Show that $\Pi$ meets the axes $x = 0$, $y = 0$ and $z = 0$ altogether in three points, and find the equations of the four tangents to $\Pi$ at these three points, including the two at the double point. Show that $\Sigma$ has an asymptote and write down its equation.

Show that the affine view $\Gamma$, for $y = 1$, of the projective curve $\Pi$ is a graph. Find its asymptotes. Make a rough sketch of $\Gamma$ indicating the asymptotes and marking the tangent where it crosses the axis $z = 0$. (You may assume that the projective curve $\Pi$ has an inflexion at the point where it meets the axis $z = 0$.)

**16.17.** The projective curve $\Pi$, given by

$$x^3 + 3xy^2 + x^2z - y^2z = 0,$$

has as affine view for $z = 1$, the folium of Descartes $\Sigma$, given by

$$x^3 + 3xy^2 + x^2 - y^2 = 0.$$

Prove that the origin is a double point of the affine curve $\Sigma$ and find the tangent lines to the curve there. Show that $\Pi$ meets the axes $x = 0$, $y = 0$ and $z = 0$ altogether in three points, and find the equations of the four tangents to $\Pi$ at these three points, including the two at the double point. Show that $\Sigma$ has an asymptote and write down its equation.

Make a rough sketch of the affine view of the projective curve $\Pi$ for $y = 1$, marking the tangent where it crosses the axis $z = 0$, and the asymptotes for this view. (You may assume that the projective curve $\Pi$ has an inflexion at the point where it meets the axis $z = 0$.)

$$[y = \pm x; y = \pm x, 3x = z, x + z = 0; 3x = 1; x = \pm 1, x + z = 0]$$

**16.18.** Show that the projective curve $P$ with equation

$$x^3 - x^2 y - 3x^2 z + 7xz^2 - 3yz^2 - 9z^3 = 0$$

has only one singular point and find it. The cubic curve $\Gamma$ is obtained from $P$ by putting $z = 1$.

Find the equation of the tangent line to $\Gamma$ at $(0, -3)$ and, by considering its intersection multiplicity with the curve, deduce that the point is a point of inflexion. Similarly show that $(3, 1)$ is a point of inflexion. Deduce that $(-3, -7)$ is the third point of inflexion and determine the slope of the tangent line at this point. Find the asymptote to the curve.

(You may assume that the singular point of $P$ is isolated.)

# 17

# Practical work

We now include projects for seven two-hour practical classes involving drawing curves by a variety of methods. They are designed for students to work in groups of six. The group draws a portfolio of curves, with each student contributing two or three curves in each practical class. Generally the students in the group are drawing different, though related, curves at any given time.

## Practical 1

Each person has three curves to draw. For Project 1, which everyone must do, all that is required is a rough sketch after some lines have been drawn precisely.

---

**1.** The rectangular hyperbola $3x^2 + 8xy - 3y^2 + 38x - 16y - 18 = 0$ has centre $(-1, -4)$, its transverse axis is parallel to the vector $(2, 1)$, and (the length of) the semi-transverse axis is 1. Choose and draw coordinate axes parallel to the edges so that the centre of the hyperbola is near the centre of the paper and take 1 unit equal to 1 cm. Use a ruler to draw the transverse axis (recall that a line is determined by two points on it – choose the points reasonably far apart for greater accuracy). Use the angle bisection method to draw the conjugate axis and the asymptotes. Mark the intersections of the hyperbola with its transverse axis, mark the foci, and make a rough sketch of the hyperbola.

---

The second project involves plotting points.

**Table of points.** It is suggested that initially you draw up a table of points, determining the values taken by $x$ and $y$ at successive values of the parameter. Use either the algebraic equation with $x$ or $y$ as parameter, or the standard parametric equation.

**Distance between plotted points.** Choose increments in the parameter so that successive points are about 1 cm apart. Where the curve bends sharply (e.g., near the vertices), choose the points closer together.

**Scale and position of origin.** Use the same scale for the $x$ and $y$ directions. Adopt this as a general policy throughout this course. The position of the origin and the scale for a given example may only be clear after an attempt at a rough sketch.

**Mode.** Paper oriented as is this sheet is in *portrait* mode: that is the long edge is vertical. For the *landscape* mode, the long edge is horizontal.

**Drawing to the edge.** Draw the axes to the edge of the paper. For unbounded curves, plot to the edge and draw the curve to the edge.

**2.** Plot one of the following conics.

i) $\dfrac{x^2}{8^2} + \dfrac{y^2}{5^2} = 1$  (landscape: algebraic equation).

ii) $\dfrac{x^2}{8^2} + \dfrac{y^2}{3^2} = 1$  (landscape: parametric equation).

iii) $x^2 - 3y^2 = 9$  ($|x| \le 10$)  (landscape: algebraic equation).

iv) $3x^2 - y^2 = 12$  ($|x| \le 7$)  (portrait: parametric equation).

v) $y^2 = 5x$  (landscape: algebraic equation).

vi)  $y^2 = 16x$   (portrait with the $y$-axis at the far left: parametric equation).

In each case take 1 unit equal 1cm, the coordinate axes parallel to the edges, and, except for vi, the origin near the centre. Where the algebraic equation is used for plotting, decide whether to take $x$ or $y$ as parameter. The standard parametric equations of conics in canonical position are given in Chapter 1. Plot at least one curve each; if you do more than one, choose one using the algebraic equation and one using the parametric equation. Your group should plot all six of these curves.

---

The third curve uses the polar equation with $\theta$ as parameter. Recall that in this case $r$ can be negative. Choose increments in $\theta$ so that plotted points again are about 1 cm apart.

**3.** Plot one of the following conics using the polar equation

$$\frac{l}{r} = 1 + e \cos \theta$$

where the semilatus rectum is $l = 2a$ for $y^2 = 4ax$, $l = \dfrac{b^2}{a}$ for $\dfrac{x^2}{a^2} + \dfrac{y^2}{b^2} = 1$
($b < a$), and $l = \dfrac{b^2}{a}$ for $\dfrac{x^2}{a^2} - \dfrac{y^2}{b^2} = 1$. (Note the different manners in which $a$ and $b$ are selected for the ellipse and for the hyperbola.) Use polar graph paper.

i)   $x^2 + 2y^2 = 36$   (landscape).
ii)  $x^2 + 4y^2 = 49$   (landscape).
iii) $x^2 - 3y^2 = 9$   (landscape).
iv)  $3x^2 - y^2 = 9$   (landscape).
v)   $y^2 = 5x$   (landscape).
vi)  $y^2 = 16x$   (portrait).

In each case take 1 unit equal to 1 cm. Plot at least one curve each. The parametrisation of the hyperbola has two open intervals as its domain. Check that your drawing shows each of these two parts. Your group should plot all six of these curves.

## Practical 2

For this practical, you are expected to choose the scale yourself. (See the notes below.) For Project 1 i-v, choose the scale so that your plotted curve is in proportion to corresponding curves given in the figures of Chapter 3; for the acnodal cubic for example, draw enough of the curve to show the inflexions indicated there.

---

**1.** Plot one of the following cubic curves, either from the algebraic equations or from the parametric equations.

   i) Semi-cubical parabola: $y^2 = x^3$; or $x = t^2$, $y = tx$.

   ii) Crunodal cubic I : $y^2 = x^3 + x^2$; or $x = t^2 - 1$, $y = tx$.

   iii) Crunodal cubic II : $3y^2 = x^3 + x^2$; or $x = 3t^2 - 1$, $y = tx$.

   iv) Acnodal cubic: $y^2 = x^3 - x^2$; or $x = t^2 + 1$, $y = tx$.

   v) Tachnodal quartic: $x^3 + x^2 = y^4$; or $x = t$, $y = \pm \sqrt[4]{t^2(1+t)}$ $(t \geq -1)$.

   vi) Witch of Agnesi: $(y^2 + 1)x = 1$; or $x = (1 + y^2)^{-1}$.

Your group should plot all six of these curves.

**Scale, area, axes, mode, and increments.** Before drawing a curve you need to decide the scale that you will use, where you will place the axes, whether to use landscape or portrait mode, and what increments you will take in the values of the primary variable or parameter that you use. In most of the cases considered in Project 1, the values which $x$ can take are automatically restricted by the fact that $y^2 \geq 0$. Begin in each case by noting the relevant range of $x$ and $y$ so that you can determine that the curve will fit in the area which you propose to allot for it; in general cases this may involve some trial and error. It is suggested that, in the examples here in Project 1, you use graph paper in the portrait mode and start with a scale of about 8 cm to 1 unit.

**Table of points.** First draw up a table of points $(x, y)$ taking for example increments for the $x$ values of 0.2 units in the first instance and using a calculator to compute $y$, or possibly use $y$ as a parameter instead. Alternatively use the parametric representation. Try each method in your group, and compare the results.

**Same scale.** Remember to use the same scale in each of the vertical and horizontal directions.

**Special increments.** Close to certain points, such as nodes and cusps, it may be necessary to choose smaller increments in the values of the parameter or of $x$ etc. than the ones that you have chosen for the curve as a whole, in order more precisely to indicate the local shape of the curve.

**Partial parametrisations.** Some parametrisations are only valid over part of the curve. In particular a parametric representation will not give isolated points of an algebraic curve. This is relevant to iv of Project 1 for example.

---

**2.** Plot one of the following curves on polar graph paper.

i)  $r = e^{a\theta}$ (logarithmic spiral: Descartes c. 1638). Choose $a$ and the scale so that your drawing looks like a larger version of that drawn in Chapter 3.

ii)  $r = a\theta$ (Archimedean spiral: c. 225 BC). Choose $a$ and the scale so that your drawing looks like a larger version of that drawn in Chapter 3, with at least three circuits showing.

iii)  $r^2 = a^2\theta$ (Fermat's spiral: c. 1636). Choose $a$ and the scale so that your drawing looks roughly like a larger version of the Archimedean spiral drawn in Chapter 3, with at least three circuits showing.

iv)  $r = \dfrac{\cos 2\theta}{\cos \theta}$ (strophoid – Barrow: c. 1670). Choose the scale so that your drawing is in similar proportion to the crunodal cubic drawn in Chapter 3.

v)  $r = \tan\theta \sin\theta$ (cissoid – Diocles: c. 100 BC). Choose the scale so that your drawing is in similar proportion to the cuspidal cubic drawn in Chapter 3.

vi)  $r^2 = (4 - 3\sin^2\theta)$ (hippopede – Proclus: c. 75 BC). Choose the scale so that this 'prodded ellipse' fits the graph paper.

Your group should plot all six of these curves.

## Practical 3

In this practical you should each draw three curves: two from Project 1, one of these on rectangular graph paper, and the other as a conchoid (see below), and one from Project 2.

---

**1. Limaçons.** These curves have the polar equation of the form:

$$r = a + b\cos\theta \quad (a, b \geq 0).$$

Plot the curve in one of the following cases. The shape of these curves is indicated in Figure 10.1 and Figure 13.7. Choose the scale for each one carefully (see below).

|       |              |                |                             |
|-------|--------------|----------------|-----------------------------|
| i)    | $a = 6$,     | $b = 0$        | Draw this curve freehand    |
| ii)   | $a = 5$,     | $b = 1$        |                             |
| iii)  | $a = 4$,     | $b = 2$        |                             |
| iv)   | $a = 3\frac{1}{2}$, | $b = 2\frac{1}{2}$ |                   |
| v)    | $a = 3$,     | $b = 3$        |                             |
| vi)   | $a = 2$,     | $b = 4$        |                             |
| vii)  | $a = 1$,     | $b = 5$        |                             |
| viii) | $a = 0$,     | $b = 6$        | Draw this curve freehand    |

**Type of graph paper.** Since the curves are given in polar coordinates, it would be possible to plot them on polar graph paper. However, after using your calculator to draw up a table of points using $x = r\cos\theta$ and $y = r\sin\theta$, it is also easy to draw them on rectangular graph paper. Do at least one on rectangular graph paper. Recall that $r$ defined by the polar equation can take negative values.

**Conchoid.** Your second limaçon (choose one with $5 \geq b \geq 2$) should be drawn as a conchoid of a circle as follows. Draw a circle of diameter $b$, and mark a point $Q$ on its circumference such that $Q$ lies on a diameter parallel to the long edge of the paper. A line through $Q$ will in general meet the circle at a second point $T$. Mark the two points $P$ and $P'$ on this line distant $a$ from $T$. As the line through $Q$ varies, $P$ and $P'$ will describe a limaçon. Choose the scale and the position of the circle by calculation or by trial and error. Do not choose the first or last in the above table for this method. You only need to draw parts of the line so that you can make accurate measurements. Use your compass to measure $a$. Use plain paper (not graph paper) for the conchoid.

**Complete set.** Each person should try one of the cases about half-way down the table (i.e., iv, v, or vi), and each group should have a complete set.

**Origin, axes, mode, and scale.** For curves in general choose the position of the origin, the mode, and the axes, and a scale so that your drawing will be as large as possible. In this case you could as an exercise in the calculus formally calculate the (local) maximum values taken by $y = r\sin\theta$, and similarly by $x = r \cos \theta$. However drawing up a table of points and working from it will suffice. Try increments of 10 degrees. At special points, such as cusps, use smaller increments to get the detail needed. The position of the origin and the scale will differ for each curve.

For ordinary graph paper the scale for the first one of 1 cm as a unit is about right. For polar graph paper the scale chosen will be different in general since the reference point is at the centre. Write the scale you use clearly on each of your graphs.

**Polar coordinates.** In many situations where polar coodinates are used we have $r \geq 0$. However in the case of curves given by equations such as these we allow $r$ to take negative values. The case where $r$ is negative corresponds to the same point of the graph given by increasing the polar angle by $\pi$ and taking the absolute value of $r$.

**Same scale.** Remember to use the same scale in each of the vertical and horizontal directions.

---

## 2. Cycloids, curtate cycloids, and prolate cycloids: rolling a circle on a line.

These curves are given by

$$\varphi \mapsto a\varphi + ai + ie^{-i\varphi}(q - a) = (a\varphi - (a - q)\sin\varphi, a - (a - q)\cos\varphi).$$

Using rectangular graph paper, plot the curve in one of the following cases

i)    $q = 0$,
ii)   $q = \frac{1}{4}a$,
iii)  $q = \frac{1}{2}a$,
iv)   $q = -\frac{1}{2}a$,
v)    $q = -\frac{3}{4}a$,
vi)   $q = -a$.

**Scale.** Use the landscape position. Find the maximum and minimum values taken by the $y$–coordinate (this should be obvious – you do not need to differentiate). You need to draw the curve over a range of values for the parameter of a little more than $0 \leq \varphi \leq 2\pi$. Now adjust the scale, or equivalently $a$, and the vertical position of the $x$–axis accordingly, so that your curve fills up as much as possible of the page. The $x$–axis will have a different position for each of these curves.

Your group should plot all six of these curves.

## Practical 4

For this practical you should each draw at least two curves.

**1. Centres of curvature.** Plot the following curves on the same diagram

$$z(t) = (\tfrac{1}{2}t^2, t) \quad \text{and} \quad z_*(t) = (\tfrac{3}{2}t^2 + 1, -t^3), \quad \text{where} \quad -1 \le t \le 1,$$

using increments of 0.1 in the values of $t$. Take 6 cm as one unit, the landscape position and the $y$–axis on the far left. Draw lines through each pair of points on the two curves having the same parameter values, and extend each line slightly beyond the two points.

(The point $z_*(t)$ is the centre of curvature of the parabola $t \mapsto z(t)$ at the point $z(t)$, and $t \mapsto z_*(t)$ is the evolute of $t \mapsto z(t)$).

Draw part of the circle of curvature to the parabola at the point $z(t)$ in one of the following cases. Mark and label clearly the centre of the circle of curvature and the point of contact.

    i)      $t = 0$,
    ii)    $t = 0.2$,
    iii)   $t = 0.4$,
    iv)   $t = 0.5$,
    v)    $t = 0.6$
    vi)   $t = 0.8$.

Your group should plot all six of these curves.

**2. Cusps.** Plot one of the following curves. Initially try $-1.1 \le t \le 1.1$. Each curve will need a different scale and/or a different position of the axes. Suggested scales are given, but you may prefer to choose a different scale. Remember always to use the same scale for the $x$ and the $y$ coordinates. Plot to the edge of the paper. Draw the cuspidal tangent line at each cusp. Some of the curves have two cusps; in other cases the cusp is rhamphoid. When your group has completed these, you should, as in all other practicals, each look at the whole collection. Which of the cusps are rhamphoid for example?

i)   $r(t) = (\tfrac{3}{2}t^2 - t^3, \tfrac{10}{9}t^2 - t^4)$,      20 cm = 1 unit    $-0.6 \le t \le 1.2$

ii)   $r(t) = (t^2 - t^3 + \tfrac{1}{4}t^4, t^2 - \tfrac{1}{3}t^3)$,   10 cm = 1 unit    $-1 \le t \le 3$

iii)   $r(t) = (\tfrac{1}{2}t^2, \tfrac{1}{4}t^4 + \tfrac{1}{5}t^5)$,           20 cm = 1 unit

iv)   $r(t) = (\tfrac{1}{2}t^2 + \tfrac{1}{3}t^3, \tfrac{1}{4}t^4)$,         10 cm = 1 unit

v)   $r(t) = (3t^2 + 2t^3, 2t^2 - t^4)$,         10 cm = 1 unit

vi)   $r(t) = (t^2 - t^3, 2t^2 - t^4)$,            4 cm  = 1 unit,   $-\sqrt{2} \le t \le \sqrt{2}$.

Your group should plot all six of these curves.

## Practical 5

For this practical you should each draw at least 2 curves: one from Projects 1-3, and one from Projects 4-9. Each group should present two each of Projects 1-3.

---

### A. Evolutes (as envelopes) and parallels: conics.

The evolute of a curve $\Gamma$ can be drawn, for example, by finding the equation of the evolute and plotting points in the usual way. A second method is to draw normals to the (oriented) curve $\Gamma$ and to measure off a distance from $\Gamma$ along each normal, equal to the radius of curvature. In this practical we consider a third method, namely to draw the evolute as the envelope of the normals. You may wish to satisfy yourself, before or after the practical class, that the constructions and formulae used below do indeed give the normal lines. Once the normals have been drawn, the parallels can also be drawn by marking off the fixed (oriented) distance, from the curve, along each normal.

In using this method, it is necessary to find a construction for drawing the normals. Often this is done by finding two points on the normal (one of these may be on the curve) and simply drawing the line through them.

When drawing the envelope first continue each normal line to the points where it meets the 'adjacent' normals.

When drawing the envelope remember that the evolute has each normal as a tangent, and that the tangency will often take place at a point on the normal between its points of intersection with the two 'adjacent' normals.

When drawing envelopes of lines, always draw the lines faintly; the drawn curves should 'stand out' from the lines.

### 1. The ellipse $\dfrac{x^2}{a^2} + \dfrac{y^2}{b^2} = 1$.

Use the drawn ellipse supplied in which the foci are already marked. Draw the normals as follows.

Consider circles with centres at successive points on the minor axis at distances 0, 1, 2, 3, 4, 6, and 12 cm from the origin in each direction (13 points in all), and which pass through the two foci; do not draw the whole circle but only the intersections needed. The normal to the ellipse at a point where one of these circles meets the ellipse is the line given by joining that point to one of the two points where the circle meets the minor axis (using the other of these will give the tangent). Continue each normal line until it meets 'adjacent' normals. In each case note any points at which the curve does not appear to be regular.

Outline the evolute of the ellipse as the curve touched by these normals. Calculate the radius of curvature at those points where the ellipse meets its major and its minor axes, and mark on the diagram the centres of curvature at each of these four points. Draw parallels to the ellipse at distances of 2 cm and 4 cm measured along the inward normals.

### 2. The parabola $y^2 = 4ax$.

Use the drawn parabola supplied in which the foci are already marked. Draw the normals as follows.

Consider circles with centre at the focus $(a, 0)$; do not draw the whole circle but only the intersections needed. The normal to the parabola at a point where one of these circles meets the parabola is the line given by joining that point to one of the two points where the circle meets the $x$-axis (decide which one of these). In each case choose the normal closest to the origin. Continue each normal line until it meets 'adjacent' normals. Choose a sufficient number of normals to enable you to draw the following curve. (A simple arithmetic sequence in the radii may not give a sufficient number.) In each case note any points where the curve does not appear to be regular.

Outline the evolute of the parabola as the curve touched by these normals. Calculate the radius of curvature at the vertex of the parabola, and mark on the diagram the centre of curvature at the vertex. Draw parallels to the parabola at distances of 1 cm and 4 cm measured along the inward normals.

Alternatively it is possible to draw the normals using the fact that the normal at $(x, y)$ passes through $(x + 2a, 0)$.

### 3. The rectangular hyperbola $x^2 - y^2 = 4a^2$.

Use the drawn hyperbola supplied in which the foci are already marked. Draw the normals as follows. Consider circles through the foci with centres on the $y$-axis at the points $(0, \pm 4)$ and at 1 cm increments between them; do not draw the whole circle but only the intersections needed. The normal to the hyperbola at a point where one of these circles meets the hyperbola is the line given by joining that point to one of the two points where the circle meets the $y$-axis (decide which one of these). In each case choose the normal closest to the origin. Continue each normal line until it meets 'adjacent' normals. In each case note any points where the curve does not appear to be regular.

Outline the evolute of the hyperbola as the curve touched by these normals. Calculate the radius of curvature at the vertex of the hyperbola, and mark on the diagram the centre of curvature at the vertex. Draw parallels to the hyperbola at distances of 1 cm and 4 cm measured along the inward normals.

---

### B. Some other curves: drawn as conchoids, strophoids, cissoids, etc.

### 4. The conchoid of Nicomedes (225 BC):   $r = a + b \sec \varphi$.

This is the conchoid of a straight line. Use a plain sheet of paper and the standard method for drawing conchoids (see Practical 3, Question 1). Choose the pole at distance $c = 2$ cm from the line, and the distance to be measured from the intersection to be $d = 6.5$ cm.

**5. The conchoid of Nicomedes (225 BC):** $r = a + b \sec \varphi$.
As in question 4, but with $c = 2$ cm and $d = 8$ cm.

**6. The right strophoid (Barrow 1670).** On a piece of plain paper, draw a line $L$ parallel to the long sides and midway between them. Mark the midpoint $P$ of this line, and a further point $Q$ distant 4 cm from $P$ and such that $PQ$ is perpendicular to $L$. Each (finely drawn) line through $Q$ will meet $L$ at a point $T$. Mark on it two points $R$ and $S$ such that $RT = TS = PT$. Repeat this procedure for a number of such lines, and draw the locus of the points $R$ and $S$.

**7. Gutschoven's curve (c. 1662):** $r = a \cot \varphi$.
On a plain piece of paper draw the coordinate axes through its centre $O$ with the $x$–axis parallel to its long edge. The line $\Lambda$ is the line $y = 5$ cm: it meets the $y$–axis in the point $C$. For a finely-drawn line $L$ through $O$ which meets the line $\Lambda$ in the point $Q$, mark on $L$ the two points $P$ and $P'$ which satisfy $OP = OP' = CQ$. Repeat this for a suitable collection of lines $L$ through the origin, and draw the curve by joining these plotted points in the usual way.

**8. The nephroid of Freeth (1879):** $r = 1 + 2 \sin \left( \dfrac{\theta}{2} \right)$ $\quad 0 \leq \theta \leq 4\pi$.

Plot this curve using as the strophoid of a circle with the centre of the circle as pole and with the fixed point on the circumference of the circle. Use polar graph paper in the landscape mode with the circle being that radius 3 cm centred at the origin. The point $T$ is where this circle meets the initial half-line. Consider the 18 points $Q$ on the same circle given by increments of 20 degrees (thus $\theta = 0, 20, 40, \ldots$). Mark points $P$ and $P'$ (in colour for clarity) on the line $OQ$ such that $P'Q = QP = QT$. Draw the strophoid in the usual way as the locus of the points $P$ and $P'$ after including any additional points $Q$ needed for accuracy.

**9. The cissoid of Diocles (200 BC).** On a piece of plain paper, draw a circle radius 4 cm with its centre at the centre of the paper. Draw a tangent line $L$ to this circle parallel to the long side of the paper. Let $A$ be the point on the circle at the opposite extremity of the diameter through the point of tangency. Each (finely drawn) line through $A$ will meet the the circle and the line $L$ in points $Q$ and $R$ respectively. Mark the point $P$ on the line such that $AP = QR$. Repeat this procedure for a number of such lines, and draw the locus of the point $P$.

Your group should draw all six of the curves 4-9.

## Practical 6

For this practical you should draw three curves: one from Project 1, one from Project 8, and one from Projects 2-7.

---

### A. Evolutes as envelope of normals: limaçons
As in Practical 5, first draw a number of normals, and then sketch in the evolute remembering that each normal line is a tangent to the evolute.

**Fine pencil.** Remember to use a fine pencil (H and sharp). This gives greater accuracy. The lines which you draw should not dominate the picture, but should be clearly visible; the curve itself and the evolute should stand out from the normals.

The evolutes in this question are reflection caustics.

---

**1.** Using the technique that follows, draw the curve

$$z = ae^{2it} + 2e^{it}$$

for one of the cases $a = 2$, $\frac{5}{4}$, $1$, $\frac{3}{4}$, $\frac{1}{2}$, and $\frac{1}{4}$, and then draw normals to the curve using sufficiently small increments so that the evolute $z_*$ can be 'seen' as the envelope of the normals. (Note that $t$ is not the polar angle.) The case $a = \frac{1}{4}$ is drawn as a special case – see below.
The evolute should have one cusp for each vertex of the limaçon, and becomes unbounded (tends to infinity) near an inflexion.

**Technique for drawing the curves and the normals.** Draw two concentric circles on polar paper, of radii 1 unit and $a$ units. (If you need advice on scale see below.) Taking increments in $t$ of, for example, 20 degrees, mark the pairs of points $e^{it}$ and $-ae^{2it}$ on the two circles, and draw the line joining the two points of each pair; this line is the normal to the curve at the point $z(t)$. The point $z(t)$ lies on this line and is the same distance from $e^{it}$ as $-ae^{2it}$ but is on the opposite side of $e^{it}$. Having drawn the segment from $-ae^{2it}$ to $z(t)$, continue it on one side only, just beyond the point where it meets adjacent normals.

Try to devise a quick method, using compasses, to find the successive pairs of points required. If you cannot, ask for advice.

Note that the equations giving the limaçons here are different from those in Practical 3 since the origin has been translated to a different position.

**Advice on scale.** The maximum value of $|z|$ is evidently $2 + a$, and is attained at $t = 0$ (note that $|z| \leq |ae^{2it}| + |2e^{it}| = a + 2$). It is suggested that you adjust the scale so that the smallest distance from the origin of your paper (9 cm on the polar paper) is about $(2 + a)x$ where 1 unit $=$ $x$ cm; thus $x = 2\frac{1}{2}$ or 2 for $a = 2$, $x = 3$ for $a = \frac{5}{4}$, $x = 4$ for $a = \frac{1}{2}$, etc. The key concern is that the points with *polar* values $\theta = 0$ and $\theta = \frac{\pi}{2}$ lie on the paper and that the curve is fairly large. '(Recall that $t$ is not the polar angle.)

**Proof.** To show that the above construction works, note that $z'(t) = 2iae^{2it} + 2ie^{it}$. Thus $iz'$ is the normal vector, and

$$e^{it} = \frac{1}{2}(z(t) + (-ae^{2it}))$$

is the midpoint of the normal line which joins $z(t)$ to $-ae^{2it}$.

**Special case:** $a = \frac{1}{4}$. In this case do not draw the limaçon. Draw the large circle as large as possible, and draw the family of normal lines as given by the above construction.

---

## B. More on envelopes

The following examples are envelopes of families of curves. An *envelope* of a one-parameter family $\{\mathbf{z}_\lambda\}$ of curves $t \mapsto \mathbf{z}_\lambda(t)$ is a curve $\lambda \mapsto \mathbf{e}(\lambda)$ which is tangent to the curve $\mathbf{z}_\lambda$ at $\mathbf{e}(\lambda)$ for each $\lambda$; thus for each $\lambda$ there is a $t$ for which $\mathbf{e}(\lambda) = \mathbf{z}_\lambda(t)$ and $\mathbf{e}'(\lambda)$ is parallel to $\mathbf{z}'_\lambda(t)$.

Usually we shall consider families of straight lines or circles, since these are easy to draw. In the examples below, draw a large number of the curves (normally about 20 will do), and the envelope will then become apparent. The curves of the family must be drawn *very accurately*.

Remember to draw the curves of the family lightly but visibly using a sharp pencil. The envelope should stand out clearly from the curves of the family, though you should never draw a thick curve. Never use the same diagram for drawing two different families of curves.

The most important parts of the curves of the family are close to the intersection of 'adjacent' curves in the sequence which you are drawing, since the envelope will be tangent to both near this point of intersection.

---

**2.** Draw the envelope of the family of lines $\{\ell_t\}$, where $\ell_t$ joins $(t, 0)$ to $(0, 1 - t)$.
(Take the origin at the centre of the page, 1 unit = 6 cm, and draw some lines with $t < 0$ and some with $t > 1$.)
This curve is a parabola.

**3.** Draw the envelope of the family of lines $\{\ell_t\}$, where $\ell_t$ joins a point $P_t$ on the $x$–axis to a point $Q_t$ on the $y$–axis which lies at a fixed distance $c$ from $P_t$.
(Take the origin at the centre of the page, and $c = 6$ cm.)
This curve is an astroid. Part of the curve is the envelope of an up-and-over garage door.

**4.** Draw enough members of the family of circles having centres on the supplied ellipse ($a = 6$ cm and $b = 4$ cm) and having radius $d$ to enable you to draw the envelope. Do this in the cases where $d = 2$, 4, and 6 cm. For $d = 4$ and 6 cm draw only that part of the envelope which is determined by the inner normals (i.e., lies inside the ellipse).

**5.** Draw the envelope of the family of circles $C_t$ which have centres on the unit circle and which touch the $x$−axis. (Take 1 unit $= 5$ cm, and use the portrait mode.)

Note that the envelope consists of part of the $x$−axis and a nephroid. The nephroid is the locus of a point on the edge of a circle radius $\frac{1}{2}$ as it is rolled around the outside of the unit circle.

---

**Orthotomics as envelopes.** The orthotomic of a curve $\Gamma$ with respect to a fixed point $P$ can be considered as the envelope of the family of circles $C_t$ which have centres on $\Gamma$ and which pass through $P$. The point $P$ is called the pole. Usually we shall consider parametrised curves. As usual rather more circles should be drawn to indicate points of the envelope such as cusps.

**6.** Using 3 copies of the parabola supplied (its vertex is near the centre of the paper), construct the orthotomic of $y^2 = 4x$ for each of the poles $(0,0)$, $(1,0)$, and $(-1,0)$. (Draw more circles having centres near to the vertex of the parabola.)

**7.** Construct the orthotomic of the ellipse supplied ($a = 4\frac{1}{2}$ cm and $b = 3$ cm) taking the pole as one of the foci.

---

Your group should draw all six of the curves 2-7.

---

**8.** Construct the orthotomic of the unit circle centre $(0,0)$ with respect to one of the poles $(-\frac{1}{4},0)$, $(-\frac{1}{2},0)$, $(-\frac{3}{4},0)$, $(-1,0)$, $(-\frac{5}{4},0)$, and $(-2,0)$. (Start with rough sketches in order to determine the scales.)

Note that these envelopes are limaçons.

Your group should draw all six of the curves of Project 8.

## Practical 7 (Anti-orthotomics and reflection caustics)

For this practical you should each draw at least 2 curves: one from Project 1, and one from Project 2.

---

### A. Anti-orthotomics

*Definition.* The anti-orthotomic of a parametrised curve $\Omega$ with respect to a *pole* $P$ (a fixed point) is the envelope of the perpendicular bisectors of the lines joining $P$ to the points of the curve.

Note that if $\Omega$ is the orthotomic of $\Gamma$ for the pole $P$, then $\Gamma$ is the anti-orthotomic of $\Omega$ for the same pole $P$.

**1.** Construct the anti-orthotomic of
   i)   a line, with respect to a pole not on the line,
   ii)  a unit circle with respect to a point distant $\frac{1}{2}$ from the centre,
   iii) a unit circle with respect to a point distant 2 from the centre,
   iv) a parabola with respect to the vertex, and
   v)  a parabola with respect to the focus.
(Use plain paper for i–iii.)

**Note:** Here and elsewhere, draw enough curves so that the consecutive intersections are about 1cm apart or less – less where the envelope has a cusp, large curvature, or a node.

---

### B. Reflection caustics

Let $t \mapsto \mathbf{m}(t)$ be a curve (called the *mirror*), and let $P$ be a point (called the *light source*). Light rays from $P$ are reflected in the mirror. (The ray and the reflected ray make equal angles with the normal at the point of reflection.)

*Definition.* The *reflection caustic* of $t \mapsto \mathbf{m}(t)$ with respect to $P$ is the envelope of the reflected rays.

**2.** Draw the reflection caustics for a circle of radius $r$ where $P$ lies at one of the distances $2r$, $r$, $\frac{3}{4}r$, $\frac{1}{2}r$, $\frac{1}{4}r$, and $\infty$ from its centre.

Draw only the reflected rays and include the whole line $z_t$, where $z_t$ is the line, on both sides of the mirror, of the reflected ray through the point $\mathbf{m}(t)$. Do not draw the rays issuing from $P$. By including the whole line $z_t$ we obtain also the 'virtual caustic', determined by the whole lines of those reflected light rays where the reflected ray is directed to the outside of the circle.

### A geometric construction for the lines $z_t$.

For each point $\mathbf{m}(t)$ on the circle mark the point $Q_t$ where the line from $P$ to $\mathbf{m}(t)$ meets the circle again. Then mark the point $R_t$ on the circle where distance $(\mathbf{m}(t),Q_t)=$ distance $(\mathbf{m}(t),R_t)$.

The line which joins $\mathbf{m}(t)$ to $R_t$ is $z_t$.

In the case where $P$ is at infinity along the $x-$axis, consider parallels to the $x-$axis through $\mathbf{m}(t)$, instead of the lines from $P$ to $\mathbf{m}(t)$. Use lightly ruled graph paper to give these parallel lines.

# 18

# Drawn curves

We now give computer programs for use with MATLAB for drawing standard curves that are needed in the practical classes and for some of the curve-plotting exercises given in the book. The programs given here and their usage in the practical classes are as follows.

Ellipse 1: Practical 5, question 1.

Ellipse 2: Practical 6, question 4.

Ellipse 3: Practical 6, question 7.

Parabola 1: Practical 5, and Practical 7, question 1, parts iv and v.

Parabola 2: Practical 6, question 6.

Parabola 3.

Hyperbola: Practical 5, question 3.

Semicubical parabola.

Polar graph paper: available for all practicals.

## 18.1 Personalising MATLAB for metric printing

Metric measurements are used throughout. You need to create a folder 'matlab' on the 'disc' from which you will be running MATLAB. Store all your MATLAB '.m' files in this folder or a subfolder. Include in the folder 'matlab' a 'startup.m' file which will give you the appropriate default page setup for printing. This 'startup.m' file needs to be created within MATLAB and/or connected to MATLAB by a 'path' in the usual way. Choose the appropriate one of the following 'startup.m' files.

### startup.m file for UK A4 paper

```
set(0,'DefaultFigurePaperType','a4letter')
set(0,'DefaultFigurePaperUnits','centimeters')
```

**startup.m file for US Letter paper**

```
set(0,'DefaultFigurePaperUnits','centimeters')
```

## Procedure for printing a sized MATLAB figure

To print the figure after running MATLAB:

On 'File' select 'Page Position'.
Select 'Select Paper Position Explicitly'.
Paper units should already be 'centimeters' using your 'startup.m' file.
Paper size should already be that of A4 or US Letter as appropriate.
Select 'landscape' or 'portrait', and type in the 'Paper Position' as indicated in the program for the curve you are drawing. The 'Paper Position' entry involves four dimensions enclosed in square brackets: a space should be left between each dimension.
Press 'Done'.
The 'Page Position' display should remain on your screen. If this does not happen repeat the above procedure, again pressing 'Done'. The 'Page Position' display should remain on your screen.
From 'Page Position', press 'Print'.

The programs are all designed so that the printed output is based on a rectangle 22 cm × 18 cm. You should check that the output has the correct dimensions. For the curves the printed axes should be 22 cm and 18 cm long, except that part of the 22 cm long axis may be off the page when the portrait position is used. For the polar graph paper the printed bounding rectangle should be 22 cm × 18 cm.

## 18.2 Ellipse 1 and Ellipse 2

```
% MATLAB program for Ellipse 1 and Ellipse 2
  set (gcf, 'position', [360 514 560 460])
  axes('position', [0 0 1 1]);
  plot([0,0],[-9,9]);
  hold on
  axis([-11 11 -9 9]);
  plot([-11,11],[0,0]);
  plot([-.1,.1],[4.47,4.47]);
  plot([-0.1,0.1],[-4.47,-4.47]);
  plot([-11,11],[0,0]);
  t= 0:0.01:2*pi;
  x=4*cos(t);
  y=6*sin(t);
  plot(x,y);
```

```
axis('equal');
axis('off')
% see 'Procedure for printing' in §18.1
% Ellipse 1: for UK A4 paper select 'landscape'
% and type '[3.75 1.5 22 18]' in Paper Position,
% for US letter paper select 'landscape'
% and type '[3 1.25 22 18]' in Paper Position.
% Ellipse 2: for UK A4 paper select 'portrait'
% and type '[-0.5 5.7 22 18]' in Paper Position,
% for US letter paper select 'portrait'
% and type '[-0.2 5 22 18]' in Paper Position.
```

## 18.3 Ellipse 3

```
% MATLAB program for Ellipse 3
set (gcf, 'position', [360 514 560 460])
axes('position', [0 0 1 1]);
plot([0,0],[-9,9]);
hold on
axis([-11 11 -9 9]);
plot([-11,11],[0,0]);
plot([-.1,.1],[3.354,3.354]);
plot([-0.1,0.1],[-3.354,-3.354]);
plot([-11,11],[0,0]);
t= 0:0.01:2*pi;
x=3*cos(t);
y=4.5*sin(t);
plot(x,y);
axis('equal');
axis('off')
% see 'Procedure for printing' in §18.1
% For UK A4 paper select 'portrait'
% and type '[-0.5 5.7 22 18]' in Paper Position.
% For US letter paper select 'portrait'
% and type '[-0.2 5 22 18]' in Paper Position.
```

## 18.4 Parabola 1

```
% MATLAB program for Parabola 1
set (gcf, 'position', [360 514 560 460])
axes('position', [0 0 1 1]);
plot([0,0],[-1,17]);
hold on
axis([-11 11 -1 17]);
```

```
plot([-11 11],[0,0]);
plot([-.1,.1],[1,1]);
t= -3.47:0.01:3.47;
y=t.^2;
x=2*t;
plot(x,y);
axis('equal');
axis('off')
% see 'Procedure for printing' in §18.1
% For UK A4 paper select 'landscape'
% and type '[3.75 0.5 22 18]' in Paper Position.
% For US letter paper select 'landscape'
% and type '[3 0.5 22 18]' in Paper Position.
```

## 18.5  Parabola 2

```
% MATLAB program for Parabola 2
  set (gcf, 'position', [360 514 560 460])
  axes('position', [0 0 1 1]);
  plot([0,0],[-6,12]);
  hold on
  axis([-11 11 -6,12]);
  plot([-11 11],[0,0]);
  plot([-.1,.1],[1,1]);
  t= -3.47:0.01:3.47;
  y=t.^2;
  x=2*t;
  plot(x,y);
  axis('equal');
  axis('off')
% see 'Procedure for printing' in §18.1
% For UK A4 paper select 'landscape'
% and type '[3.75 1.5 22 18]' in Paper Position.
% For US letter paper select 'landscape'
% and type '[3 1.75 22 18]' in Paper Position.
```

## 18.6  Parabola 3

```
% MATLAB program for Parabola 3
  set (gcf, 'position', [360 514 560 460])
  axes('position', [0 0 1 1]);
  plot([0,0],[-0.5,17.5]);
  hold on
```

```
axis([-11 11 -0.5 17.5]);
plot([-11 11],[0,0]);
plot([-.1,.1],[0.75,0.75]);
t= -3.1:0.01:3.1;
y=.75*t.^2;
x=1.5*t;
plot(x,y);
axis('equal');
axis('off')
% see 'Procedure for printing' in §18.1
% For UK A4 paper select 'landscape'
% and type '[3.75 0.5 22 18]' in Paper Position.
% For US letter paper select 'landscape'
% and type '[3 0.5 22 18]' in Paper Position.
```

## 18.7 Hyperbola

```
% MATLAB program for Hyperbola
set (gcf, 'position', [360 514 560 460])
axes('position', [0 0 1 1]);
plot([0,0],[-6,12]);
hold on
axis([-11 11 -6,12]);
plot([-11 11],[0,0]);
plot([-0.1,0.1],[2.828,2.828]);
plot([-0.1,0.1],[-2.828,-2.828]);
t= -1.32:0.01:1.32;
y=2*sec(t);
x=2*tan(t);
plot(x,y);
t= -1.15:0.01:1.15;
y=-2*sec(t);
x=2*tan(t);
plot(x,y);
axis('equal');
axis('off')
% see 'Procedure for printing' in §18.1
% For UK A4 paper select 'landscape'
% and type '[3.75 1.5 22 18]' in Paper Position.
% For US letter paper select 'landscape'
% and type '[3 1.25 22 18]' in Paper Position.
```

## 18.8 Semicubical parabola

```
% MATLAB program for Semicubical-parabola
  set (gcf, 'position', [360 514 560 460])
  axes('position', [0 0 1 1]);
  plot([0,0],[-0.5,17.5]);
  hold on
  axis([-11 11 -0.5 17.5]);
  plot([-11 11],[0,0]);
  t= -1.8:0.01:1.8;
  y=3*t.^2;
  x=2*t.^3;
  plot(x,y);
  axis('equal');
  axis('off')
% see 'Procedure for printing' in §18.1
% For UK A4 paper select 'landscape'
% and type '[3.75 0.5 22 18]' in Paper Position.
% For US letter paper select 'landscape'
% and type '[3 0.5 22 18]' in Paper Position.
```

## 18.9 Polar graph paper

Commercial coloured centimetre polar graph paper is recommended. Plotted curves appear much clearer where the circles and lines printed on the paper are a light colour. However the following MATLAB program will draw polar graph paper. You may prefer to adjust the three Linewidth commands in the program for the thicker lines and/or the thinner lines. The size of the printed enclosing box is 22 cm × 18 cm. A good laser printer of at least 600 dpi and with memory is recommended. An old computer can take several minutes to print the polar graph paper.

```
% MATLAB program for Polar paper
% To print the figure after running MATLAB:
% on 'File' select 'Page Position'
% select 'Select Paper Position Explicitly'.
% Paper units should already read 'centimeters'
% because of your 'startup.m' file.
% Paper size should already be that of
% A4 or US Letter as appropriate.
% For UK A4 paper select 'landscape'
% and type '[4 2 22 18]' in Paper Position.
% For US letter paper select 'landscape'
% and type '[3 2 22 18]' in Paper Position.
```

```
% Select 'Done'.
% From 'Page Position', select 'Print'.

set (gcf, 'position', [360 514 560 460])
axes('position', [0 0 1 1]);
plot([0,0],[-9,-0.6]);
plot([0,0],[0.6,9]);
plot([0,0],[-0.1 0.1]);
hold on
plot([-11,-0.2],[0,0]);
plot([-0.1,0.1],[0,0]);
plot([0.2,11],[0,0]);
set(gca,'Xticklabel',{});
set(gca,'Yticklabel',{});
axis([-11 11 -9 9]);
set(gca,'Ticklength',[0 0]) ;

% multiplot A1 : CIRCLES
for b=1:15,
t= 0:0.01:2*pi;
x=b.*cos(t),y=b.*sin(t);
plot(x,y), end;

% multiplot A2 : CIRCLES
for b=1:75,
t= 0:0.01:2*pi;
x=(b/5).*cos(t),y=(b/5).*sin(t);
plot(x,y,'LineWidth', 0.05), end;

% multiplot B1 : RADIAL LINES
da=pi/9;
a=0;
while a<2*pi
s= 0.2:14.8:15;
x=s.*cos(a),y=s.*sin(a);
plot(x,y);
a=a+da;
end;

% multiplot B2 : RADIAL LINES
da=pi/9;
a=0;
while a<2*pi
s= 0.6:14.4:15;
x=s.*cos(a+(pi/18)),y=s.*sin(a+(pi/18));
plot(x,y);
```

```
a=a+da;
end;

% multiplot B3 : RADIAL LINES
da=pi/(18);
a=0;
while a<2*pi
s= 1:1:2;
x=s.*cos(a+(pi/36)),y=s.*sin(a+(pi/36));
plot(x,y,'LineWidth', 0.05);
a=a+da;
end;

% multiplot B4 : RADIAL LINES
da=pi/(90);
a=0;
while a<2*pi
s= 2:13:15;
x=s.*cos(a),y=s.*sin(a);
plot(x,y,'LineWidth', 0.05);
a=a+da;
end;
axis('equal');
```

# 19

# Further reading

Abhyankar S. S., *Algebraic Geometry for Scientists and Engineers*, American Mathematical Society Providence 1990.
  Contains mostly advanced material.

Archbold J. W., *Introduction to Algebraic Geometry of a Plane*, Edward Arnold London 1948.
  Chapters VI and VII contain relevant material.

Archibald R. C. and Court N. A., Curves, special, *Encyclopædia Brittanica*, William Benton Chicago.
  Lists some 60 special curves, with properties.

Basset A. B., *Cubic and Quartic Curves*, Deighton Bell Cambridge 1901.
  Contains relevant material though sometimes from a different viewpoint.

Brieskorn E., and Knorrer H., *Plane Algebraic Curves*, Birkhäuser Verlag Basel 1986.
  Generally an advanced text; however, the first part gives a substantial history of algebraic curves.

Coolidge J. L., *A Treatise on Algebraic Plane Curves*, Oxford University Press 1931, reprinted by Dover New York 1959.
  Much of this older text is advanced.

*Encyclopædia of Mathematics*, Reidel Dordrecht 1987–1994.
  Contains articles on some special curves.

Fowler R. H., *The Elementary Differential Geometry of Plane Curves*, Cambridge University Press 1929, reprinted by Stechert-Hafner New York 1964.
  Contains articles on some special curves.

Frost P., *Curve Tracing*, Macmillan London 1872, reprinted by Chelsea New York 1961.

Includes an unrivalled collection of sketches of algebraic and some algebraic/trigonometric curves illustrating aspects of the theory, such as cusps, nodes, and asymptotes. The variations of forms and combinations of these are given in an encyclopædic manner. The theory included is generally elementary and many classes of curves are considered.

Hilton H., *Plane Algebraic Curves*, Oxford University Press 1920.

Contains advanced and some elementary theory.

Hunt K. H., *Kinematic Geometry of Mechanisms*, Oxford University Press 1978.

For curves drawn by linkages.

Lawrence J. Dennis, *A Catalog of Special Plane Curves*, Dover New York 1972.

Contains elementary theory, and details of some 60 curves, including some transcendental curves, plus properties of each.

Lockwood E. H., *A Book of Curves*, Cambridge University Press 1961.

Contains a large number of drawing techniques for, and properties of, special curves.

Loria G., *Spezielle Algebraische und Transzendente Ebene Curven: Theorie und Geschichte I, II*, Teubner Verlag Leipzig and Berlin 1910, 1911.

Large number of special curves with historical notes.

Primrose E. J. F., *Plane Algebraic Curves*, Macmillan London and St. Martin's Press New York 1955.

Contains elementary theory.

Semple J. G. and Kneebone G. T., *Algebraic Curves*, Oxford University Press 1959.

Generally an advanced book. Includes information on Puiseux fractional power series.

von Seggern D., *CRC Standard Curves and Surfaces*, CRC Press Boca Raton 1990.

Contains many drawn standard curves (graphs, algebraic and other) including data on how to get a particular shape.

Teixeira F. G., *Traité des Courbes spéciales remarqueables planes et gauche*, Coïmbre 1908–1915, reprinted by Chelsea New York 1971.

Large number of special curves with historical notes.

Walker Robert John, *Algebraic Curves*, Princeton University Press 1950, reprinted by Springer Verlag Heidelberg 1978.

Covers quite a lot of theory.

# Index

absolute curvature, 134
acnodal cubic, 63–64, 87, 88, 91, 140,
    152, 159, 171, 279, 318, 332
acnode, 63, 278, 292
affine
    line, 298
    section, 304
    view, 304
affinely equivalent plane curves, 318
algebraic curve, 59, 88–89, 157–160,
    271–296
analytic function, 98
angle, 5–6
    between lines, 5
    between vectors, 5
    non-oriented, 6
    oriented, 6
Apollonius, 26
applications
    Archimedean spiral, 73
    conics, 26–29
    cubics, 66
    cycloid, 217–218
    trochoid, 228–229
arc-length, 94, 134, 211
Archimedean spiral, 71–73, 83, 84, 87,
    145, 147, 149, 333
Archimedes, 73
Argand diagram, 4, 84, 87, 146–147
astroid, 89, 95, 145, 194, 226, 250,
    264, 341
asymptote, 318–320
    hyperbola, 20, 43, 45, 313

plane curve, 312
    procedure for determining, 313–314
axis
    central conic, 42, 45
    conjugate, see conjugate axis
    major, see major axis
    minor, see minor axis
    projective plane, 301–302
    transverse, see transverse axis

Barrow, 327, 333, 339
Bernoulli
    Jakob, 71, 228
    Johann, 218, 228
Bernoulli's lemniscate, 76, 145
bicorn, 89
blunt point, 130
bounded curve, 309
brachistochrone, 218
branch, 59, 61, 62, 66, 281–284,
    286–288
    near a cusp, 123

canonical form, see conic, canonical
    form
cardioid, 69, 84, 87, 95, 145, 147, 177,
    223, 263, 291
    cusp, 180
Cartesian coordinates, 1–2
catenary, 95, 141, 207
caustic, 265
    evolute of the orthotomic, 266

of a circle, 268

central conic, *see* conic, central

centre

  conic, *see* conic, centre

  ellipse, *see* ellipse, centre

  hyperbola, *see* hyperbola, centre

  of curvature, 167, 336

    algebraic curve, 173

    parameter $x$, 171

ceratoid cusp, 160

characteristic

  equation, 36

  polynomial, 36

circle, 13–15, 79, 88, 90, 145, 147, 207, 276, 277

  great, *see* great circle

  negatively oriented, 167

  of curvature, 168, 251

  positively described, 136

  positively oriented, 167

  standard parametrisation, 167

cissoid, 89, 93, 290, 327, 333, 339

closed curve, 78

cocked hat, 89

column

  matrix, 34

  vector, 34

comparable parametrisation, 210

concave side of the curve, 170

concentrate, 243

conchoid, 229, 280, 285, 291, 326, 334, 338, 339

cone, 15, 306

conic, 15–16, 306

  canonical form, 15, 31, 40

  central, 40–45

  centre, 40–45

  discriminant, 43–45

  general equation, 15, 31, 37

  general position, 31–58

conjugate axis, 19, 42

contact, 103–108

  between circle and curve, 173–175

corner, 82

critical point, 116

crunodal cubic, 62–63, 79, 86, 88, 91, 140, 171, 281, 284, 296, 332

cubic, 61–64, 291–295

  classification, 293

  curve, 61

  degenerate, 292

  non-degenerate, 292

  rational parametrisation, 293

curtate cycloid, 217, 234, 335

curvature, 134, 230

  absolute, *see* absolute curvature

  algebraic curve, 157–160

  at singular point, 295

  limiting, 160–163

  negative, 135

  of a circle, 148

  parametrised by $x$, 137–142

  polar equation, 142–145

  positive, 135

  zero, 135, 149

curve of concentration, 243, 244

cusp, 59, 61, 123, 124, 292, 336

  coincident branches, 124

  locus, 255

  ordinary, 123, 288–291

  projective curve, 314

cuspidal cubic, 61–62, 79, 87, 88, 91, 93, 140, 261, 281

cuspidal tangent, 61, 125, 165

cycloid, 140, 175, 192, 202, 215–218, 234, 335

  curtate, *see* curtate cycloid

  prolate, *see* prolate cycloid

  standard parametrisation, 217

Dürer, 228

deltoid, 226

Descartes, 26, 66, 71, 327, 333

determinant, 35

diagonalising quadratic forms

  algebraic method, 37–38

  geometrical method, 31–33

differentiability class, 80

differentiable curve, 78

Diocles, 327, 333, 339

direction of a curve, 83, 85, 133

directrix, 22

discriminant, 255

  conic, *see* conic, discriminant

envelope, 255
domain of a function, 78
double point, 59, 62, 66, 69
duality, 301

eccentricity
    ellipse, *see* ellipse, eccentricity
    hyperbola, *see* hyperbola,
        eccentricity
    parabola, *see* parabola, eccentricity
eigenvalue, 36
eigenvector, 36
ellipse, 49, 53, 58, 86, 140, 141, 159,
    172, 175, 192, 201–203, 233, 250,
    261, 262, 264, 306, 317
    canonical position, 16–18
    centre, 17, 40–45
    eccentricity, 17
    focus, 17
end point, 78
envelope
    degenerate, 255
    different definitions, 259–260
    discriminant, 255
    geometrical, 243
    limiting-positions, *see*
        limiting-positions envelope
    of circles of curvature, 251
    singular-set, *see* singular, set,
        envelope
    singularity of, 259–260
epicycloid, 194, 220, 229, 234
epitrochoid, 219–223
    standard parametrisation, 220
equi-angular spiral, 69–71, 79, 84, 87,
    144, 147, 149, 172, 194, 333
equivalent projective curves, 318
Euclid, 26
Euler's theorem, 100, 314
evolute, 183
    at a cusp, 185
    astroid, 189
    at a vertex, 186
    envelope of normals, 184, 244–245,
        340
    epicycloid, 189
    is circle, 193

matrix method, 186
non-regular point, 186
of a cardioid, 189–190
of a cycloid, 187–189
of a hyperbola, 338
of a parabola, 338
of an ellipse, 337

family of parametrised curves, 246
Fermat, 26
Fermat's spiral, 333
final point, 78
fixed
    curve, 209
    plane, 209
focus
    ellipse, *see* ellipse, focus
    hyperbola, *see* hyperbola, focus
    parabola, *see* parabola, focus
folium, 285, 291, 327
four bar linkage, 73–76
four vertex theorem, 156, 164
Freeth, 339
fundamental theorem of Algebra, 274

Galileo, 28, 218
geometric curve, 78
geometrical envelope, 243
global positioning, 29
great circle, 301
Gutschoven's curve, 89, 93, 325, 339

half-line, 1, 2
Hessian, 157
higher
    cusp, 127
    inflexion, 115
    singularity, 128
    undulation, 115
hippopede, 231, 280, 333
history
    Archimedean spiral, 73
    conics, 26–29
    cubics, 66
    cycloid, 217–218
    equi-angular spiral, 71

semicubical parabola, 62
trochoid, 228–229
homogeneous
coordinates, 299
function, 100, 302
degree of, 302
polynomial, 273–274, 302
degree of, 302
Huygens, 218, 228
hyperbola, 46, 53, 58, 93, 140, 142,
158, 193, 261, 306, 312, 317, 323
canonical position, 18–20
centre, 19, 40–45
eccentricity, 19
focus, 19
rectangular, 20, 43, 48, 140, 145
hypocycloid, 189, 194, 224, 234
hypotrochoid, 223–228
standard parametrisation, 224

implicit function theorem, 89
inflexion, 112–120, 151–153, 238–240
circle, 239
general definition, 118
parametrised by $x$, 116
projective curve, 316
initial
half-line, 2
point, 78
inner product, 6, 34
interior point, 78
intersection multiplicity, 272
interval, 78
invariance of point-contact order, 108
inverse
function, 100
function theorem, 100
involute, 203–207
equation of, 205
geometrical interpretation, 204–205
wrap/unwrap, 204–205
irreducible
quadratic factor, 274
isochrone, 218
isolated
point, 63
projective curve, 314

zero, 97
iterated line, 40

Jordan curve, 82
theorem, 82

Kempe, 73
Kepler, 26, 291

L'Hôpital's theorem, 100, 125, 129,
141, 163, 200
La Hire, 228
latus rectum, 23
lemniscate, *see* Bernoulli' lemniscate
length
of vector, 34
limaçon, 68–69, 84, 108, 163, 164,
177–182, 220, 234, 267, 280, 285,
334
alternative equation, 182
curvature, 179
cusp, 180
five classes, 181–182
inflexion, 180
node, 179
non-regular point, 179–180
one loop, 177
undulation, 181
vertex, 180–181
limiting
curvature, *see* curvature, limiting
positions envelope, 260
tangent direction, 164
line, 9–13
linear point, 113
linearly dependent vectors, 6
linearly independent vectors, 6
linkage, 73–76
local parametrisation, 89, 288
logarithmic spiral, *see* equi-angular,
spiral
loop, 69

Mênaechmus, 26
MacLaurin, 326
major axis, 17, 42

matrix, 34
  diagonal, 36
  identity, 34
  inverse, 35
  invertible, 35
  non-singular, 35
  orthogonal, 36
  product, 35
  rotation, 36
  symmetric, 35
  transpose, 35
minor axis, 17, 42
moving
  curve, 209
  plane, 209
multiple point, 80
multiplicity
  non-singular point, 275
  of a point, 275

negative curvature, *see* curvature,
  negative
nephroid, 89, 223, 227, 263, 342
  of Freeth, 339
Newton, 26, 66
Newton's
  formula, 141
  polygon, 291
Nicomedes, 326, 338, 339
nodal
  point, 284
    projective curve, 314
node, 59, 62, 69, 284
  2-node, 292
non-singular point, 88
  projective curve, 307
normal
  direction, 85
  line, 85, 87
    algebraic curve, 92
  vector, 85, 87
    algebraic curve, 91

order
  of a cusp, 127
  of a zero, 98
  of contact, 104
  of point contact, 104
  of tangency, 104
ordinary
  branch, 281
  cusp, 288–291, *see* cusp, ordinary
orientation
  along a curve, 83, 133
  of a circle, 169
oriented curve, 133
orthogonal vectors, 6–7
orthonormal basis, 37
orthotomic, 264, 342
  of a circle, 266–267
osculate, 174
osculating
  circle, 170, 173–175
  curve, 174

Pappus, 26
parabola, 41, 51, 54, 58, 95, 140, 145,
  171, 172, 191, 201, 249, 261, 263,
  277, 286, 306, 317
  axis, 22
  canonical position, 21–22
  eccentricity, 22
  focus, 22
  vertex, 22
parallel, 298
  at inflexion, 199
  at non-regular point, 196
  line, 196
  non-regular point of, 198
  of a curve, 195, 252–253
  of ellipse, 199–203, 337
  of hyperbola, 338
  of involute, 203
  of parabola, 338
  same centre of curvature, 197
  same tangent direction, 197
  vectors, 6–7, 87
parameter, 2
parametric
  curve, 78
  equation, 2, 5
parametrisation
  comparable, *see* comparable
    parametrisation

local, *see* local parametrisation
polynomial, 55–58, 104
rational, 55–58
Pascal, 218
Peano curve, 81
pear-shaped quartic, 291
pencil, 275
piecewise smooth curve, 78
piriform curve, 291
Plücker, 66
plane curve, 78
   view of projective curve, 305–306
point at infinity, 24, 298, 300
point contact
   analytical interpretation, 121
   geometrical interpretation, 120
   infinite, 286
polar
   angle, 2
   coordinates, 2–4
   equation, 4–5
pole, 2, 17
polynomial parametrisation, *see*
   parametrisation, polynomial
positive curvature, *see* curvature,
   positive
Proclus, 231
product of vectors, 34
projective curve, 302
   affine view, 304
   associated to a plane curve, 303
   degree of, 302
   determined by a plane curve, 303
   non-singular point, 307
   singular point, 307
   tangent line, 308
projective line, 297, 302
projective plane, 298
   axis, 301–302
   diagrammatic representation,
     301–302
   geometrical interpretation, 299–300
projectively equivalent
   affine curves, 318
   plane curves, 316
prolate cycloid, 217, 335

quadratic cusp, 123
quadrifolium, 226
quartic, 66–68

radian, 3
radius of curvature, 168
rational parametrisation, *see*
   parametrisation, rational
reflexion caustic, 265
regular, 84
   curve, 83
remainder, 101
rhamphoid cusp, 162
rhodonea, 226
rigid
   motion, 235
   self-map, 234
Roberval, 228
rolling
   curve, 211
rose curve, 226
roulette, 209
   complex equation, 213
   real equation, 213

scalar product, 6, 34
self-crossing, 80
self-intersection, 80
semi-cubical parabola, 61–62, 95, 124,
   249, 263, 286, 287, 290, 332
sign of the curvature, 135
signed curvature, 134
simple
   closed curve, 82, 156
   inflexion, 112
   intersection, 104
   tangency, 106
   undulation, 114
singular
   envelope, 253–259
   line, 275–278
   point, 88, 271–296
     at infinity, 314
     isolated, 278
     projective curve, 307
   set, 246, 247
     envelope, 244, 246–250, 256

using complex multiplication, 250–252
sinusoidal spiral, 145
smooth
  curve, 78
  function, 98
space-filling curve, 77, 81
speed, 83, 84
spiral
  Archimedean, *see* Archimedean spiral
  equi-angular, *see* equi-angular spiral
  logarithmic, *see* equi-angular spiral
  sinusoidal, *see* sinusoidal spiral
St. Laurent, 229
standard parametrisation of a circle, *see* circle, standard parametristion
stationary point, 116
  of inflexion, 117
strophoid, 285, 324, 327, 333, 339

tachnodal quartic, 66–68, 89, 91, 93, 332
tachnode, 66, 286, 287, 324
tangency, 106
tangent, 85, 87
  algebraic curve, 92, 276–278
  cuspidal, *see* cuspidal tangent
  direction, 85
    limiting, *see* limiting tangent direction
  projective curve, 308
  singular point, 280–282
  vector, 85, 87
    algebraic curve, 91
tautochrone, 218
Taylor's theorem, 97, 101
time $t$, 212
transverse
  axis, 19, 42
  intersection, 104
trifolium, 226
trigonometry, 7
trisectrix, 69, 223, 285, 326
trochoid, 218

unbounded, 64

undulation, 112–120, 153–157
  general definition, 118
  parametrised by $x$, 116
unit
  speed, 96
    curve, 83, 84
  vector, 6

vector, 34
velocity vector, 83, 84
vertex, 153–157

Wallis, 26
Watt, 73
Watt's curves, 73–76
witch of Agnesi, 318, 332

zero curvature, *see* curvature, zero
zero of homogeneous polynomial, 273